Advanced Materials Science and Engineering of Carbon

Advanced Materials Science and Engineering of Carbon

Michio INAGAKI

Professor Emeritus of Hokkaido University, 228-7399 Nakagawa, Hosoe-cho, Kita-ku, Hamamatsu 431-1304, Japan

Feiyu KANG

Dean and Professor, Graduate School at Shenzhen, Tsinghua University, University Town, Shenzhen City, Guangdong Province 518055, China

Masahiro TOYODA

Professor of Oita University, 700 Dannoharu, Oita 870-1192, Japan

Hidetaka KONNO

Professor Emeritus of Hokkaido University, 702, 1-1 Nishi-10, Minami-15, Chuou-ku, Sapporo 064-0915, Japan

AMSTERDAM • BOSTON • HEIDELBERG • LONDON
NEW YORK • OXFORD • PARIS • SAN DIEGO
SAN FRANCISCO • SINGAPORE • SYDNEY • TOKYO

Butterworth-Heinemann is an imprint of Elsevier

Butterworth-Heinemann is an imprint of Elsevier
The Boulevard, Langford Lane, Kidlington, Oxford, OX5 1GB, UK
225 Wyman Street, Waltham, MA 02451, USA

First published 2014

British Library Cataloguing in Publication Data
A catalogue record for this book is available from the British Library

Library of Congress Cataloguing in Publication Data
A catalog record for this book is available from the Library of Congress

ISBN: 978-0-12-407789-8

For information on all Butterworth-Heinemann publications visit
our website at **store.elsevier.com**

Working together
to grow libraries in
developing countries

www.elsevier.com • www.bookaid.org

Contents

Preface

Carbon materials, the targeted materials of the present book, are very important in many fields of science, engineering, and technology, and so papers reporting on "carbon material" are published in journals in a wide range of specialties. Even focusing on a specific subject—for example, carbon nanotubes, template carbonization, anode materials for lithium-ion batteries, and so on—huge numbers of scientific papers are published. Therefore, the search of related references published in journals without omission is an onerous and time-consuming task. Naturally, it is not easy to provide a comprehensive overview of a particular subject within the science of carbon and cover the whole range of material released. That is what makes it so challenging; and above all, comprehensive summary and review of the published results are remarkably helpful to many people and vital to further development of the field.

In the present book, the authors attempt to give summaries and reviews on selected themes concerning carbon materials, based on the material and information, as much as is obtainable to us. Principal results in advanced materials science and engineering of carbon materials are reviewed with reference to a vast number of papers published in scientific journals. The book is organized into 17 chapters, including the introduction in Chapter 1. Chapters 2 to 10 are focused on issues of formation and preparation of carbon materials, and Chapters 11 to 17 cover different applications. In Chapters 2 and 3, carbon nanotubes and graphene are reviewed, with emphasis on their formation. Processes with specific procedures and the resultant carbon materials are reviewed in Chapters 4 to 10: they cover carbonization under pressure; graphitization under high pressure, including stress graphitization; glass-like carbons, with special attention to their activation and graphitization; template carbonization to control morphology and pore structure; carbon nanofibers prepared via electrospinning; carbon foams creating new applications; and nanoporous carbon membranes including carbon fiber webs. In Chapters 11 to 17, carbon materials used in specific fields are reviewed: electrochemical capacitors; lithium-ion rechargeable batteries; photocatalysis; spilled-oil recovery; adsorption of hydrogen, methane, volatile organic compounds, and metal ions; highly-oriented and highly-crystalline graphite, emphasizing its high thermal conductivity; and isotropic high-density graphite, emphasizing its nuclear applications.

To understand the advanced science and engineering of carbon materials, a wide range of fundamental knowledge in the field of carbon materials is essential; that is, knowledge of aspects such as carbonization, graphitization, intercalation, and so on, in addition to basic knowledge of chemistry, physics, biology, and other subjects. For readers' convenience, it is recommended to consult *Carbon Materials Science and Engineering: From Fundamentals to Applications*, published by Tsinghua University Press. The book will supply fundamental knowledge on carbon materials and help in the understanding of the broad range of topics in the present book.

It would give great pleasure to the authors if the content of this book can provide useful information which may be used to inspire the readers to new research directions.

Acknowledgments

The authors would like to express their sincere thanks to the people who kindly provided the data and figures for this book. They also thank all of the people who have taken care of this book in Tsinghua University Press and also in Elsevier.

Introduction

1

Carbon materials have always played important roles for human beings; for example, charcoals as a heat source and adsorbent since prehistoric times, flaky natural graphite powder as pencil lead and soot in black ink in the development of communication techniques, graphite electrodes in steel production, carbon blacks for reinforcing tires in the development of motorization, graphite membrane switches making computers and control panels thinner and lighter, carbon fibers for reinforcing plastics, high-purity graphite blocks in nuclear reactors, compounds of graphite with fluorine in lithium primary batteries, graphite in lithium-ion secondary batteries. Many carbon materials have been developed and more will be developed in the future. They are widely used from the home to the industrial setting.

In Figure 1.1, some examples of applications of carbon materials are illustrated, in order to show how widely they are used, although listing every application is not possible here. In the aircraft and aerospace fields, carbon-fiber-reinforced plastics are used in body parts. In automobiles, carbon-fiber-reinforced carbons are used in brakes; carbon/metal composites in brushes; carbon blacks in tires; and activated carbons to create comfortable space in the car, and also in the canister to save gasoline and to avoid air contamination. In the building and civil engineering fields, carbon-fiber-reinforced concrete is successfully used in buildings and bridges exposed to sea water to avoid erosion by salts. Carbon fibers are used for reinforcing the piers of expressways. In electronic devices such as computers and mobile phones, carbon materials are used in power sources as electrodes of primary and secondary batteries, and conductive graphite sheets printed on polymer films are used as switches and conducting leads, helping to make the devices greatly lighter and thinner. To produce semiconductors for electronic devices, like silicon single crystals, carbon/carbon composites for heaters and high-density graphite for crucibles and susceptors are essential. Carbon materials also find their way into sustainable energy development: the blades of windmills consist of carbon-fiber-reinforced composites. To stabilize the varying electricity produced by windmills and solar cells, lithium-ion rechargeable batteries and electrochemical capacitors are essential devices, both using carbon materials as electrodes. Electric conductive carbon rods and carbon blacks support the development of primary batteries. Compounds of graphite with fluorine, graphite fluorides, improve the performance of primary batteries, and the reaction of lithium intercalation/deintercalation into the galleries of graphite is greatly furthering the development of lithium-ion rechargeable batteries. In addition, carbon nanotubes and fullerenes are promoting the development of nanotechnology in various fields of science and engineering.

Aircraft and aerospace

Automobiles

- Carbon laminate
- Carbon sandwich
- Fiberglass
- Aluminum
- Aluminum/steel/titanium pylons

Other 5%
Steel 10%
Titanium 15%
Aluminum 20%
Composites 50%

Civil engineering

Electronics

Sustainable energy

Industrial equipment

Pressure bombs

FIGURE 1.1 Carbon Materials Supporting Our Lives

Carbon materials are predominantly composed of carbon atoms, only the single element, but they have widely diverse structures and properties. Diamond has a three-dimensional structure, graphite has a two-dimensional nature, while carbon nanotubes are one-dimensional, and buckminsterfullerene, C_{60}, is zero-dimensional. Fullerenes behave as molecules, although other carbon materials do not. Graphite is an electrical conductor and its conductivity is strongly enhanced by AsF_5 intercalation, becoming almost comparable to that of metallic copper, whereas diamond is completely insulating. Diamond, the hardest material, is used for cutting tools, and graphite is so soft that it can be used as a lubricant.

1.1 Classification of carbon materials

Classification of carbon materials has been done on various different bases; for example, the chemical nature of the carbon-carbon bonds (structures), the production procedure, structural change at high temperatures, nanotexture, and time of

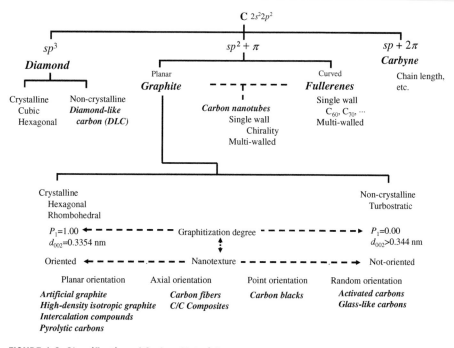

FIGURE 1.2 Classification of Carbon Materials

appearance [1,2]. Carbon materials have been named from different viewpoints. Structurally, they can be divided into diamond, graphite, fullerenes, carbynes, glass-like carbons, etc., and by production procedure into artificial graphite, intercalation compounds, activated carbons, carbon/carbon composites, etc. In Figure 1.2, some representative carbon materials are shown. Those in bold italics are listed in relation to carbon families classified on the basis of carbon-carbon bonds, together with some information on the diversity in each family.

Based on the nature of the carbon-carbon bonding, four carbon families have been defined and named after representative carbon materials: the diamond family constructed by C-C bonds based on sp^3 orbitals, the graphite family with bonds based on planar sp^2 orbitals, fullerenes on curved sp^2 orbitals, and carbynes on sp orbitals. In the graphite and fullerene families, the two π electrons per one carbon atom have a pronounced influence on the properties.

Commonly known carbon materials are shown in Figure 1.2. Most carbon materials that have been produced on an industrial scale belong to the graphite family. The fundamental structural unit is a stack of layers of carbon hexagons, i.e. graphite-like hexagonal carbon layers. These hexagonal carbon layers have strong anisotropy because of strong covalent bonding due to sp^2 orbitals in the layer, but weak bonding of van der Waals force between π electron clouds of stacked layers.

When these layers are large enough, they stack with specific regularity: ABAB stacking results in graphite crystals belonging to the hexagonal crystal system [3] and ABCABC stacking results in graphite crystals belonging to the rhombohedral crystal system [4]; the latter crystals being in metastable phase under atmospheric pressure. When the layers are too small, there is no stacking regularity even though they stack in parallel. The structure, where layers are just stacked without regularity, cannot be called graphite and has been named *turbostratic* [5]. It is known to be common in various layered compounds, such as clays. Turbostratic stacking is also metastable under normal conditions, around room temperature under atmospheric pressure, and it is thought to be stabilized by the presence of hydrogen and other foreign atoms, which bond to carbon atoms located at the edges of the layers and also by dangling bonds with neighboring layers.

The spacing between two layers that stack without regularity (turbostratic stacking), is larger than that of graphitic stacking. Interlayer spacing of graphite crystals has been accurately determined to be 0.3354 nm [6]. Interlayer spacing of turbostratic stacking is reported to be 0.344 nm [7], but now it is understood not to be a unique value, because of the presence of foreign atoms at the edges of layers. A turbostratic structure is commonly observed after the carbonization of many carbon precursors, because carbonization occurs at temperatures as low as 700–1300 °C. By heat treatment at higher temperatures, carbon layers grow in both directions, parallel to the layers (increase in the crystallite size La) and perpendicular to the layers (increase in the thickness of parallel stacking, which is measured as crystallite size Lc). This crystallite growth was known to depend strongly on carbons prepared at low temperatures, and it was proposed to classify carbons into two groups, graphitizing and non-graphitizing carbons [7], later named graphitizable and non-graphitizable carbons [8] or soft and hard carbons [9]. However, this classification into graphitizing and non-graphitizing is not critical, mainly because carbon materials with a variety of nanotextures have been developed and they show very different behaviors under high-temperature treatment.

During crystallite growth, the average interlayer spacing, usually measured by X-ray powder diffraction analysis, increases gradually with increase in heat treatment temperature, so that graphitic stacking with the spacing of 0.3354 nm occurs randomly in the crystallite with turbostratic stacking, as shown schematically in Figure 1.3 for a certain moment during heat treatment.

The development of graphitic stacking (graphitization degree) can be determined exactly by the probability of incidence in neighboring layers P_1, which can be done by Fourier analysis of the X-ray powder pattern [10]. Since the procedure to determine P_1 is not simple, however, the average interlayer spacing, d_{002}, is often used as a convenient parameter for the graphitization degree, which can be measured from the diffraction angle of $00l$ diffraction lines. In Figure 1.4, change in the average d_{002} with heat treatment temperature is shown for different carbon materials, demonstrating how widely the behavior of carbon materials varies.

From measured d_{002}, different parameters, such as p and g, have been proposed to evaluate the graphitization degree, by making various assumptions [7,11,12]. However, the value d_{002} itself is now used as a measure of graphitization degree, often coupled with the crystallite sizes of Lc, La, and $Lc(112)$. The specifications for

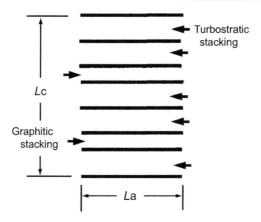

FIGURE 1.3 Structural Model of a Partially-graphitized Carbon

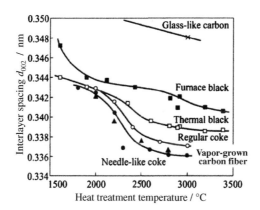

FIGURE 1.4 Changes in the Interlayer Spacing, d_{002}, with Heat Treatment Temperature for Various Carbon Materials

the analysis of these X-ray diffraction parameters, d_{002}, c_0 ($=2d_{002}$), a_0, Lc, La, and $Lc(112)$, have been proposed in the international journal *Carbon* [13]. The relationship between d_{002} and P_1 was reported to be linear [10,14], but it is now known not to be a simple linear relationship, as shown in Figure 1.5, on the basis of the measurement of various carbon materials [13].

1.2 Nanotexture of carbon materials

The preferred orientation of the anisotropic carbon layers in a carbon material is known to be important to understand the properties and structural changes at high temperatures. Based on a scheme of preferred orientation of carbon layers, the nanotexture of carbon materials has been used to classify them into two groups, random

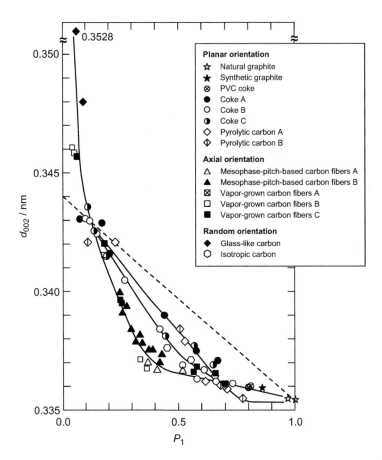

FIGURE 1.5 The Relationship Between Interlayer Spacing, d_{002}, and Degree of Graphitization, P_1

From [13]

and oriented, and the latter into a further three: planar, axial, and point orientations [15,16]. This classification is shown in Figure 1.6, together with some examples of carbon materials for each nanotexture.

Nanotexture with a planar orientation scheme can be represented by graphite, flaky natural graphite, kish graphite, and highly oriented pyrolytic graphite (HOPG), in which the orientation degree is almost 100%. In pyrolytic carbons, which have been prepared by chemical vapor deposition (CVD) of gaseous hydrocarbon precursors, such as methane, propane, and benzene, various degrees of planar orientation are obtained, depending on the preparation conditions and also on the heat treatment temperature. In cokes, particularly needle-like coke, planar and axial orientation schemes are mixed and the orientation degree depends strongly on the heat treatment temperature. In the axial orientation scheme, co-axial and radial modes have to be differentiated. The extreme of co-axial orientation (orientation degree of 100%) is

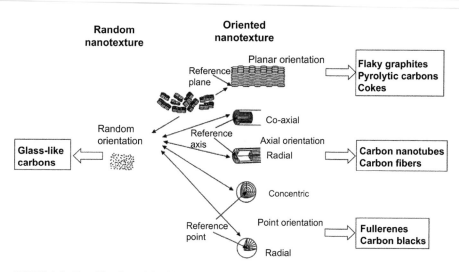

FIGURE 1.6 Classification of the Nanotexture of Carbon Materials, with Examples

found in carbon nanotubes (CNTs). Both co-axial and radial modes, in addition to various degrees of axial orientation, can be found in the cross-section of carbon fibers. The co-axial mode associated with a central hollow tube is found in vapor-grown carbon fibers. In mesophase-pitch-based carbon fibers, both co-axial and radial modes, and even intermediate modes, are formed. Isotropic-pitch-based carbon fibers, however, have a uniquely random orientation. Polyacrylonitrite (PAN) -based and phenol-based carbon fibers have random orientation along the fiber axis, as well as in the cross-section. In the point orientation scheme, concentric and radial modes are found. All fullerene molecules are the extremes of concentric point orientation. In carbon blacks, various degrees of concentric point orientation are found, depending on the size of primary particles and the heat treatment temperature. Radial point orientation is realized in carbon spherules, which have been synthesized from the mixture of polyethylene and poly(vinyl chloride) under pressure [17]. In mesocarbon microbeads (MCMBs), which are mesophase spheres separated from the isotropic pitch matrix, radial point orientation can be found near the surface, but not at the center [18]. Even in random orientation schemes, the fundamental structure is based on hexagonal carbon layers, though they form minute closed shells as found in glass-like carbons.

These nanotextures depend strongly on the precursors; in general, thermoplastic precursors tending to give oriented nanotexture but thermosetting precursors random nanotexture. However, carbonization conditions, not only temperature and residence time but also heating rate, pressure, etc., have an additional influence on the resultant nanotextures. Carbonization under high pressure is known to give a specific nanotexture, different from that obtained under normal pressure [19].

These nanotextures established in the process of carbonization govern the structural change at high temperatures from 1500 to 3000 °C, as shown in Figure 1.4. In carbons where the nanotexture consists of planar orientation, such as needle-like coke, the transformation of turbostratic stacking to graphitic stacking and the crystallite growth both parallel and perpendicular to the carbon layers occur easily; in other words, graphitization proceeds more easily with increasing temperature, and can reach high graphitization degrees. In carbons with nanotextures of a random orientation scheme, on the other hand, the development of graphite structure is strongly depressed and limited to a low graphitization degree even at a temperature as high as 3000 °C, as in glass-like carbon. Carbons with point orientation show intermediate behavior, depending on the size of the primary particles (Figure 1.4); thermal black having a diameter of a few hundreds of nanometers and furnace black about 30 nm size. Vapor-grown carbon fibers with axial orientation show very similar behavior to needle-like coke (Figure 1.4).

In order to overcome the depression in the development of graphite structure due to the nanotexture established through carbonization, heat treatment under severe conditions, much more severe than necessary to modify the nanotexture during carbonization, is known to be required, in such a way as to melt at a high temperature above 3400 °C under moderate pressure [20] or to heat-treat at a temperature above 1600 °C under high pressure at 0.3–1 GPa [21].

1.3 Microtexture of carbon materials

Most particles with planar and axial orientation, such as cokes, carbon fibers, and carbon nanotubes, are also anisotropic and by agglomeration they can create further variety in texture. In order to understand the properties of different carbon materials, therefore, it is necessary to take into consideration the texture formed by the preferred orientation of these anisotropic particles, in addition to the nanotexture and graphitization degree of each particle. The texture due to the preferred orientation of anisotropic particles may be called the *microtexture*, because the particles are often of micrometer or millimeter size. The microtexture is usually created during the forming process of bulky carbon materials. In large-sized graphite electrodes for metal refining, for example, the particles of needle-like coke tend to be oriented along the rod axis during the forming process by extrusion with pitch binder. To prepare carbon-fiber-reinforced plastics (composites), different microtextures based on the orientation of carbon fibers have been applied in order to get high strength and high modulus of the composites. Some examples are shown in Figure 1.7.

Two methods have been employed for realizing the isotropy of the bulk of carbon materials that fundamentally consist of anisotropic structural units or crystallites: (1) the random aggregation of micrometer- or millimeter-sized particles, even though those particles are anisotropic, and (2) the random agglomeration of nanometer-sized crystallites in the bulk. The former is realized in so-called isotropic high-density graphite blocks, where small coke particles are formed by using binder pitch under isostatic pressing. The latter, random aggregation of nanometer-sized carbon layers, has been realized in glass-like carbons, which are isotropic and have a non-graphitizing nature.

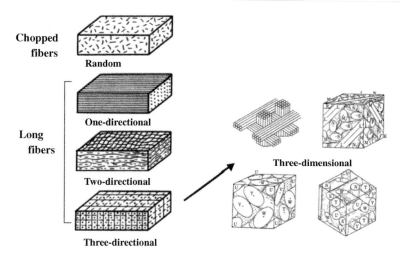

FIGURE 1.7 Microtextures Formed by Carbon Fibers

FIGURE 1.8 Pores Observed in Isotropic High-density Graphite Blocks. Bulk density increasing from A to F

From [22]

Pores in the formed carbons influence the bulk properties of the carbon materials. In such a case, the microtexture, including the shape and size of pores in addition to that due to the orientation of anisotropic particles, has to be taken into account. Optical microscopy images of polished sections of six isotropic high-density graphite blocks with different bulk densities from 1.735 to 1.848 g/cm^3 are shown in Figure 1.8 [22]. Although the difference in bulk density looks rather small, a marked difference is seen in the micrographs, showing different shapes, sizes, and distributions of pores in the cross-sections. Different pore parameters of these carbon materials, such as density, average cross-sectional area, roundness, fractal dimension, etc., have been determined with the help of image analysis. The mechanical properties, such as elastic modulus, bending strength, and fracture toughness (K_{Ic}), showed close dependence on the pore parameters.

1.4 Specification of carbon materials

Carbon materials are used in many different fields of science and engineering. They appear in various scientific and technological reports, describing their preparation and applications, not only as functional materials but also as parts of devices. In those cases, sufficient information on the carbon materials has to be presented. Commonly used names of carbon materials, such as coke, carbon fiber, and carbon nanotube, give *a priori* some information: the carbon family to which they belong and even the nanotexture they have. In many cases, however, more information is required to understand correctly the carbon materials used.

Carbon fibers are prepared from different precursors by varying processes, and consequently they have different nanotextures and structures. The precursor used enables estimation of the nanotexture of the resultant carbon fiber: PAN, isotropic-pitch, and phenol are known to give random nanotexture in the cross-section, but mesophase-pitch-based carbon fibers can have a wide range of nanotextures (straight radial, corrugate radial, concentric, etc.), and vapor-grown carbon fibers have concentric axial nanotexture. The heat treatment temperature of the carbon fiber is also important information for understanding the structure and properties of the fibers; vapor-grown carbon fibers can have the interlayer spacing, d_{002}, close to graphite (0.3354 nm) after heat treatment at 3000 °C, but isotropic-pitch-based carbon fibers are far from graphite values even after 3000 °C heat treatment, as shown in Figure 1.9A. Crystallite magnetoresistance $(\Delta\rho/\rho)_{cr}$ shows very marked difference among carbon fibers prepared from different precursors and under different conditions, as shown in Figure 1.9B. In addition, it has to be pointed out that the word "graphitized," which has been used, for example, in expressions like "graphitized carbon fibers," does not always mean the development of graphitic structure in the carbon fibers, but that a high temperature of 2500–3000 °C was applied. In order to specify the carbon fiber correctly, therefore, not only the precursor and heat treatment temperature but also some structural parameters, such as d_{002} or $(\Delta\rho/\rho)_{cr}$, and nanotexture in the cross-section have to be presented, particularly in the case of mesophase-pitch-based carbon fibers.

In the case of carbon nanotubes, the situation is more complicated, mainly because there is no consistent definition of carbon nanotube and also because many people want to use "carbon nanotube" in publications even without unambiguous identification of the structure and nanotexture. It is regrettable to say that the terms "carbon nanotube" and "graphene" have not, in many papers published, been used in a precise sense according to the definitions. The present authors have proposed that a carbon nanotube has to consist of straight layers, as shown by its 002 lattice fringes in the transmission electron microscopy (TEM) image in Figure 1.10A, as this is the reason why the name "carbon nanotube" was first proposed. According to this definition, the fibrous carbon consisting of small carbon layers preferentially oriented along its fiber axis, as in the example shown in Figure 1.10B, should not be called carbon nanotube. These two fibrous carbons have quite different properties, particularly their electronic, mechanical, and chemical properties. In reports where the term "carbon nanotube" is used, therefore, it is strongly recommended to show how the carbon layers are extended on the wall of tubes, by TEM imaging or other techniques.

Carbon blacks are named by the production process during gas-phase carbonization, such as thermal black, furnace black, acetylene black, which suggests to us some information on the size of primary particles and the aggregation (so-called *structure*): thermal black has large particle size and almost no structure, but furnace

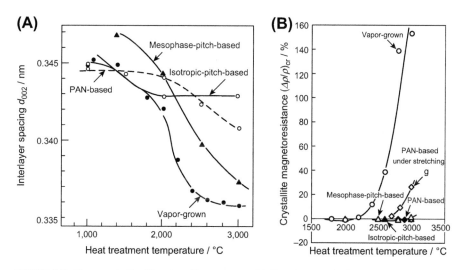

FIGURE 1.9 Changes in the Structure Parameters with Heat Treatment Temperature for Carbon Fibers

(A) Interlayer spacing, d_{002} and (B) crystallite magnetoresistance, $(\Delta\rho/\rho)_{cr}$.

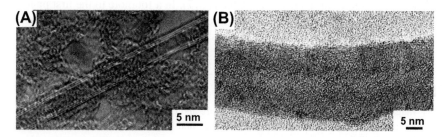

FIGURE 1.10 002 Lattice Fringe Images of Fibrous Carbons

(A) Carbon nanotube and (B) carbon nanofiber.

black has small particle size and well-developed structure. In furnace black, the size distribution of primary particles, ranging from 5 to 100 nm, and the degree of structure development are known to be very important for reinforcing rubber.

1.5 Construction of the present book

In 2006, the authors (M.I. and F.K.) published a book entitled *Carbon Materials Science and Engineering: From Fundamentals to Applications* from Tsinghua University Press. The book was written with the aim of giving the fundamentals in science and engineering on all carbon materials, in relation to their applications. The principal readers of the book were assumed to be beginners in carbon science and engineering, such as young researchers, scientists, and graduate students who were studying or intending to study carbon materials.

In the present book, advanced materials science and engineering of carbon materials are reviewed over 17 chapters, including this introduction. Chapters 2 to 10 are focused on the issues of formation and preparation of carbon materials, and Chapters 11 to 17 cover different applications. In Chapters 2 and 3, carbon nanotubes and graphene are reviewed, with emphasis on their formation. Since fundamental carbonization and graphitization processes were described in our previous book, processes containing additional and/or specific procedures and the formed carbon materials are reviewed in Chapters 4 to 10: carbonization under pressure (Chapter 4); graphitization under high pressure (Chapter 5); glass-like carbons, with special attention given to the activation and graphitization (Chapter 6); template carbonization to control morphology and pore structure (Chapter 7); carbon nanofibers prepared via electrospinning (Chapter 8); carbon foams creating new applications (Chapter 9); and nanoporous carbon membranes, including carbon-fiber webs (Chapter 10). In Chapters 11 to 17, carbon materials used in specific fields are reviewed: electrochemical capacitors (Chapter 11); lithium-ion rechargeable batteries (Chapter 12); photocatalysis (Chapter 13); spilled-oil recovery (Chapter 14); adsorption of hydrogen, methane, volatile organic compounds (VOCs), and metal ions (Chapter 15), highly-oriented and highly-crystallized graphite, with emphasis on its high thermal conductivity (Chapter 16); and isotropic high-density graphite, emphasizing its nuclear applications (Chapter 17). In a footnote to the first page of each chapter, the location in our previous book of fundamental information on each topic is indicated, as prerequisite for readers.

References

[1] Inagaki M. New Carbons. Control of Structure and Functions 2000:1–3, Elsevier.
[2] Inagaki M, Kang F. Carbon Materials Science and Engineering 2006:23–31, Tsinghua University Press.
[3] Bernal JD. Proc Roy Soc A 1924;106:749–73.
[4] Lipson H, Stokes AR. Nature 1942;149:328.

[5] Warren BE. J Chem Phys 1934;2:551–6.
[6] Nelson JB, Riley DP. Proc Phys Soc A 1945;57:477–85.
[7] Franklin RE. Acta Cryst 1951;4:253–61.
[8] International Committee for Characterization and Terminology of Carbon. Carbon 1982;20:445–9.
[9] Mrozowski S. Proceedings of the First and Second Conferences on Carbon. University of Buffalo: Waverly Press; 1956;31–45.
[10] Houska SR, Warren BE. J Appl Phys 1954;25:1503–10.
[11] Bacon GE. Acta Cryst 1951;4:558–61.
[12] Maire J, Mering J. Chem Phys Carbon 1979;6:125–90.
[13] Iwashita N, Inagaki M. Carbon 1993;31:1107–13.
[14] Noda T, Iwatsuki M, Inagaki M. TANSO 1966; No. 47: 14–22 [in Japanese].
[15] Inagaki M. TANSO 1985; No. 122: 114–122 [in Japanese].
[16] Inagaki M. New Carbon Mater 1999;14:1–3.
[17] Hishiyama Y, Yoshida A, Inagaki M. Carbon 1982;20:79–84.
[18] Augie D, Oberlin M, Oberlin A, et al. Carbon 1980;18:337–46.
[19] Inagaki M, Park KC, Endo M. New Carbon Mater 2010;25:409–20.
[20] Noda T, Inagaki M. Bull Chem Soc Jpn 1964;37:1709–10.
[21] Inagaki M, Meyer RA. Chem Phys Carbon 1999;26:149–244.
[22] Oshida K, Ekinaga N, Endo M, Inagaki M. TANSO 1996; No. 173: 142–147 [in Japanese].

Carbon Nanotubes: Synthesis and Formation

2

Carbon materials can be classified into three categories on the basis of their period of development: classic carbons, new carbons, and nanocarbons [1]. Classic carbons include synthetic graphite blocks mainly used as electrodes, carbon blacks, and activated carbons, for which production procedures were developed before the 1960s. In the 1960s, carbon materials different from these classic carbons were invented: carbon fibers from various precursors, including vapor-grown carbon fibers; pyrolytic carbons produced via chemical vapor deposition processes; glass-like carbons with high hardness and gas impermeability; high-density isotropic carbons produced by isostatic pressing; intercalation compounds with different functionalities, such as high electrical conductivity; and diamond-like carbons as transparent carbon sheets. These newly developed carbon materials are classified as new carbons. Since the 1990s, various fullerenes with closed-shell structure, carbon nanotubes with nanometer diameters, and graphene flakes of only a few atoms' thickness have attracted attention from nanotechnology; these are classified as nanocarbons.

If these carbon materials are considered from the point of view of their texture, however, they may be classified into two groups: nanotextured and nano-sized carbons [2]. Most carbon materials in the new carbon category are classified as nanotextured carbon, because their nanotexture is controlled via different processes in their production, in addition to the structural control. On the other hand, fullerenes, carbon nanotubes, and graphene can be classified as nano-sized carbon, because the shell size of fullerenes, diameter of carbon nanotubes, and thickness of graphene flakes are on the nanometer scale. Carbon blacks in classic carbon are composed of nano-sized particles, but they are not usually classified as nanocarbons because they have various applications as a mass, not as individual nano-sized particles.

These two classifications for carbon materials are summarized in Figure 2.1.

In this chapter, synthesis of carbon nanotubes (CNTs), representative nanocarbons, and their formation into yarns, sheets, and sponges are summarized. The formation of CNTs is very important for developing their applications in different engineering fields.

Prerequisite for readers: Chapters 2.3 (Nanotexture development in carbon materials) and 3.4 (Fibrous carbons) in *Carbon Materials Science and Engineering: From Fundamentals to Applications*, Tsinghua University Press.

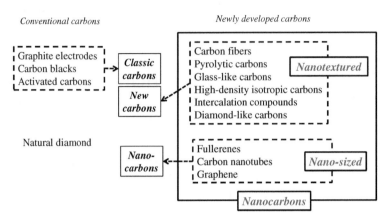

FIGURE 2.1 Classification of Carbon Materials

2.1 Synthesis of carbon nanotubes

Carbon nanotubes (CNTs) are synthesized at the first stage of growth of so-called vapor-grown carbon fibers [3–6], as shown in Figure 2.2, and are found in the carbon deposits on graphite anodes during arc discharge [7–10]. Under similar arc discharge conditions, fibrous carbon composed of a scroll of carbon layers was obtained in 1960, and was called "graphite whisker" [11].

Arc discharge between graphite electrodes forms carbon nanotubes together with other forms of amorphous carbon, such as carbon blacks and pyrolytic carbons, but with difficulty in controlling the structure of CNTs, i.e. thickness (single-, double-, or multi-walled), diameter, and length. In order to obtain CNTs of high purity, different techniques such as laser-abrasion, modified arc discharge, and catalytic chemical vapor deposition have been proposed [12]. Since catalytic chemical vapor deposition (CCVD) was thought to be practical for large-scale production, it has been applied using various carbon precursors and various catalytic metals, either supported on a substrate or floating as fine particles [13–21]. The so-called HiPco process, using the disproportionation reaction of high-pressure and high-temperature CO with iron pentacarbonyl as a catalyst precursor, is producing single-walled CNTs (SWCNTs) almost free from amorphous carbon [22–24]. Double-walled CNTs (DWCNTs) were successfully obtained in a high yield by the catalytic CVD method by placing a Mo catalyst on Al_2O_3 at the end of the furnace and an Fe catalyst on MgO at the center of the furnace [25,26]. It was thought that the Mo catalyst worked as the moderator for enhancing the active carbon species, and the Fe catalyst was the catalyst for nanotube formation. A CH_4/Ar gas mixture was fed onto the catalysts, heated at 875 °C for 10 min. The yield of DWCNTs was more than 95% after purification. Long DWCNTs have been synthesized with a relatively high production rate of 0.5 g/h in a quartz tube (45 mm in diameter and 180 cm in length) heated to a temperature of 1100 °C, by introducing a xylene solution of ferrocene with a small amount of sulfur at a rate of 0.05–0.15 cm^3/min under a flow of argon and hydrogen at rates of 2500–3500 and 500 cm^3/min, respectively [27].

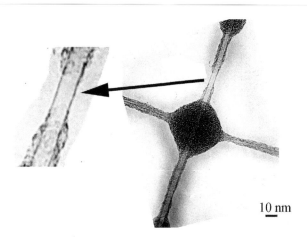

10 nm

FIGURE 2.2 Single-walled Carbon Nanotube Formed at the First Stage of Growth of Vapor-grown Carbon Fibers

Courtesy of Prof. M. Endo of Shinshu University, Japan

Although the yield and growth rate of CNTs with uniform structure, either single-, double-, or multi-walled, were markedly improved by using catalysts under controlled conditions, the resultant CNTs usually contained the minute metal particles used as a catalyst at their tips. Therefore, opening of the tips by oxidation and removal of the metal particles through either dissolution by acid or evaporation as halides were essential for getting chemically pure CNTs.

CNTs aligned in parallel with each other have been prepared from different hydrocarbon gases on different metal catalysts. By CVD of 2-amino-4,6-dichloro-*s*-triazine at 950 °C on a Co film (10–100 nm thick) deposited on a silica substrate by laser etching, well-aligned CNTs were synthesized, of lengths up to about 50 μm and fairly uniform diameters (30–50 nm); these were produced in high yield without noticeable amounts of other forms of carbon [28]. CVD of acetylene on iron/silica substrates produced CNTs (20–40 nm diameter) aligned perpendicularly to the substrate surfaces [29]. The lengths of CNTs increased with growth time, and reached about 2 mm after 48 h. Well-aligned CNTs over areas up to several square centimeters were grown on nickel-coated glass at 666 °C by plasma-enhanced hot-filament CVD of acetylene containing NH_3 [30]. NH_3 gas was thought to etch the surface of the nickel catalyst. Nanotubes with controllable diameters from 20 to 400 nm and lengths from 0.1 to 50 μm were obtained.

CNTs aligned perpendicularly to the substrate (called "arrays") with well-defined patterns were synthesized by CVD at 700 °C under a flow of ethylene at 1000 standard cm^3/min (SCCM) on a P-doped porous Si(100) wafer, on which an Fe thin film (5 nm thick) was patterned by electron beam evaporation, as shown in Figure 2.3 [31]. The resultant CNTs were multi-walled and had diameters of 16 ± 2 nm. Growth rate of CNTs was very high; the length of CNTs, i.e. the height of the arrays, reached

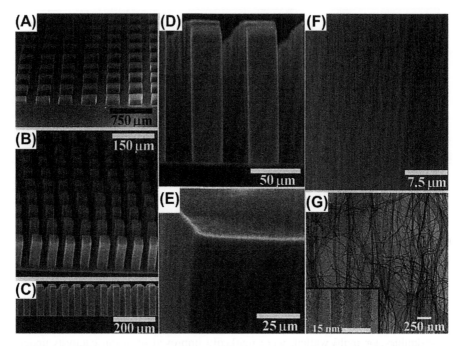

FIGURE 2.3 Carbon Nanotube Arrays. (A-F) SEM images with different magnifications, (G) TEM images of carbon nanotubes in the array.

From [31]

30 and 240 μm after 5 and 60 min CVD, respectively. In the arrays, CNTs aligned almost perfectly perpendicularly to the substrate, as shown in Figure 2.3F, although CNTs formed on a porous Si wafer without an Fe film were not aligned (Figure 2.3G). The CNT arrays can be formed with a well-defined pattern (Figures 2.3A–C), of which the edges and corners are very sharp (Figures 2.3D and E). Instant addition of inert gas (Ar) into the supply of the precursor gas (acetylene), as shown in Figure 2.4D, was found to give a distinct straight line on the scanning electron microscopy (SEM) image of the as-grown CNT array (Figure 2.4E), and so the growth kinetics and mechanism of the multi-walled CNT (MWCNT) array were discussed with relation to these marks [32]. Growth rate of CNTs by CVD of acetylene was almost constant at 600 °C, c. 3 μm/min, but at 680 °C it increased from 17 to 23 μm/min during the first 15 min and then tended to be saturated. The growth process was well approximated by the first order reaction $E_a = 159 \pm 5$ kJ/mol. The diameter corresponding to its maximum population tended to shift to larger sizes with increasing thickness of Fe catalyst: 6.2 nm (3–4 walls) to 9.2 nm (6–9 walls) with increase in Fe thickness from 0.2 to 5.0 nm [33]. The presence of a buffer layer of Al_2O_3 or SiO_2 in between the substrate and catalyst layer was essential for the synthesis of CNT arrays, no formation of CNT array being observed on a flexible stainless steel foil either with or without a thin film of iron [34]. Preheating of precursor gas (C_2H_4/H_2)

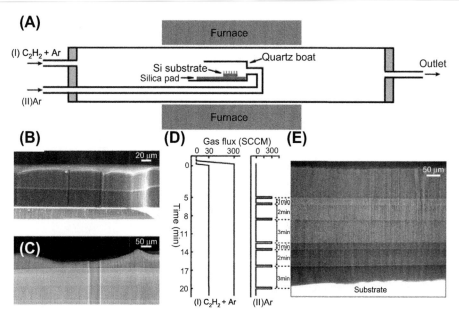

FIGURE 2.4 Determination of Growth Rate of Carbon Nanotubes Forming in an Array.
(A) Illustration of experimental setup, (B) SEM image of a CNT array grown on a silicon substrate, with white line in the middledue to turning off C_2H_2 inlet, (C) SEM image of a CNT array with a series of line marks, (D) plot of gas flux versus time during CNT growth, (E) image of CNT array captured by a digital camera on an optical microscope.

From [32]

to 770 °C lowered the deposition temperature for CNT arrays to 500 °C, but did not appreciably change the diameters of the resultant CNTs [35,36].

Carbon nanofibers with 10–140 nm diameters have been synthesized by using H_2/ methane gas and a 0.5–5 nm thick Ni or Fe catalyst film through low-power microwave plasma-assisted CVD [37].

Accelerated growth of SWCNTs and DWCNTs can be achieved by either using an alcohol as the carbon precursor or adding water vapor to the precursor gas [38–42]. High-purity SWCNTs with a diameter of about 1 nm have been synthesized from alcohol by using Y-type zeolite powder loaded with Fe/Co nanoparticles; SWCNTs were formed at 550 °C from methanol vapor [38]. Sheets of vertically aligned SWCNTs with a few micrometers' length have been grown by CVD of ethanol vapor by using Co-Mo catalyst particles of about 1.0–2.0 nm size, which were dispersed densely on a quartz substrate by a dip-coating method [40]. Continuous reduction of catalysts with Ar/H_2 (3% H_2) during CVD was essential for generating dense SWCNTs with vertical alignment. The diameter of SWCNTs decreased with increasing acetonitrile content in ethanol, showing a drastic decrease in diameter from 2.1 to 0.7 nm by the addition of 0.1% acetonitrile, although the nitrogen content saturated at about 1 at%, as shown in Figure 2.5 [42].

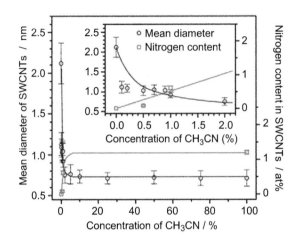

FIGURE 2.5 Changes in Diameter and Nitrogen Content of SWCNTs with Concentration of Acetonitrile in Ethanol

The insert is an enlarged presentation for a region of low CH_3CN concentration.

From [42]

FIGURE 2.6 Single-walled Carbon Nanotube Forest

From [43]

Vertically aligned SWCNTs (called "forests") have been grown from ethylene on various metal catalysts in either Ar/H_2 or He/H_2 containing a small amount of water vapor, as shown in Figure 2.6 [43]; this process was called water-assisted growth because of the acceleration by water vapor, and also named "supergrowth" because of the superior growth rate of CNTs. Dense and vertically aligned SWCNTs with 2.5 mm length could be grown in 10 min by using thin catalyst films of Al_2O_3 and Fe sputtered on a Si wafer. The ratio of ethylene and water was crucial for the production. The as-grown SWCNT forest had a very low content of metal catalyst, less than 2×10^{-5} mass%, and was free from both amorphous carbon and multi-walled carbon nanotubes. The SWCNT forest obtained could be easily removed from the substrate

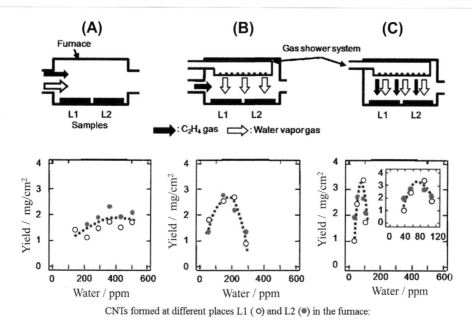

CNTs formed at different places L1 (○) and L2 (●) in the furnace:

FIGURE 2.7 Yield of Single-walled Carbon Nanotubes Depending on the Water Vapor Concentration and Method of Gas Delivery

(A) Both gases injected from the side, (B) ethylene from the side and water vapor from the top, and (C) both gases from the top.

From [47]

with a razor blade and the substrate could be re-used for the growth of CNT forests. By using a lithographically patterned catalyst on the substrate, a well-defined vertically standing structure of SWCNT forests was produced. In the forest, SWCNTs with a diameter of 3.0 ± 0.07 nm were formed, occupying the surface area of the catalyst by about 84 mass%, showing efficient usage of catalyst surface [44].

The structure of CNTs and the characteristics of their forests synthesized via the supergrowth process were explored in detail. High growth rate of CNTs was obtained by using an Fe catalyst layer on an Al_2O_3 buffer layer; it was possible to obtain an SWCNT forest with a height of 500–1000 μm after 10 min [45,46]. The diameter of CNTs grown by this process depended linearly on the thickness of the Fe catalyst layer; with Fe thicknesses of 1.6, 3, and 5 nm, SWCNTs, DWCNTs, and MWCNTs, respectively, were formed. The method of gas delivery to the catalyst and the water vapor content in the gas have a strong influence on the yield of SWCNTs as forests [47]. As shown in Figure 2.7, three different ways of supplying carbon precursor gas (ethylene) and water vapor were tested with different water contents. By supplying both gases from the top (Figure 2.7C), a relatively high yield, more than 3 mg/cm², was obtained, with a low water content of about 80 ppm. The length of SWCNTs, i.e. the height of the forest, increased with time when both gases were supplied from the top, more than 3 μm after 80 min, although growth of SWCNTs was terminated at around 1 μm after 20 min when the two gases were supplied from the side (Figure 2.7A) and when a water supply from the top (Figure 2.7B) was employed.

The formation of patterned CNT forests (or arrays, i.e. vertically aligned CNTs) has been reviewed [48].

2.2 Formation of carbon nanotubes

2.2.1 Formation into yarns

For many applications, ultrathin CNTs are required to be much thicker yarns (called sometime fibers or strands) to make their handling easier. The formation of CNTs into a yarn has been done by three methods: (1) in situ spinning in a CVD furnace, (2) spinning from the suspension of as-prepared CNTs after purification, and (3) spinning from a CNT forest.

SWCNTs synthesized by laser a vaporization technique were known to be bundled into a rope with a diameter of 5–20 nm [12,49]. Long and thick SWCNTs yarns have been synthesized by CVD of either benzene or n-hexane with floating catalyst particles formed from ferrocene and thiophene at 1100–1200 °C [18,50,51]. By either inserting a rod or winding up in the lower part of the furnace, as schematically shown in Figures 2.8A and 2.8B, yarns can be prepared from SWCNTs synthesized in the hot zone by the floating catalyst method; an example of the twisted yarn obtained being shown in Figure 2.8C [52]. Concentration of additive (thiophene), flow rate of carrier gas (H_2), and CVD temperature governed the structure of CNTs, whether

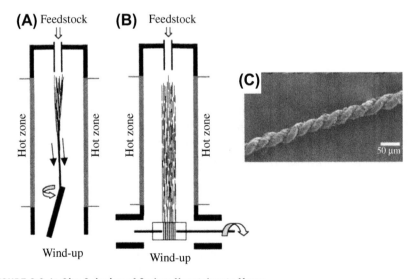

FIGURE 2.8 In Situ Spinning of Carbon Nanotubes to Yarns

(A and B) Schematic illustrations, and (C) spun yarn.

From [52]

multi-walled or single-walled. The rate of winding-up (5.5–20 m/min) influenced the density and strength of the resultant yarns owing to the alignment of CNTs, which improved with a higher rate [53]. Yarns prepared from hexane had a high tensile strength (1.46 GPa) and a high Young's modulus (30 GPa), higher than those from ethanol and ethylene glycol [54].

Long yarns, the authors claiming of several kilometers, have been prepared by assembling multiple layers of CNTs concentrically, which were produced by CVD of mixed gases of acetone and ethanol with ferrocene and thiophene in flowing hydrogen, followed by spinning and water-densification [55]. The CNT yield from the mixed gases was about 240 mg/h, while that from ethanol alone was about 110 mg/h. The stable spinning could be conducted with a velocity in the range of 5 to 20 m/min.

CNTs synthesized through arc-discharge and laser-ablation methods have been formed into yarns via their suspension. By immersing a graphitized carbon fiber (CF) bundle into a SWCNT/DMF (N,N-dimethylformamide) suspension and applying a voltage of 1–2 V between the CF and the suspension, a cylindrical cloud of the SWCNT was formed around the positive CF, resulting in a yarn with 2–10 μm diameter by slow withdrawal from the suspension (electrophoretic drawing) [56]. The diameter of the yarns depended on the time allowed for the SWCNT cloud assembly before withdrawal and the rate of withdrawal. SWCNT yarns with a diameter of 16–42 μm and length of more than 3 cm have been obtained by a withdrawal rate of 0.85 μm/s [57]. SWCNT yarns have been prepared by injecting a SWCNT suspension in a sodium dodecyl sulfate (SDS) aqueous solution into a PVA (poly(vinyl alcohol)) aqueous solution through a syringe, followed by taking out the yarn in gel, washing out the PVA and SDS with water, and drying [58–60]. Drying the gel yarns under tension was effective to improve the alignment of CNTs and Young's modulus [59]. Spinning from CNT suspension in SDS solution was performed by injecting into either an ethanol/glycerol or an ethanol/glycol mixture and then washed out the remaining SDS and glycerol or glycol with water and ethanol to get yarns of pure CNTs [61]. SWCNT yarns have also been prepared by adding a SWCNT suspension of lithium dodecyl sulfate aqueous solution into HCl aqueous solution at a rate of 0.25 cm^3/min under rotation at 33 rpm, followed by excluding the HCl in methanol [62]. SWCNTs suspended in concentrated sulfuric acid have been formed into yarns via a similar procedure [63,64]. Alignment of CNTs in the yarns is improved by using a smaller nozzle [63]. The resultant yarns had a bulk density of 0.87–1.11 g/cm^3 [64]. Nematic liquid crystal solution prepared by suspending <0.5 mass% SWCNTs in 102% sulfuric acid was able to give needle-like yarns with a length of c. 20 mm after addition of a small amount of water [65]. In the yarn, SWCNTs were aligned well.

From forests where CNTs were vertically aligned, long yarns or ribbons have been continuously spun at room temperature [66]. In Figure 2.9, a yarn 30 cm long and 200 μm wide is spun from an MWCNT forest 100 μm high (Figures 2.9A and 2.9B); the alignment of CNTs in the forest (Figure 2.9C) and spun yarn (Figure 2.9D) are also shown. A yarn of about 10 m length could be spun from a forest with an area of 1 cm^2. Passing through a volatile liquid was effective to densify the spun yarns [67]. Introduction of twist during spinning was also effective to improve the

(A)

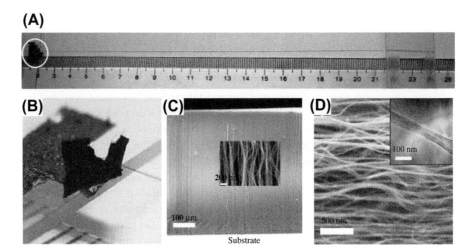

FIGURE 2.9 Spinning of Carbon Nanotube Yarn From a Forest

(A) Yarn of 30 cm length and 200 μm width, spun from (B) an MWCNT forest 100 μm high. (C) Alignment of the CNTs in the forest, (D) alignment of CNTs in the spun yarn.

From [66].

mechanical properties of the resultant yarns [68–71]. As-spun yarn with 4 and 13 μm diameter had tensile strength of 0.85 and 0.17 GPa, respectively, but after twisting the diameter of the yarn decreased to 3 and 10 μm and the tensile strength increased to 1.91 and 0.41 GPa, respectively [69]. Application of DC voltage of 70 V for 3 h under high vacuum increased the electrical conductivity of the yarn by 13%, as well as its tensile strength [66]. By applying tension to a CNT web, which had been prepared from a forest by compression and then twisted, CNT yarns with high tensile strength were prepared [72].

2.2.2 Formation into sheets

The formation of CNTs into a sheet (film) has been performed in various ways, which may be classified into three methods: (1) filtration of a CNT suspension, (2) control of CVD condition, and (3) formation of the CNT forest.

CNTs prepared through the arc-discharge, laser abrasion, and catalytic CVD methods were suspended in a solution and then recovered on a filter as a thin sheet (called "buckypaper") [25,26,73]. A buckypaper with a thickness of about 30 μm is shown in Figure 2.10A, composed of high-purity DWCNTs fabricated through the catalytic CVD method (Figure 2.10D), oriented with their tube axes parallel to the sheet (Figure 2.10C), but randomly within the sheet (Figure 2.10B) [25,26]. This process can be further classified into three on the basis of how the stable suspension of CNTs is accomplished: without any pretreatment, with modification of the CNT surface in advance, and using a surfactant.

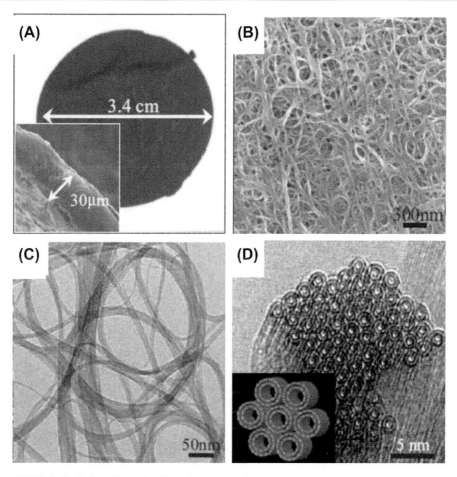

FIGURE 2.10 Carbon Nanotube Sheet (Buckypaper)

(A and B) Appearance of the sheet, (C and D) transmission electron microscopy images of constituent double-walled carbon nanotubes, with a model.

Courtesy of Prof. M. Endo of Shinshu University, Japan

    Suspension of CNTs without any treatment can be done only at very low concentration, of 2–3 mg/L [74,75]. By cutting CNTs into lengths of 0.5–2 μm through oxidation in H_2SO_4/HNO_3 mixed acid, washing, and annealing at 200 °C, a suspension of 1 g/L was possible [76]. Applying a magnetic field parallel to the filter surface was tried to orientate CNTs [77–79]. An SWCNT sheet could be produced with a relatively high bulk density by applying a magnetic field of 25 T; a sheet with thickness of 1.3 μm was 1.33 g/cm^3 and one of 7 μm was 1.21 g/cm^3 [77]. During filtration of an SWCNT-suspending solution of dimethylformamide under a 25 T field, dense, thin ribbons were obtained [78]. Also, by applying an electric field of 5 MHz at 2–10

V (electrophoresis), an oriented CNT sheet was prepared [80–83]. A spin-coating technique was successfully applied to coat a substrate with a CNT sheet, the surface of the substrate being modified by either amino- or phenyl-terminated silane [84,85].

Oxidation treatment of a CNT surface in order to obtain their stable suspension has often been carried out by using H_2SO_4/HNO_3 mixed acid [86–89]. From an SWCNT suspension of chloroform after oxidation in H_2O_2/H_2SO_4, the sheet was prepared by the repetition of pulling-up the substrate by the Langmuir-Blodgett method [90]. On the interface between an SWCNT/oleum suspension and air, a thin CNT sheet was formed upon absorbing water vapor in air [91]. Modification of a CNT surface by DNA [92,93], tetraoctylammonium bromide [94], and thionyl [95] have also been proposed.

Various surfactants have been employed to suspend CNTs in a solution: Triton X-100 [96–104], sodium dodecyl sulfate (SDS) [105–112], and sodium dodecylbenzene sulfonate (SDBS) [113,114]. A CNT suspension with about 1 g/L was easily prepared. From an SWCNT/X-100 suspension, transparent, electrically conductive CNT sheets (50–150 nm thick) have been prepared through filtrating under vacuum, washing, and dissolving out the filter [98]; transparency of the CNT sheet being comparable to Indium tin oxide (ITO) sheet. Transparent and flexible CNT sheets have been prepared by dipping a polyethylene terephthalate (PET) substrate into an SWCNT/X-100 suspension after repetitive washing under ultrasonic vibration [99]. The optoelectronic performance of conductive transparent coatings of different SW-, DW-, and MWCNTs on glass substrates depended strongly on the length, diameter, chirality, wall number, structural defects, and metallic-to-semiconducting tube ratio of the CNTs and also on the properties of the substrates [115].

Sheets composed of well-aligned MWCNTs were pulled out from the side face of a CNT forest, as shown in Figure 2.11 [116]. A strong sheet with 3.4 cm width could be pulled out from the forest (Figure 2.11A) with a rate as high as 1 m/min, of which side views are shown in Figures 2.11B and 2.11C. The sheets are transparent, as shown in Figure 2.11A where a logo mark under the sheet can be seen, and in Figure 2.11D where four overlapping sheets with different directions can be easily recognized. The thickness of the sheets depended on the height of the forest: a sheet about 18 μm thick was obtained from a forest 247 μm high. The bulk density of the sheet was very low at 0.0015 g/cm^3, but by passing through a vaporizable liquid it increased to 0.5 g/cm^3. Densification of CNT sheets by passing through a volatile liquid was found to be effective in order to improve various physical properties of the sheets [117]. The mechanism of the sheet formation has been discussed [118]. On MWCNT sheets drawn from forests of different diameters of CNTs prepared by changing the thickness of the Fe catalyst sheet, electrical resistance and optical transmittance have been measured [33].

SWCNT forests synthesized through water-assisted CVD [43] have been densified by adding a drop of liquid, such as water, alcohol, acetone, hexane, or liquid nitrogen, and called "SWNT solid" [119]. The forest, with a bulk density of 0.029 g/cm^3, was densified to 0.57 g/cm^3 in two steps, by surface tension after the addition of the liquid and van der Waals forces during vaporization of the liquid. Overhead and side

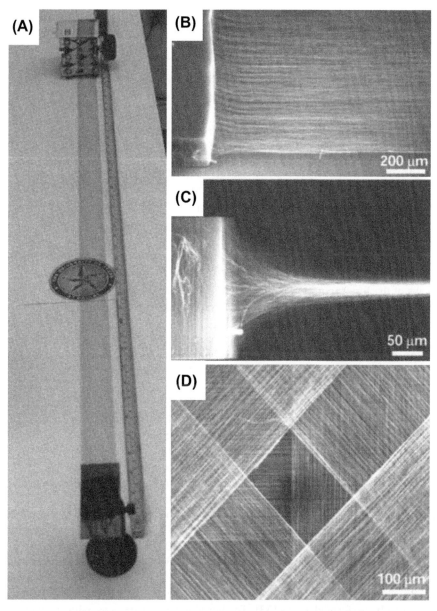

FIGURE 2.11 Formation of a Transparent Carbon Nanotube Sheet from a Forest. (A) MWCNT sheet (3.4 cm wide and meter long) pulled out of the forest, **(B)** SEM image at the pulling site, **(C)** SEM image of pulling with 90° rotation, **(D)** SEM image of the sheet overlaying four CNT sheets with 45° rotation

From [116]

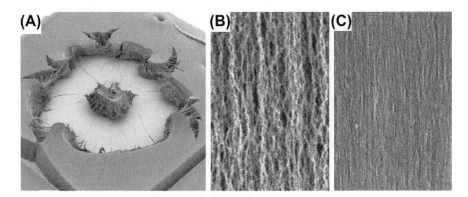

FIGURE 2.12 Shrinkage of a Carbon Nanotube Forest

(A) After shrinkage, (B and C) SEM images of side view before and after shrinkage.

From [119]

views of the forest, composed of SWCNTs with about 2.8 nm diameter, are shown in Figure 2.12A, and SWCNT alignment before and after densification is shown in Figures 2.12B and 2.12C, respectively.

By applying a water-assisted CVD process, it was possible to form a forest consisting of more than 95% SWCNT on conducting metal foils of a wide composition range of Ni-Cr-Fe alloys [120]. On Ni, Fe, and Fe/Cr alloy, MWCNTs were formed together with SWCNTs.

SWCNTs have been synthesized without using any metal catalyst through CVD of methane at 900 °C on a SiO_2 film deposited on a Si or Si/SiO_2 substrate [121]. The formation of SWCNTs was confirmed by radial breathing mode (RBM) of Raman spectrum and transmission electron microscopy (TEM). The patterned growth of SWCNTs was shown to be possible by scratching the Si/SiO_2 substrate.

Horizontally aligned SWCNTs have been uniformly grown on various crystallographic planes of sapphire (α-Al_2O_3) [122,123] and on quartz (SiO_2) [124–127]. Oriented growth of SWCNTs along [1 10 1] direction of sapphire crystal was observed by using ferritin as a catalyst [123]. Films of long SWCNTs have been prepared on quartz using a Cu catalyst by a CVD process at 900 °C with a flow of H_2/Ar gas through an ethanol bubbler, as shown in Figure 2.13 [125]. SWCNTs thus obtained were semiconducting (>95%) [126].

By controlling gas flow in a CVD furnace, oriented growth of CNTs has been successfully performed [128–133]. By introducing CH_4/H_2 gas with a relatively high flow rate (800/700 standard (SCCM)) onto Si wafers with mono-dispersed Cu nanoparticles, horizontally-oriented dense SWCNT films were prepared at 925 °C [128]. By using an extremely low feeding rate of methane at 1.5 SCCM, horizontally aligned SWCNTs with diameter about 2 nm were formed on a SiO_2/Si substrate deposited with Fe-Mo catalyst nanoparticles [130]. Cross-orientation of SWCNTs has been carried out through two-step CVD by using a Cu catalyst [131].

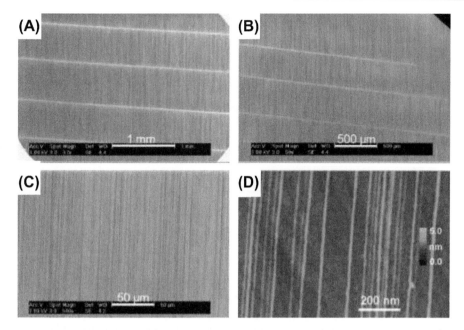

FIGURE 2.13 Horizontally Aligned, High-density SWCNTs Grown on a Quartz Substrate

(A and B) SEM images of the film, the bright stripes corresponding to the Cu catalyst, (C) high-magnification SEM, and (D) atomic force microscopy (AFM). image of a $1 \times 0.75 \ \mu m^2$ area.

From [125]

SWCNTs synthesized by the arc-discharge process have been formed into films by passing through a magnetic field of 0.56 T on different substrates, such as glass, Si, poly(methyl methacrylate (PMMA), and PET [134]. SWCNTs deposited randomly on a SiO_2/Si substrate have been transferred to another substrate by mechanical force with shear stress to create well-oriented, dense films [135].

The formation of CNT sheets has been reviewed [136,137].

2.2.3 Formation into sponges

Agglomerates of MWCNTs have been prepared by compression, with bulk density of 0.4–0.5 g/cm^3 and excellent resilience up to 250 MPa [138]. By controlling the injection rate of the source gases (dichlorobenzene and ferrocene), sponge-like agglomerates of MWCNTs with densities of 0.005–0.025 g/cm^3 have been prepared by catalytic CVD that showed high compressibility up to 0.1 MPa and high thickness recovery of 93% [139]. CNT sponges have shown reversible stress-strain curves with marked hysteresis in air and ethanol, and rapid sorption of oil floating on water [140]. Highly resilient composites have been prepared by compressing a mixture of natural graphite and MWCNTs [141].

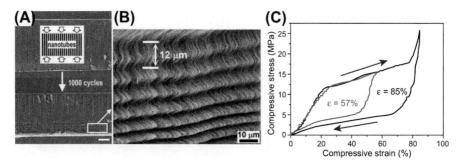

FIGURE 2.14 Compression of Vertically Aligned MWCNTs

(A) Before and after compression of 1000 cycles (SEM images), (B) lower part under high magnification (SEM image), and (C) stress-strain curves under compression.

From [142]

Vertically aligned MWCNT arrays, produced by catalytic CVD with xylene and ferrocene, show excellent mechanical behavior under compression, and high compressive strength and recovery rate, by keeping their open-cell nature, as shown in Figure 2.14C [142]. Their high resiliency is due to the formation of zigzag buckles of MWCNTs, as shown in Figures 2.14A and 2.14B. On vertically aligned MWCNT mats (arrays), mechanical behavior under tension and compression has been measured, showing very low stiffness under compression, due to buckling of the CNTs, but being considerably stiff under tension [143].

2.3 Applications of carbon nanotubes

CNTs are known to have excellent electrical, thermal, and mechanical properties. However, it has also been pointed out that it is very difficult to show these intrinsic properties of individual CNTs on their aggregates. In order to address this point, various problems have to be solved: finding efficient methods for preparation of well-characterized CNTs and purification of the prepared CNTs, establishment of characterization methods, etc. Many of these problems have not yet been solved completely and considerable efforts are being made to do this.

As-prepared CNTs are usually accompanied by various impurities, metal catalysts, carbonaceous impurities with amorphous and fullerene-like structures, and CNT structure variations, such as different wall thickness (single-, double-, and multi-walled), capped and open tubes, and various diameters and lengths of tubes. For the purification of these CNTs, various processes—chemical oxidation, physical separation, and a combination of chemical and physical processes—have been proposed. A comprehensive review on the purification of CNTs has been published [144].

In as-grown SWCNTs, nanotubes having metallic and semiconducting properties are usually mixed. Semiconducting SWCNTs have been extracted from as-grown CNTs without detectable remaining metallic SWCNTs or impurities by

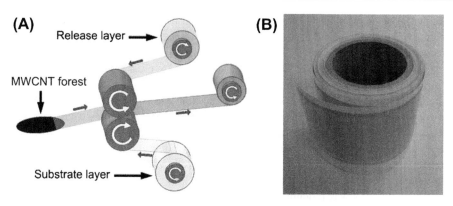

FIGURE 2.15 Preparation of a CNT/PE Composite

(A) Roll-to-roll setup, and (B) the CNT/PE composite tape.

From [153]

using polyfluorene as an extracting agent in toluene, assisted by ultracentrifugation [145]. Continuous separation of metallic and semiconducting SWCNTs has been performed by passing an SWCNT/sodium sulfate dispersion through a column containing agarose gel beads [146]: the latter was trapped by the beads and the former passed through the column. Metallic SWCNTs have been separated by using SDS as surfactant and an allyldextran-based gel with multicolumn chromatography [147].

Most CNTs are bundled by strong van der Waals forces, which can be troublesome for some applications; for example, bundled SWCNTs have a smaller surface area than that theoretically possible. Debundling of SWCNTs has been performed through intercalation of lithium with solvent molecules between the tubes in dimethyl sulfoxide [148]. This process was confirmed not to give any change in the quality of SWCNTs by TEM observation and Raman spectroscopy. The effective dispersion of CNTs in solvents and polymers has been reviewed, focusing on the surface modification of CNTs, to give some guidelines [149].

MWCNT sheets prepared from a forest by simple drawing have been tested in various applications, such as grids for high-resolution transmission microscopy [150], loudspeaker cones [151], anodes for lithium-ion batteries by loading SnO_2 nanoparticles [152], heating elements for incandescent display [153], stretchable touch panels [154]. A transparent conducting flexible sheet of well-aligned MWCNTs prepared from a forest has been transferred to a polyethylene (PE) sheet by a roll-to-roll technique (Figure 2.15). The MWCNT sheets were reported to possess performance of sheet resistances and transmittances comparable to ITO films.

MWCNT yarns prepared by spinning from a forest have been tested as field electron emitters in high vacuum [155,156]. Electron emission I-V curves were measured at the bent part of the yarn, with diameter about 20–30 μm and 2 cm length at 1500–2200 K, giving a work function of 4.54–4.64 eV and thermoionic emission constant of 228–824 A/cm^2K^2, larger than that for conventional tungsten cathodes [155]. On the cross-section of the yarn, which was densified by passing through acetone and had a diameter of about 50 μm, electron emission performance was measured; the electron emission started at the relatively low voltage of 400 V, current at 1000 V reached 2.1 mA, and current density was 100 A/cm^2 [156].

Sheets have been prepared from SWCNT forests by pressing after the densification through a liquid collapsing process, in which SWCNTs were aligned in one direction along the surface of the sheet (called SWCNT solids) [119]. These sheets were successfully applied to the electrodes of supercapacitors, giving capacitance as high as 80 F/g and energy density of 69.4 Wh/kg in 1 mol/L Et$_4$NBF$_4$/PC (propylene carbonate). electrolyte solution [120,157,158]. Their electrochemical properties are unique, characterized by a butterfly-shaped cyclic voltammogram measured in a three-electrode cell, as shown in Figure 2.16 [157]. Electrochemical doping in semiconducting CNTs was thought to occur at the interface between the electrolyte and the nanotube surface, together with electric double layer formation on the surface of CNTs. By controlled oxidation, the SWCNT forest was converted to a material with high surface area of 2240 m^2/g, where 85% of carbon atoms were thought to constitute a surface [159]. SWCNT films prepared via filtration have been treated with pyrrole after the arylsulfonic acid-functionalization, which gave high values of capacitance (350 F/g), power density (4.8 kW/kg), and energy density (3.3 kJ/kg) in 6 mol/L KOH [160]. SWCNT films prepared by spray-drying were reported to give high energy and power densities: 23 and 70 kW/kg for aqueous gel electrolytes

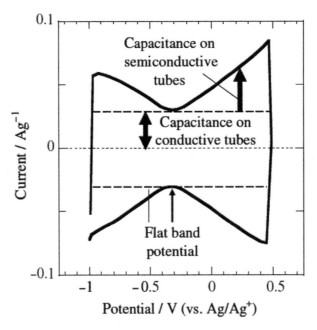

FIGURE 2.16 Cyclic Voltammogram of Single-walled Carbon Nanotube Sheet

Courtesy of Dr H. Hatori of AIST, Japan

(PVA/H$_3$PO$_4$ and liquid H$_3$PO$_4$) and an organic electrolyte (LiPF$_6$/EC + DEC), respectively, and 6 Wh/kg for both aqueous and non-aqueous electrolytes [161].

SWCNT yarns and sheets have been tested as high-performance components in electronic devices, sensors, and other applications, such as thin film transistors [126,162–165], high-current field-effect transistors and sensors in nano-size [124], transparent electrodes of organic solar-cells [166], and transparent flexible anodes for organic light-emitting diodes [167].

In order to improve the electrical conductivity of CNTs, doping of either K, Br, or I has been proposed [168–174]. By the intercalation of Br and K into SWCNTs prepared via the laser-ablation method, their electrical resistivity at 300 K decreased by a factor of 30 [169]. Doping of polyiodide chains into the interstitial channels in SWCNT ropes was found to be effective to improve the electrical properties, resistance of I$_2$-saturated ropes being reduced by almost two orders of magnitude at 300 K [170]. Doped I$_2$ reached 25 mass% and was stable even in Na/THF solution at 300 °C after removing I$_2$ adsorbed on the exterior [172]. The presence of poly-iodine anions and the charge transfer between iodine and DWCNTs were confirmed by Raman spectroscopy [171]. Iodine-doped DWCNT cables (yarns) gave low electrical resistivity, reaching around 10^{-7} Ωm, and their specific conductivity (conductivity/ weight) was reported to be higher than that of copper and aluminum [174]. Electrical conductivity improvement by I$_2$ doping has also been reported in SWCNTs [175].

Large amounts of CNTs, up to 100 tons/year, have been used as one of the fillers in the anode of commercial lithium-ion rechargeable batteries, in which the resilience and electrical properties of CNTs are believed to play an important role. CNTs have also been successfully used for reinforcement of the cathode of lithium-ion rechargeable batteries [176–180]. In Figure 2.17, cyclic performance is compared for

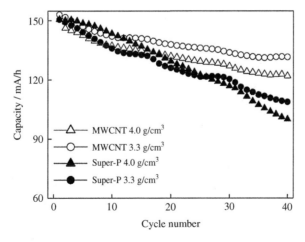

FIGURE 2.17 Cycle Performance of the LiCoO$_2$ Cathode Formed by the Addition of Different Amounts of Conductive Additive (Super P) and MWCNT

From [179].

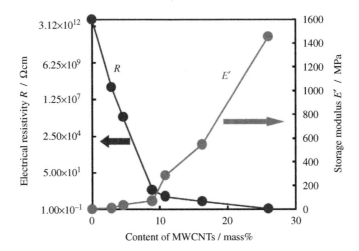

FIGURE 2.18 Changes in Electrical Resistivity, *R*, at Room Temperature and Storage Modulus, *E′*, with the Content of MWCNTs in Fluorine Rubber

Courtesy of Prof. M. Endo of Shinshu University, Japan

the $LiCoO_2$ cathode formed using a conventional conductive additive (furnace black, Super-P) and multi-walled carbon nanotube (MWCNT) [179]. The MWCNTs used were prepared by catalytic CVD and had an average size of 10–15 nm in diameter and 10–20 μm in length. The addition of 3.3 g/cm^3 of MWCNTs into the cathode was effective to improve the cyclic performance [179]. The use of 0.5 mass% MWCNTs after graphitization, together with 1 mass% acetylene black, gave cyclic performance much better than that of 1 mass% acetylene black and a little better than that of 3 mass% acetylene black [180]. The addition of CNTs to the powder of insulating cathode materials, like $LiCoO_2$, was found to give greater structural integrity, higher thermal conductivity, higher density, and shortened electrolyte absorption time, in addition to better electrochemical performance and higher electrical conductivity. The effectiveness of CNT addition into the anode has been reviewed [181].

The addition of a small amount of MWCNTs into conventional carbon fiber/phenol resin composites was found to improve thermal conductivity of the composites [182]. Without addition of MWCNTs, the thermal conductivity was relatively high, at about 250 W/m·K, because the carbon fibers used had a high conductivity along the fiber axis. By the addition of 7 mass% MWCNTs, thermal conductivity increased to 393 W/m·K. This result may encourage the use of CNTs as one of hybrid fillers in conventional composites in order to improve both thermal and mechanical properties. Effective reinforcement of rubber has been obtained by using MWCNTs prepared by catalytic CVD after heat treatment at 2800 °C [183]. The key issues in this reinforcement were to use MWCNTs modified at their surface with fluorinated rubber to generate durable sealants that could be operated satisfactorily at high temperatures up to 260 °C and high pressure up to 239 MPa. Electrical resistivity, *R*, and storage modulus, *E′*, are shown as a function of the content of MWCNTs in fluorinated rubber in Figure 2.18. The marked increase in

E' with increasing MWCNT content is thought to be due to the formation of a cellular structure of MWCNTs. The composites are expected to work as sealants for the probing and production of oil in deeper wells.

CNTs have also been tested as an adsorbent for H_2 [184–186]. Hydrogen storage in CNTs was revisited and reported to be less than 1.7 mass% at pressures up to 12 MPa at room temperature [186]. The results suggest that the hydrogen storage capacity of CNTs is far below the benchmark settled by Department of Energy, USA (DOE). Interaction between carbon atoms in SWCNTs and atomic hydrogen has been studied by using in situ synchrotron-radiation-based core-level photoelectron spectroscopy and Raman spectroscopy [187].

2.4 Concluding remarks

The synthesis of CNTs as forests and sheets by CVD using a catalyst deposited on an insulating buffer layer of SiO_2 or Al_2O_3 is advantageous in its high growth rate, lack of included catalyst metal, high purity of CNTs, and homogeneous structure, as well as high alignment vertically to the substrate in the forests and parallel to the substrate in the sheets, respectively. The growth rate of CNTs reaches 250 μm/min in the case of the water-assisted synthesis [43], although it is 3–23 μm/min without water assistance [32]. In most cases, catalyst particles stay on the insulating layer, not at the top of CNTs, and so there is no possibility of including metallic catalyst particles during removal of the forest from the substrate by cutting. The remaining catalyst can be used repeatedly for the synthesis of CNT forests [43]. The catalyst works efficiently, the surface of the catalyst being covered by SWCNTs up to 84% [44]. The forests and sheets thus synthesized contain negligibly small amounts of other forms of carbon, no carbon blacks, and no fullerene-like particles. The CNTs in the forests and sheets possess homogeneous structure, either single-, double-, or multi-walled, which can be controlled by the thickness of catalyst film: 1.6–2 nm thickness of catalyst Fe giving SWCNTs, 3 nm Fe giving DWCNTs, and more than 4 nm Fe MWCNTs [46]. These advantages of the synthesis process might be a breakthrough for the development of practical applications for CNTs.

Catalysts for the synthesis of the forests are made from transition metals and their alloys. Most of these metals can dissolve carbon atoms at high temperatures, which seems to be an important factor for employing them as catalysts for CNT synthesis, as well as for the graphitization of carbon materials. For the same reason, they are not recommended to be employed as a substrate for the synthesis of graphene via CVD [188]. Even using a transition metal as catalyst for CNT synthesis, however, it remains at the surface of the substrate (insulating layer)—in other words, at the bottom of CNTs during the growth of CNT forests [31,40,43]—but it exists at the top of the CNTs during the floating catalyst process for CNT synthesis [18]. This difference in the location of catalyst particles in the tube is certainly related to the mechanism of catalytic action of the metal, but unfortunately is not understood at all well. The ratio of Co/Mo bimetallic catalyst governed the formation of SWCNTs from CO at 700 °C [189]. On the catalyst $Mo_xCo_yMg_{1-x-y}O$ ($x = 0$–0.09, $y = 0.05$–0.09), the addition of Mo was effective in increasing the yield of CNTs, and addition of Co in improving the quality of SWCNTs from methane [16,190]. Tungsten acted as a promoter for the synthesis of CNT from methane, as Mo does, although the selectivity was less [191]. The composition in Ni_xFe_{1-x} bimetallic catalysts with a constant mean diameter of 2.0 nm had an effect on the yield of SWCNT via a floating catalyst method: the SWCNT fraction reducing with increasing Fe content in the catalysts, although the yield of as-grown CNTs was almost constant at about 0.65 mg/h [192]. These results were explained by higher carbon solubility of Fe compared to Ni, which results in a larger amount of carbon precipitation and the formation of MWCNTs.

In contrast to these transition metals, Cu film is recommended for the CVD deposition of graphene layers, mainly because of no dissolution of carbon and, for the same reason, it is thought not to be suitable for the acceleration of graphitization. However, Cu was reported to be an effective catalyst for the synthesis of sheets in which CNTs aligned in parallel to the substrate surface [73,118,126]. The effect of Cu catalysts for the synthesis of CNTs is not yet clearly explained. Undoubtedly, transparent CNT sheets are very interesting materials. These sheets have to be seriously compared with other transparent conductive sheets, not only CNT sheets prepared from forests, but also ITO films, which are commonly used in various devices, and graphene membranes, which will be explained later in this book (Chapter 3).

References

[1] Inagaki M. New Carbons: Control of Structure and Functions. Elsevier; 2000.
[2] Inagaki M, Radovic LR. Carbon 2002;40:2279–82.
[3] Oberlin A, Endo M, Koyama T. J Cryst Growth 1976;32:335–49.
[4] Endo M. CHEMTECH 1988:568–76.
[5] Endo M, Hayashi T, Kim YA, et al. Jpn J Appl Phys 2006;45:4883–92.
[6] Endo M. Jpn J Appl Phys 2012;51; 040001.
[7] Iijima S. Nature 1991;354:56–7.
[8] Ebbesen TW, Ajayan PM. Nature 1992;358:220–2.
[9] Iijima S, Ichihara T. Nature 1993;363:603–5.
[10] Bethune DS, Kiang CH, deVries MS, et al. Nature 1993;363:605–7.
[11] Bacon RJ. Appl Phys 1960;31:283–91.
[12] Thess A, Lee R, Nikolaev P, et al. Science 1996;273:483–7.
[13] Dai H, Rinzler AG, Nikolaev P, et al. Chem Phys Lett 1996;260:471–5.
[14] Kong J, Cassell AM, Dai H. Chem Phys Lett 1998;292:567–74.
[15] Colomer JF, Benoit JM, Stephan C, et al. Chem Phys Lett 2001;345:11–7.
[16] Tang S, Zhong Z, Xiong Z, et al. Chem Phys Lett 2001;350:19–26.
[17] Flahaut E, Govindaraj A, Peigney A, et al. Chem Phys Lett 1999;300:236–42.
[18] Cheng HM, Li F, Sun X, et al. Chem Phys Lett 1998;289:602–10.
[19] Satishkumar BC, Govindaraj A, Sen R, et al. Chem Phys Lett 1998;293:47–52.
[20] Ci L, Xie S, Tang D, et al. Chem Phys Lett 2001;349:191–5.
[21] Zhang Y, Chang A, Cao J, et al. Appl Phys Lett 2001;79:3155–7.
[22] Nikolaev P, Bronikowski MJ, Bradley RK, et al. Chem Phys Lett 1999;313:91–7.

[23] Bronikowski MJ, Willis PA, Colbert DT, et al. J Vac Sci Technol A 1800–1805;2001:19.
[24] Zhou W, Ooi YH, Russo R, et al. Chem Phys Lett 2001;350:6–14.
[25] Endo M, Muramatsu H, Hayashi T, et al. Nature 2005;433:476.
[26] Kim YA, Muramatsu H, Hayashi T, et al. Chem Vap Deposition 2006;12:327–30.
[27] Wei J, Jiang B, Wu D, et al. J Phys Chem B 2004;108:8844–7.
[28] Terrones M, Grobert N, Olivares J, et al. Nature 1997;388:52–5.
[29] Pan ZW, Xie SS, Chang BH, et al. Nature 1998;394:631–2.
[30] Ren ZF, Huang ZP, Xu JW, et al. Science 1998;282:1105–7.
[31] Fan SS, Chapline MG, Franklin NR, et al. Science 1999;283:512–4.
[32] Liu K, Jiang K, Feng C, et al. Carbon 2005;43:2850–6.
[33] Liu K, Sun YH, Chen L, et al. Nano Lett 2008;8:700–5.
[34] Lepro X, Lima MD, Baughman RH. Carbon 2010;48:3621–7.
[35] Nessim GD, Seita M, O'Brien KP, et al. Nano Lett 2009;9:3398–405.
[36] Meshot ER, Plata DL, Tawfick S, et al. ACS Nano 2009;3:2477–86.
[37] Zhong G, Tachiki M, Umezawa H, et al. Chem Vapor Dep 2004;10:125–8.
[38] Maruyama S, Kojima R, Miyauchi Y, et al. Chem Phys Lett 2002;360:229–34.
[39] Murakami Y, Miyauchi Y, Chiashi S, et al. Chem Phys Lett 2003;377:49–54.
[40] Murakami Y, Chiashi S, Miyauchi Y, et al. Chem Phys Lett 2004;385:298–303.
[41] Murakami Y, Maruyama S. Chem Phys Lett 2006;422:575–80.
[42] Thurakitseree T, Kramberger C, Zhao P, et al. Carbon 2012;50:2635–40.
[43] Hata K, Futaba DN, Mizuno K, et al. Science 2004;306:1362–4.
[44] Futaba DN, Hata K, Namai T, et al. J Phys Chem B 2006;110:8035–8.
[45] Yamada T, Namai T, Hata K, et al. Nat. Nanotechnol 2006;1:131–6.
[46] Zhao B, Futaba DN, Yasuda S, et al. ACS Nano 2009;3:108–14.
[47] Yasuda S, Futaba DN, Yamada T, et al. ACS Nano 2009;3:4164–70.
[48] Hahm MG, Hashim DP, Vajtai R, et al. Carbon Lett 2011;12:185–93.
[49] Gennett T, Dillon AC, Alleman JL, et al. Chem Mater 2000;12:599–601.
[50] Zhu H, Xu CL, Wu DH, et al. Science 2002;296:884–6.
[51] Zhu H, Jiang B, Xu C, et al. Chem Commun 2002:1858–9.
[52] Li Y-L, Kinloch IA, Windle AM. Science 2004;304:276–8.
[53] Koziol K, Vilatela J, Moisala A, et al. Science 2007;318:1892–5.
[54] Motta M, Li YL, Kinloch I, et al. Nano Lett 2005;5:1529–33.
[55] Zhong XH, Li Y-L, Liu Y-K, et al. Adv Mater 2010;22:692–6.
[56] Gommans HH, Alldredge JW, Tashiro H, et al. J Appl Phys 2000;88:2509–14.
[57] Annamalai R, West JD, Luscher A, et al. J Appl Phys 2005;98:114307.
[58] Vigolo B, Penicaud A, Coulon C, et al. Science 2000;290:1331–4.
[59] Vigolo B, Poulin P, Lucas M, et al. Appl Phys Lett 2002;81:1210–2.
[60] Poulin P, Vigolo B, Launois P. Carbon 2002;40:1741–9.
[61] Steinmetz J, Glerup M, Paillet M, et al. Carbon 2005;43:2397–9.
[62] Kozlov ME, Capps RC, Sampson WM, et al. Adv Mater 2005;17:614–7.
[63] Zhou W, Vavro J, Guthy C, et al. J Appl Phys 2004;95:649–55.
[64] Ericson LM, Fan H, Peng H, et al. Science 2004;305:1447–50.
[65] Davis VA, Ericson LM, Parra-Vasquez ANG, et al. Macromolecules 2004;37:154–60.
[66] Jiang KL, Li QQ, Fan SS. Nature 2002;419:801.
[67] Zhang X, Jiang K, Feng C, et al. Adv Mater 2006;18:1505–10.
[68] Zhang M, Atkinson KR, Baughman RH. Science 2004;306:1358–61.
[69] Zhang X, Li Q, Tu Y, et al. Small 2007;3:244–8.

[70] Zhang X, Li Q, Holesinger TG, et al. Adv Mater 2007;19:4198–201.
[71] Liu K, Sun Y, Zhou R, et al. Nanotechnology 2010:21; 045708.
[72] Tran CD, Humphries W, Smith SM, et al. Carbon 2009;47:2662–70.
[73] Zhou Y, Hu L, Gruener A. Appl Phys Lett 2006;88:123109.
[74] Armitage NP, Gabriel J, Gruner G. J Appl Phys 2004;95:3228–30.
[75] Hu L, Hecht DS, Gruener G. Nano Lett 2004;4:2513–7.
[76] Shimoda H, Oh SJ, Geng HZ, et al. Adv Mater 2002;14:899–901.
[77] Smith BW, Benes Z, Luzzi DE, et al. Appl Phys Lett 2000;77:663–5.
[78] Walters DA, Casavant MJ, Qin XC, et al. Chem Phys Lett 2001;338:14–20.
[79] Fischer JE, Zhou W, Vavro J, et al. J Appl Phys 2003;93:2157–63.
[80] Chen XQ, Saito T, Yamada H, et al. Appl Phys Lett 2001;78:3714–6.
[81] Seo HW, Han CS, Choi DG, et al. Microelectron Eng 2005;81:83–9.
[82] Boccaccini AR, Cho J, Roether JA, et al. Carbon 2006;44:3149–60.
[83] Banerjee S, White BE, Huang LM, et al. J Vac Sci Technol B 2006;24:3173–8.
[84] LeMieux MC, Roberts M, Barman S, et al. Science 2008;321:101–4.
[85] LeMieux MC, Sok S, Roberts ME, et al. ACS Nano 2009;3:4089–97.
[86] Shaffer MSP, Fan X, Windle AH. Carbon 1998;36:1603–12.
[87] Li Y-H, Xu C, Wei B, et al. Chem Mater 2002;14:483–5.
[88] Davis VA, Parra-Vasquez ANG, Green M, et al. Nat Nanotech 2009;4:830–4.
[89] Li P, Xue W. Nanoscale Res Lett 2010;5:1072–8.
[90] Kim Y, Minami N, Zhu W, et al. Jpn J Appl Phys 2003;42:7629–34.
[91] Sreekumar TV, Liu T, Kumar S. Chem Mater 2003;15:175–8.
[92] Badaire S, Zakri C, Maugey M, et al. Adv Mater 2005;17:1673–6.
[93] Zamora-Ledezma C, Blanc C, Maugey M, et al. Nano Lett 2008;8:4103–7.
[94] Kamat PV, Thomas KG, Barazzouk S, et al. J Am Chem Soc 2004;126:10757–62.
[95] Choi SW, Kang WS, Lee JH, et al. Langmuir 2010;26:15680–5.
[96] Hennrich F, Lebedkin S, Malik S, et al. Phys Chem Chem Phys 2002;4:2273–7.
[97] Casavant MJ, Walters DA, Schimdt JJ, et al. J Appl Phys 2003;93:2153–6.
[98] Wu Z, Chen Z, Du X, et al. Science 2004;305:1273–6.
[99] Saran N, Parikh K, Suh DS, et al. J Am Chem Soc 2004;126:4462–3.
[100] Ko H, Peleshanko S, Tsukruk VV. J Phys Chem B 2004;108:4385–93.
[101] Pasquier AD, Unalan HE, Kanwal A, et al. Appl Phys Lett 2005;87:203511.
[102] Ko H, Tsukruk VV. Nano Lett 2006;6:1443–8.
[103] Engel M, Small JP, Steiner M, et al. ACS Nano 2008;2:2445–52.
[104] Vichchulada P, Zhang QH, Duncan A, et al. ACS Appl Mater Interfaces 2010;2:467–73.
[105] Meitl M, Zhou Y, Gaur A, et al. Nano Lett 2004;4:1643–7.
[106] Kaempgen M, Duesberg GS, Roth S. Appl Surf Sci 2005;252:425–9.
[107] Artukovic E, Kaempgen M, Hecht DS, et al. Nano Lett 2005;5:757–60.
[108] Zhou YX, Hu LB, Gruener G. Appl Phys Lett 2006;88:123109.
[109] Zhang DH, Ryu K, Liu XL, et al. Nano Lett 2006;6:1880–6.
[110] Yoon YH, Song JW, Kim D, et al. Adv Mater 2007;19:4284–7.
[111] Geng HZ, Kim KK, So KP, et al. J Am Chem Soc 2007;129:7758–9.
[112] Paula S, Kimb DW. Carbon 2009;47:2436–41.
[113] Shaver J, Parra-Vasquez ANG, Hansel S, et al. ACS Nano 2009;3:131–8.
[114] Zhang S, Li Q, Kinloch IA, et al. Langmuir 2010;26:2107–12.
[115] Li Z, Kandel HR, Dervishi E, et al. Langmuir 2008;24:2655–62.
[116] Zhang M, Fang SL, Zakhidov AA, et al. Science 2005;309:1215–7.

[117] Wei Y, Jiang KL, Feng XF, et al. Phys Rev B 2007:76; 045423.
[118] Liu K, Sun Y, Liu P, et al. Nanotechnology 2009;20:335705.
[119] Futaba DN, Hata K, Yamada T, et al. Nat Mater 2006;5:987–94.
[120] Hiraoka T, Yamada T, Hata K, et al. J Am Chem Soc 2006;128:13338–9.
[121] Liu B, Ren W, Gao L, et al. J Am Chem Soc 2009;131:2082–3.
[122] Han S, Liu XL, Zhou CW. J Am Chem Soc 2005;127:5294–5.
[123] Yu QK, Qin GT, Li H, et al. J Phys Chem B 2006;110:22676–80.
[124] Kocabas C, Hur SH, Gaur A, et al. Small 2005;1:1110–6.
[125] Ding L, Yuan DN, Liu J. J Am Chem Soc 2008;130:5428–9.
[126] Ding L, Tselev A, Wang JY, et al. Nano Lett 2009;9:800–5.
[127] Hong SW, Banks T, Rogers JA. Adv Mater 2010;22:1826–30.
[128] Zhou WW, Han ZY, Wang JY, et al. Nano Lett 2006;6:2987–90.
[129] Hong BH, Lee JY, Beetz T, et al. J Am Chem Soc 2005;127:15336–7.
[130] Jin Z, Chu HB, Wang JY, et al. Nano Lett 2007;7:2073–9.
[131] Liu Y, Hong J, Zhang Y, et al. Nanotechnology 2009;20:185601.
[132] Wang X, Li Q, Xie J, et al. Nano Lett 2009;9:3137–41.
[133] Wen Q, Zhang R, Qian WZ, et al. Chem Mater 2010;22:1294–6.
[134] Wang B, Ma Y, Li N, et al. Adv Mater 2010;22:3067–70.
[135] Liu H, Takagi D, Chiashi S, et al. ACS Nano 2010;4:933–8.
[136] Ma Y, Wang B, Wu Y, et al. Carbon 2011;49:4098–110.
[137] Seah C-M, Chai S- P, Mohamed AR. Carbon 2011;49:4613–35.
[138] Liu Y, Qian W, Zhang Q, et al. Nano Lett 2008;8:1323–7.
[139] Gui X, Cao A, Wei J, et al. ACS Nano 2010;4:2320–6.
[140] Gui X, Wei J, Wang K, et al. Adv Mater 2010;22:617–21.
[141] Wei T, Wang K, Fan Z, et al. Carbon 2010;48:305–12.
[142] Cao A, Dickrell PL, Sawyer WG, et al. Science 2005;310:1307–10.
[143] Deck CP, Flowers J, McKee GSB, et al. J Appl Phys 2007:101; 023512.
[144] Hou P-X, Liu C, Cheng H- M. Carbon 2008;46:2003–25.
[145] Izard N, Kazaoui S, Hata K, et al. Appl Phys Lett 2008;92:243112.
[146] Tanaka T, Urabe Y, Nishide D, et al. Appl Phys Express 2009;2:125002.
[147] Liu H, Nishide D, Tanaka T, et al. Nat Commun 2011;2:309.
[148] Tanaike O, Kimizuka O, Yoshizawa N, et al. Electrochem Commun 2009;11:1441–4.
[149] Kim SW, Kim T, Kim YS, et al. Carbon 2012;50:3–3.
[150] Zhang L, Feng C, Chen Z, et al. Nano Lett 2008;8:2564–9.
[151] Xiao L, Chen Z, Feng C, et al. Nano Lett 2008;8:4539–45.
[152] Zhang HX, Feng C, Zhai YC, et al. Adv Mater 2009;21:2299.
[153] Liu P, Liu L, Wei Y, et al. Adv Mater 2009;21:3563–6.
[154] Feng C, Liu K, Wu J-S, et al. Adv Funct Mater 2010;20:885–91.
[155] Liu P, Wei Y, Jiang KL, et al. Phys Rev B 2006;73:235412.
[156] Wei Y, Weng D, Yang Y, et al. Appl Phys Lett 2006;89; 063101.
[157] Kimizuka O, Tanaike O, Yamashita J, et al. Carbon 2008;46:1999–2001.
[158] Tanaike O, Futaba DN, Hata K, et al. Carbon Lett 2009;10:90–3.
[159] Hiraoka T, Izadi-Najafabadi A, Yamada T, et al. Adv Funct Mater 2010;20:422–8.
[160] Zhou C, Kumar S. Chem Mater 2005;17:1997–2002.
[161] Kaempgen M, Chan CK, Ma J, et al. Nano Lett 2009;9:1872–6.
[162] Kong J, Franklin NR, Zhou C, et al. Science 2000;287:622–5.
[163] Someya T, Small J, Kim P, et al. Nano Lett 2003;3:877–81.
[164] Jang YT, Moon SI, Ahn JH, et al. Sens Actuators B 2004;99:118–22.

[165] Bekyarova E, Kalinina I, Itkis ME, et al. J Am Chem Soc 2007;129:10700–6.
[166] Rowell MW, Topinka MA, McGehee MD, et al. Appl Phys Lett 2006;88:233506.
[167] Li J, Hu L, Wang L, et al. Nano Lett 2006;6:2472–7.
[168] Mordkovich VZ, Baxendale M, Yoshimura S, et al. Carbon 1996;34:1301–3.
[169] Lee RS, Kim HJ, Fischer JE, et al. Nature 1997;388:255–7.
[170] Grigorian L, Williams KA, Fang S, et al. Phys Rev Lett 1998;80:5560–3.
[171] Cambedouzou J, Sauvajol J-L, Rahmani A, et al. Phys Rev B 2004;69:235422.
[172] Kissell KR, Hartman KB, Van der Heide PAW, et al. J Phys Chem B 2006;110:17425–9.
[173] Choi WI, Ihm J, Kim G. Appl Phys Lett 2008;92:193110.
[174] Zhao Y, Wei J, Vajtai R, et al. Sci Rep 2011;1:83.
[175] Khoerunnisa F, Fujimori T, Itoh T, et al. Chem Phys Lett 2011;501:485–90.
[176] Endo M, Kim YA, Hayashi T, et al. Carbon 2001;39:1287–97.
[177] Lin Q, Harb JN. J Electrochem Soc 2004;151:1115–9.
[178] Li X, Kang F, Shen W. Carbon 2006;44:1334.
[179] Sheem K, Lee YH, Lim HS. J Power Sources 2006;158:1425–30.
[180] Sotowa C, Origi G, Takeuchi M, et al. Chem Sus Chem 2008;1:911–5.
[181] de las Casas C, Li W. J Power Sources 2012;208:74–85.
[182] Kim YA, Kamio S, Tajiri T, et al. Appl Phys Lett 2007;90; 093125.
[183] Endo M, Noguchi T, Ito M, et al. Adv Funct Mater 2008;18:3403–9.
[184] Miyamoto J, Hattori Y, Noguchi D, et al. J Am Chem Soc 2006;128:12636–7.
[185] Kim DY, Yang C-M, Yamamoto M, et al. J Phys Chem C 2007;111:17448–50.
[186] Liu C, Chen Y, Wu C-Z, et al. Carbon 2010;48:452–5.
[187] Tokura A, Maeda F, Teraoka Y, et al. Carbon 2008;46:1903–8.
[188] Inagaki M, Kim YA, Endo M. J Mater Chem 2011;21:3280–94.
[189] Kitiyanan B, Alvarez WE, Harwell JH, et al. Chem Phys Lett 2000;317:497–503.
[190] Flahaut E, Peigney A, Bacsa W, et al. J Mater Chem 2004;14:646–53.
[191] Landois P, Peigney A, Ch Laurent, et al. Carbon 2009;47:789–94.
[192] Chiang W-H, Sankaran RM. Carbon 2012;50:1044–50.

Graphene: Synthesis and Preparation

3

The term "graphene" was first proposed in 1986 as the name for an isolated single two-dimensional sheet of carbon atoms, occurring in a graphite intercalation compound [1]. In the first-stage structure, a two-dimensional carbon layer has neighboring intercalate layers and is isolated from other carbon layers, as schematically shown in Figure 3.1. In structures higher than second stage, however, more than two carbon layers are stacked parallel with the same regularity as in graphite. The carbon layer occurring in the first-stage structure was proposed to be called "graphene," which comes from the suffix "-ene" for polycyclic aromatic hydrocarbons, such as naphthalene, anthracene, etc., and the prefix "graph-" from graphite.

"Graphene," therefore, was defined as an isolated single layer of carbon hexagons consisting of sp^2-hybridized C-C bonding with π-electron clouds. From the engineering point of view, thin flakes consisting of a few layers of carbon atoms, including monolayer graphene, could be very important because of their interesting structural and physical characteristics and also promising potential applications in technological fields, as reviewed from various viewpoints [2–6].

Preparation of graphene is classified into five routes: (1) mechanical cleavage of graphite crystals, (2) exfoliation of graphite through its intercalation compound, (3) chemical vapor deposition (CVD) on different substrate crystals, (4) organic synthesis processes, and (5) other processes. The preparation of the graphite intercalation compound in route (2) and CVD in route (3) include chemical reactions, but here these two routes are differentiated from the synthesis of graphene via organic synthesis processes. The pyrolysis of organics resulting mostly in the formation of nanoribbons and unzipping of carbon nanotubes by using metal nanoparticles are classified into route (5).

In this chapter, various attempts to prepare graphene are reviewed, with a discussion on the probable effectiveness of each method. There have been many papers published in this area, particularly after the Nobel Prize in physics for 2010, which was awarded to Profs A. Geim and K. Novoselov of the University of Manchester for their groundbreaking experiments on graphene [7]. In many published papers, however, the term "graphene" has not been used in its strict definition, i.e. a single layer of carbon atoms consisting of sp^2-hybridized bonds. Some authors do not pay

---

Prerequisite for readers: Chapter 3.2 (Highly oriented graphite) in *Carbon Materials Science and Engineering: From Fundamentals to Applications*, Tsinghua University Press.

Advanced Materials Science and Engineering of Carbon.

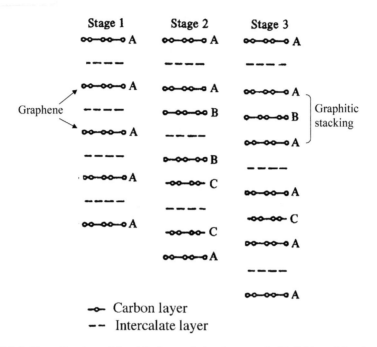

FIGURE 3.1 **Stage Structure of Graphite Intercalation Compounds (Definition of Graphene)**

enough care to how many layers are stacked in their samples, although they have called them graphene. Here, therefore, the terms either "flake(s)" or "sheet(s)" are used to designate the products of the attempts to prepare graphene, and the term "monolayer graphene" is used only for the products confirmed to be a single layer.

3.1 Preparation through the cleavage of graphite

The cleavage of highly crystalline kish graphite using double-sided adhesive tapes was repeated until the flake became transparent with a thickness of 18–108 nm [8,9]. On these thin flakes, electrical resistivity (ϱ), the Hall coefficient (R_H), and transverse magnetoresistance ($\Delta\varrho/\varrho$) were measured at temperatures between 4.2 and 300 K, as shown in Figure 3.2. The high crystallinity of the thin flakes thus prepared was confirmed from the presence of a Shubnikov–de Haas oscillation in R_H (Figure 3.2B). These results show a marked dependence of electronic properties on the thickness of the flakes, i.e. the number of layers stacked. Analysis of these experimental results shows that the overlap energy between conduction and valence bands decreases and the relaxation rate due to lattice defects increases with decreasing number of the layers stacked.

Thin flakes with a thickness of 3-100 nm and lateral size of about 2 μm have been obtained through the micromechanical cleavage technique [10,11] from the

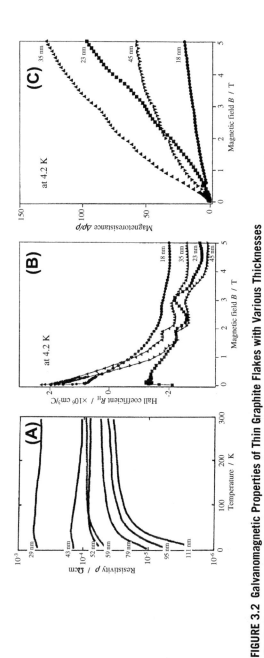

FIGURE 3.2 Galvanomagnetic Properties of Thin Graphite Flakes with Various Thicknesses

(A) Change in resistivity, ϱ, with temperature, (B) change in the Hall coefficient, R_H, with magnetic field, B, and (C) magnetoresistance, $\Delta\varrho/\varrho$, with B.

Courtesy of Prof. Y. Ohashi of Keio University, Japan

micropillars formed on highly oriented pyrolytic graphite (HOPG) by using oxygen plasma [12]. Thin flakes, including monolayer graphene, were visible on the SiO$_2$/Si substrate, although they were optically transparent [13]. Electric-field-dependent conductance measurement on these flakes showed a marked modulation as a function of gate voltage, the more markedly on the thinner flakes, as shown in Figure 3.3.

The flakes obtained through repeated micromechanical cleavage of graphite were suspended in a liquid and then attached to a micrometer-sized metallic scaffold to identify the monolayer graphene under transmission electron microscopy (TEM) [14]. The detailed observation by electron microscope (TEM) techniques showed that the flakes exhibited random microscopic out-of-plane deformations, monolayer graphene showing more marked deformations than two-layer flakes, in addition to foldings [15]. The edges of monolayer graphene were shown to have either zigzag or armchair structure through high-resolution scanning tunneling microscopy (STM) [16]. Atomically flat flakes have been obtained on the cleaved surface of mica (muscovite) [17]. The fluctuation in height on the layer surface measured by atomic force microscopy (AFM) was less than 25 pm. By peeling graphite bonded onto borosilicate glass, thin flakes with a large area consisting of a single layer or a few layers were obtained on the substrate [18].

Using sonication of exfoliated graphite for a long time in an alcohol-water mixture, it was possible to cleave the material into flakes as thin as 52 nm thick [19,20]. Dispersion of sieved graphite powder in N-methylpyrrolidone at a concentration of about 0.01 mg/cm^3 gave a grey supernatant after sonication and mild centrifugation, from which thin flakes consisting of less than five layers were recovered by pipetting and

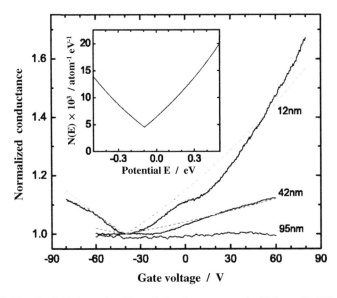

FIGURE 3.3 Electric-field-dependent Electric Conductance on the Flakes with Different Thicknesses Prepared by Peeling

From [10]

then filtration [21]. The yield of monolayer graphene was about 1 mass% and reported to be improved up to 12 mass% by repetition of the cleaving process. Other solvents, *N*,*N'*-dimethylacetamide, γ-butyrolactone, and 1,3-dimethyl-2-imidazolidinone could be used for the dispersion of thin flakes. The dispersion of sieved graphite in an aqueous solution of 5-10 mg/cm^3 with sodium dodecylbenzene sulfonate gave a supernatant after mild sonication, which contained about 3% monolayer graphene [22]. Simple sonication of graphite flakes in benzylamine in Ar atmosphere gave a suspension of flakes of few layers at a concentration of 0.5 mg/cm^3 [23]. Thin flakes with an average thickness of 1.18 nm have been prepared from graphite by sonication in cetyltrimethylammonium bromide solution in glacial acetic acid and used to prepare a composite film by a simple solution blending with poly(vinyl chloride) (PVC) [24,25].

It has to be pointed out that perfect two-dimensional atomic crystals cannot be obtained even through the cleavage of high crystalline graphite, unless in a limited size or containing many crystal defects. On the thin flakes obtained, scrolling and folding at the edges are impossible to be avoided.

3.2 **Preparation through the exfoliation of graphite**

Synthesis of graphite oxide (GO), a covalent-type intercalation compound, and its thermal exfoliation to thin flakes have been used on a large scale in industry to prepare flexible graphite sheets, of which the fundamental process is the synthesis of GO by strong oxidation, exfoliation at high temperature, and forming into thin sheets via compressing and rolling [26]. Attempts to prepare graphene through GO have been reported since 1962 in a number of papers via procedures similar to those for flexible graphite sheets, even though the word "graphene" was not used. The preparation of graphene through ionic-type intercalation compounds with H_2SO_4, HNO_3, also and potassium has also been reported. The procedures for the preparation of flexible graphite sheets and thin graphene-like flakes are compared in Figure 3.4. For graphene preparation, both thermal and chemical exfoliation are employed, and also reduction is essential in order to have high electrical conductivity. Thermal exfoliation at high temperature, where exfoliation and partial reduction are thought to occur, is used for flexible graphite sheet preparation mainly because it allows production on a large-scale. By starting from intercalation compounds, excluding graphite oxides, a chemical exfoliation is employed for graphene preparation, and reduction is not necessary, because no chemical bonding with oxygen is expected in the intercalation compounds.

3.2.1 **Preparation using graphite oxides**

In most studies on the preparation of graphene, natural graphite with various particle sizes was selected as starting material, but no information on their detailed crystalline structure (crystallinity) was presented. In some papers, so-called artificial graphite and graphite electrodes were used without a detailed characterization of their

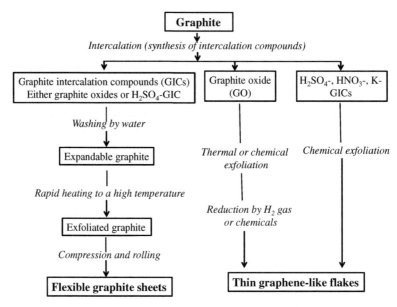

FIGURE 3.4 Processes for the Preparation of Flexible Graphite Sheet and Graphene Via Intercalation Compounds

structure, although their structure and properties are known to depend strongly on the precursor and preparation conditions.

GO has been synthesized by the so-called Hummers method [27], which was derived from the method of Staudenmaier [28], consisting of the oxidation of graphite in concentrated H_2SO_4 with $NaNO_3$ and $KMnO_4$, the exclusion of excess $KMnO_4$ by reducing to water-soluble $MnSO_4$ with H_2O_2, and then washing by methanol. The Brodie method [29] has also been used, where the oxidation of graphite is carried out in fuming HNO_3 with $KClO_3$. To synthesize GO, the electrochemical oxidation of graphite can be applied to natural graphite [30], reactor-grade graphite [31], and various carbon fibers [32,33] in either H_2SO_4 or HNO_3, and also to natural graphite in an ammonia solution [34]. The GO synthesized can have a wide range of chemical compositions, such as $C_8O_{3.5-4.3}H_{2.5-2.9}$ [35], $C_8O_{3.78-5.05}H_{2.9-4.4}$ [36], $C_8O_{2.54}H_{3.91}$, and $C_8O_{4.61}H_{6.70}$ [37], due to the presence of different oxygen-containing functional groups in different amounts.

Exfoliation of GO at a high temperature of around 1000 °C for a short time has often been used. During this high-temperature exfoliation, some of the functional groups on the surface of the GO were removed and the thin flakes tended to coalesce. The structural change from GO to graphene by thermal exfoliation by rapid heating to 1050 °C has been discussed on graphite particles of about 45 μm [38]. Thermal exfoliation of GO after being spray-dried with air at 300 °C was performed at 1050 °C for 30 s, and reported to give a high concentration of monolayer graphene, up to 80% [39]. Thin GO flakes were recovered from the supernatant of a dispersion, which was prepared in 10^{-3} mol/L NaCl aqueous solution by sonication, followed by centrifugation at 3000 rpm [40]. The resultant GO dispersion remained stable for several months.

A process of reduction of GO after exfoliation has usually been included in order to produce thinner flakes, and has been carried out by different methods. By heat treatment in a mixture of Ar and H_2 at 500 °C for 2 days, monolayer graphene with a thickness of 0.37 nm was obtained [41]. A GO film deposited on a SiO_2 substrate by spray-coating was reduced in an air flow containing hydrazine at 80 °C [42]. Reduction of GO has also been performed in liquid hydrazine, resulting in a stable dispersion of thin flakes due to the stabilization of negatively charged carbon layers surrounded by counter-ion $N_2H_4^+$ [43]. An aqueous suspension of thin flakes was obtained by adding either NaOH or KOH (8 mol/L) at 50–90 °C under mild sonication [47]. The treatment of GO flakes by phenyl-isocyanate was also reported to be effective for producing a colloidal suspension of GO flakes with a thickness of about 1 nm [48], because of the formation of hydrophobic chemical groups on the GO surface to keep the flakes separated [49]. These isocyanate-treated GO flakes were reduced by hydrazine, the resultant flakes giving a relatively high electrical conductivity after being used to make a composite with styrene: 1 vol% mixing giving about 0.1 S/m and 2.5 vol% mixing about 1 S/m [47]. The reduction of GO by using either $NaBH_4$ at a steam-bath temperature or hydroquinone under refluxing was reported to give different stacking regularities to the products after reduction, the former resulting in turbostratic but the latter in crystalline stacking of graphite layers [50]. A stable aqueous dispersion of thin flakes was obtained after the exfoliation and reduction of GO in the presence of poly(sodium 4-styrensulfonate) [45]. The reduction of GO was done in a mixed solution of N,N-dimethylacetamide (DMF) and water under microwave irradiation [51]. The photocatalytic reduction of GO has also been carried out by suspending GO with the photocatalyst TiO_2 in ethanol, but a complete reduction was not achieved [52].

It has been experimentally shown that the size and crystallinity of the starting graphite have a marked influence on the thickness of the flakes [53]. GO flakes were prepared through a modified Hummers method from different graphite materials and their thermal exfoliation, followed by reduction in a flow of H_2 at 450 °C, and dispersion in N-methylpyrrolidone by sonication. Figure 3.5 shows a histogram of the thickness distribution of 100 flakes recovered from a supernatant of the suspended solution for five kinds of starting graphite. A narrow distribution of thickness was obtained from artificial graphite, about 80% of the flakes being less than 2 nm, but a broad distribution from 4 to more than 10 layers was obtained from HOPG. This experimental result suggests that the smaller the lateral size and the lower the crystallinity of the starting graphite, the fewer the number of layers stacked in the flakes. The monolayer graphene prepared from the artificial graphite exhibited a high electrical conductivity of about 1×10^3 S/cm.

Conductive and transparent films with thicknesses of 14–86 nm have been prepared from GO by exfoliation under sonication and reduction with hydrazine in a water dispersion, followed by making films on a quartz surface and annealing up to 1100 °C [54], as shown in Figure 3.6. The optical transparency of the film with 14 nm

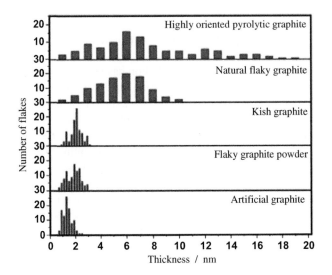

FIGURE 3.5 Distribution Histogram of Thickness of Flakes Prepared From Various Graphite Samples

Courtesy of Prof. H.-M. Cheng of the Institute of Metal Research, China.

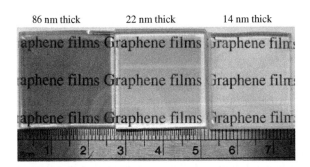

FIGURE 3.6 Thin Sheets Prepared in Different Thicknesses from Graphite Oxide Through Exfoliation, Reduction and Annealing at 1100 °C

From [54].

thickness was well over 80% in the wavelength range of 1100 to 3000 nm, and its electrical conductivity was over 200 S/cm.

Hydrogen-arc discharge on GO particles has been reported to result simultaneously in efficient exfoliation, considerable elimination of oxygen-containing functional groups, and structural annealing [55]. The sheet prepared in this way showed a high electrical conductivity of about 2×10^3 S/cm, much higher than the sheet prepared by argon arc-discharge exfoliation (about 2×10^2 S/cm) and also than that by conventional thermal exfoliation (about 80 S/cm).

A hydrosol of GO flakes prepared from natural graphite powder gave a thin membrane (0.5-20 μm thick) at the liquid/air interface by warming to 353 K, which was

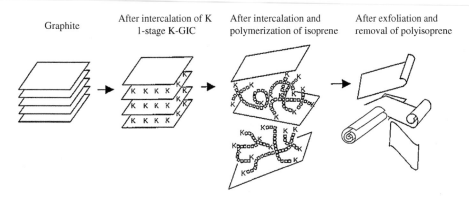

FIGURE 3.7 Process of Graphene Formation Through a First-stage Potassium Intercalation Compound

Courtesy of Dr H. Shioyama of AIST, Osaka, Japan.

flexible and self-standing [56]. It was shown that GO sheets could be deposited at any place defined by a molecular template, owing to the electrostatic attraction between the negatively charged GO flake and the positively charged template [57]. The patterned GO deposits were reduced by exposure to hydrazine vapor. Spontaneous exfoliation of graphite in chlorosulfonic acid (HSO_3Cl) gave dispersion with a relatively high concentration of thin flakes of about 2 mg/cm^3, of which 70% were identified as monolayer [58].

Exfoliated graphite prepared through the conventional process (Figure 3.4) was re-intercalated by using oleum (fuming sulfuric acid) and then tetrabutylammonium hydroxide in DMF to form ternary graphite intercalation compounds (GICs), resulting in a homogeneous suspension of thin sheets by sonication with a surfactant [59]. The thin flakes obtained were suspended again in 1,2-dichloroethane solution, after removing the surfactants and repeated centrifugation and suspension in DMF, and then formed into the large, transparent conducting films.

Electrochemical exfoliation of natural graphite and HOPG has been performed in sulfuric acid by applying a bias of 10 V [60]. Exfoliated thin flakes were dispersed in DMF and a thin sheet was obtained at the air/DMF interface. Flakes consisting of few layers were obtained by electrochemical exfoliation in $LiClO_4$/PC (propylene carbonate) electrolyte by applying a potential of -15 V, followed by sonication in a DMF/PC solution of LiCl [61].

3.2.2 Preparation using graphite intercalation compounds

A first-stage graphite intercalation compound with potassium (K-GIC, KC_8) prepared from HOPG was exfoliated by exposing to isoprene vapor and heat-treating above 500 °C [62–64]. Isoprene molecules were thought to be intercalated into the graphite gallery with potassium and polymerized in the gallery, resulting in exfoliation of host graphite. In Figure 3.7, the process is schematically illustrated. The product was composed of thin flakes, most of which were rippled and rolled up especially at the edges, together with particles that were completely rolled up (nanoscrolls).

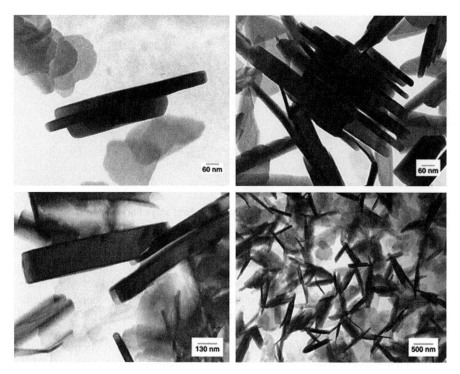

FIGURE 3.8 Nanoscrolls Formed From First-stage Graphite Intercalation Compound with Nitric Acid

From [67].

The highly exothermic reaction of KC_8 with ethanol was found to cause exfoliation of graphite, resulting in the formation of nanoscrolls [65]. Ternary GICs with potassium and tetrahydrofuran (THF), $K(THF)_xC_{24}$, exfoliated spontaneously in *N*-methylpyrrolidone to give suspended thin flakes, including monolayer graphene and nanoribbons, which were thought to be negatively charged and were air sensitive [66].

The first-stage graphite intercalation compound with nitric acid (graphite nitrate), prepared from natural graphite in red fuming nitric acid (HNO_3) by bubbling ozone, has also been used as a starting material [67]. Particles of blue-colored graphite nitrate under suspension were reacted with ethanol to exfoliate. Many nanoscrolls were formed, consisting of 15–20 layers and having a diameter of 40–120 nm and a length of up to about 750 nm, as shown in Figure 3.8.

3.3 Synthesis through chemical vapor deposition

Preparation of thin flakes of graphite through chemical vapor deposition (CVD) onto the crystal surface of different metals and metal carbides has mostly been done in

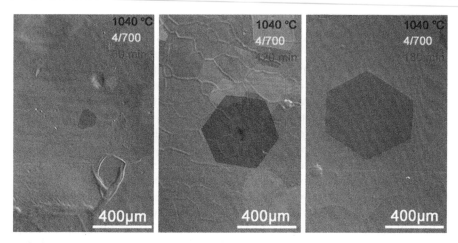

FIGURE 3.9 Synthesis of Hexagonal Single-crystal Graphene

Single-crystal graphene was synthesized by ambient-pressure CVD for different periods of time at 1040 °C under a flow of methane of 4 standard cm^3/min (SCCM) with H_2 of 700 SCCM

Courtesy of Prof. H.-M. Cheng of the Institute of Metal Research, CAS, China

a chamber under high-level vacuum. Interactions have been observed between the carbon atoms deposited and the substrate, such as epitaxial growth of thin sheets. Reviews of thin graphite films formed on the surface of various metals have been published [68–70].

CVD onto the (111) plane of a Pt crystal was carried out in a scanning tunneling microscope by exposing it to ethylene gas under a pressure of 4.4×10^{-2} Pa at 300 K and then annealing at different temperatures in the microscope [71]. The formation of monolayer graphene was concluded from the detailed analysis of STM images. A carbon layer was deposited on a clean Pt(111) surface by exposing it to about 1.3×10^{-4} Pa benzene at 1000 K [72]. A carbon layer with a $(\sqrt{19} \times \sqrt{19})R23.4°$ structure grew at the beginning of the exposure and then islands of monolayer graphene with clear hexagonal shapes were formed. Hexagonal millimeter-sized single-crystal graphene can be grown on polycrystalline Pt foil by ambient-pressure CVD with a low concentration of methane at 1040 °C, as shown in Figure 3.9 [73]. The graphene thus synthesized was separated from the substrate Pt by using it as the cathode of an electrolysis cell with an NaOH aqueous solution, after being spin-coated with poly(methyl methacrylate) (PMMA) (the bubbling method). The resultant graphene showed high crystal quality with low wrinkle height of 0.8 nm and a carrier mobility of greater than 7100 cm^2/Vs. The collision of methane molecules with a Pt(111) surface under supersonic acceleration resulted in the formation of monolayer graphene, without any carbon by-products [74].

Two mechanisms for the growth of a thin sheet composed of few layers have been suggested: CVD from hydrocarbon gas on the Ni substrate, and carbon precipitation at a low temperature from the Ni substrate, because Ni metal can absorb a large amount of carbon into its interstitial sites at a high temperature [75]. CVD in a $CH_4/H_2/Ar$ gas flow at 1000 °C was performed on a thin foil of Ni (less than 300 nm thick) to make the carbon precipitation as small as possible, resulting in the formation of thin sheets [76]. Monolayer graphene, after being transferred to a SiO_2 substrate, showed electron mobility greater than 3700 cm^2/Vs at a carrier density of about $5 \times 10^{12}/cm^2$. The resultant sheets were reported to be conducting, transparent, stretchable, and flexible, and their crystalline quality was comparable with that of mechanically cleaved flakes. In Figure 3.10, variation in resistance with bending radius is shown for a film transferred to polymer substrate; resistance parallel to the bending direction (R_y) increases with increasing bending radius, and is recovered by releasing the force, but that perpendicular to bending direction (R_x) does not change. Thin carbon layers have been obtained on a Ni(110) surface by reacting with CO at 600 K and 6.5×10^{-3} Pa [77]. A strong interaction between deposited carbon atoms with the Ni substrate was concluded from the carbon near-edge electron-energy loss and surface extended energy-loss fine-structure analyses. Thin sheets were deposited on faceted Ni(755), which consisted of the terraces of the (111) face and the steps of a high-Miller index face [78]. The vibration spectrum of monolayer graphene consisted of a single peak, revealing similar strengths of the interfacial bond between the deposited carbon and the substrate Ni on two faces.

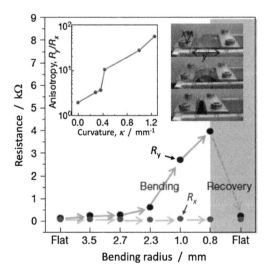

FIGURE 3.10 Variation in Resistance, R_x and R_y, Along and Across a Flake, with Bending Radius

Inserts show the relationship between anisotropy in resistance R_y/R_x and curvature of flake, and the appearance of the setup.

On the surface of polycrystalline Ni, thin carbon films consisting of 1-12 layers with a large area of about 1 cm^2 have been prepared by ambient pressure CVD [79]. The films contained domains consisting of 1-2 layers and 1-20 μm in lateral size. Composite films of monolayer hexagonal BN (*h*-BN) and monolayer graphene were prepared on a Ni(111) surface by CVD (i.e. BN/graphene/Ni and graphene/BN/Ni composite films), with both *h*-BN and graphene having a relatively strong interaction with the Ni(111) surface [80–82]. Nanoribbons consisting of a few layers have been synthesized by controlling the precipitation of carbon from Ni [181]. A Ni layer about 500 nm thickness was sputtered onto the substrate and then heated to 1300 °C in a high vacuum (1.3×10^{-4} Pa), followed by slow cooling to 800 °C (1 K/min). Upon slow cooling, the carbon atoms precipitated onto the surface of the Ni as thin sheets. High-quality transparent ribbons with thickness less than 8 nm, width up to 350 nm, length more than 3 μm, and negligibly small D band in the Raman spectrum were obtained, although many ribbons with various widths and lengths were formed at the same time.

Monolayer graphene with a large area has been grown on Cu foil by CVD of methane at a temperature up to 1000 °C [83]. The films predominantly consisted of a monolayer, probably due to a low carbon solubility of Cu (about 0.03%), in contrast to the high solubility into Ni (about 1.1%). In order to avoid the introduction of structural defects into the graphene film during its transfer to another substrate, such as SiO$_2$/Si, the deposition of PMMA onto graphene film formed on Cu foil was proposed [84]. After CVD formation of graphene on Cu, PMMA was deposited on top of the graphene film, followed by etching of Cu foil, and then the resultant PMMA/graphene was placed on a substrate. On the PMMA/graphene, an appropriate amount of PMMA solution was dropped, where the coated PMMA was partly or fully dissolved, this redissolution of the PMMA seeming to mechanically relax the underlying graphene film and to give a better contact with the substrate. In 2010, an efficient roll-to-roll transfer process of graphene formed on a flexible Cu foil was developed, which consisted of three steps, as shown schematically in Figure 3.11 [85]. Graphene film formed by CVD on a flexible Ni substrate has also been transferred to a flexible poly(ethylene terephthalate) (PET) substrate by a roll-to-roll process [86]. The graphene

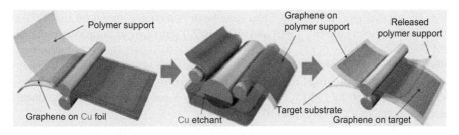

FIGURE 3.11 Scheme for a Roll-to-roll Transfer Process of Graphene Formed on a Flexible Cu Foil

From [85].

film had a sheet resistance of about 125 Ω per square and an optical transmittance of 97.4%, but four-layered films prepared by repeating the roll-to-roll process four times on the same substrate had about 30 Ω per square and about 90%. Growth of graphene has been performed on the (100) face of a high-purity Cu single crystal by CVD of methane, demonstrating the formation of defect-free graphene [87]. A polycrystalline Cu substrate has also been used for graphene synthesis, and demonstrated that the graphene grains show no definite epitaxial relationship with the Cu substrate [88]. By using surface-wave-plasma CVD of methane mixed with Ar and/or H_2 on a Cu or Al foil substrate at a low temperature of 300–400 °C, transparent conductive films with a large area (23×20 cm^2) were synthesized, which consisted of a few graphene layers [89].

Thin layers have been grown on an Ir(111) surface by CVD at a low pressure of ethylene at 1120–1320 K [90]. The flake consisted of monolayer graphene and extended over terraces and step edges of the Ir substrate. Epitaxial growth of thin flakes has been observed on a Ru(0001) surface by the precipitation of interstitial carbon in the Ru metal through slow cooling from 1150 to 825 °C [91]. The flakes were lens shaped with a size of more than 100 μm and consisted of single to few layers. A prolonged annealing of Ru(0001) in ultra-high vacuum resulted in the formation of thin layers [92]. Deposition of "graphitic" carbon and "carbidic" carbons on clean and potassium-covered Co(0001) surfaces has been reported, through adsorption of acetylene molecules followed by heating to above 420 K [93].

Monolayer graphene has been synthesized on a TiC(111) faceted surface by CVD of ethylene under a pressure of 8×10^{-3} Pa at 1100 °C [94] and at 1250 °C [95,96]. The strength of the C-C bond in the monolayer graphene was found to be weakened because of orbital hybridization between the graphene layer and the substrate TiC(111) surface [94]. On the TiC(111) surface, a sheet consisting of two crystal domains with a lateral size of about 300 nm was obtained [95]. Epitaxial growth of a graphite thin sheet [97] has also been reported.

Thin sheets were grown on the (0001) surface of 6H-SiC [98] and also of 4H-SiC [99] by CVD, and their electronic transport properties were determined. They were shown by low-energy electron diffraction (LEED) and STM measurements to grow epitaxially on the substrate and by Auger spectroscopy to consist of a few layers. The properties of these sheets could be explained by the existence of a two-dimensional electron cloud with large anisotropy, high mobility, and two-dimensional localization. By vaporization of Si, thin films have been epitaxially grown on the (0001) surface of single-crystal SiC [100] and on the (0001) surfaces of both Si-terminated and C-terminated sides of 6H- and 4H-SiC crystals [101–104]. On the C-terminated (0001) surface, the growth of carbon nanotubes due to Si vaporization has also been reported [105]. The thermal decomposition of a 6H-SiC wafer at increasing temperature from 1080 to 1320 °C led to the layer-by-layer growth of unconstrained, heteroepitaxial thin carbon films [103].

On a TaC(111) surface, mono- and bilayer flakes were formed by CVD of ethylene: the first layer was deposited at 1300 °C to make a layer with good crystalline quality, and then the second layer was formed at 1000 °C [106]. Layer-by-layer formation of the carbon film was observed by measuring the relative intensity ratio of the X-ray photoelectorn spectroscopy (XPS) C1s peak as a function of the exposure L (1 Langmuir=1×10^{-6} Torr.s), as shown in Figure 3.12. In comparison with the formation of the first layer, the second layer grew slowly and the growth rate of the third layer was much slower. The electronic states of the monolayer flake were very different from those of bulk graphite, owing to the hybridization of the π orbitals of the deposit with the d orbitals of the substrate. The lattice constant measured from the LEED pattern was 0.249 nm for the monolayer flake and for the bilayer flake was 0.247 nm, suggesting that the interaction between the flake and the substrate became weak upon the formation of the second layer.

Catalytic CVD of methane has been carried out at 900 °C on HOPG, after the deposition of Fe by an electron-beam sublimator at room temperature [107,108]. The same process has been applied to heal structural defects in the topmost layer of graphite [109]. Diamond nanoparticles prepared by electrophoretic deposition onto the HOPG substrate could be converted to monolayer graphene by heating to 1600 °C: the monolayer graphene had a lateral size of 10–15 nm and was placed epitaxially on the substrate [110]. Changes in the structure and π-electron state during the gradual conversion from nanodiamond to monolayer graphene have been discussed [111]. ZnS ribbons formed on Si have been successfully used as a substrate (template) to synthesize graphene nanoribbons consisting mainly of 10 layers by CVD of methane at 750 °C [112].

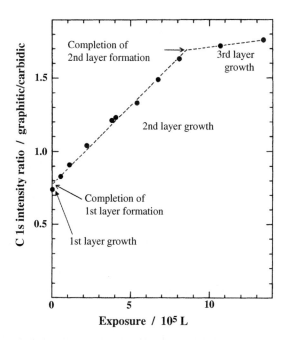

FIGURE 3.12 Change in the Intensity Ratio of the XPS C1s Peak for a Graphitic Carbon Layer to That For the Substrate TaC with Exposure to Ethylene Gas

From [106].

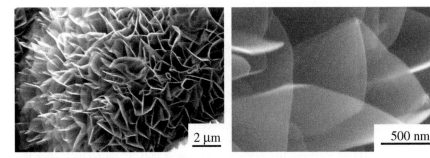

FIGURE 3.13 SEM Images of Graphene Flowers Grown by High-speed CVD Under High-pressure Ar

Courtesy of Mr K. Muramatu of Incubation Alliance Inc., Japan

CVD synthesis of thin flakes has been carried out without any substrate by passing liquid ethanol droplets into microwave-assisted Ar plasma [113]. The powder collected on the filter was easily dispersed in methanol by sonication, in which the presence of mono- and bilayer graphene was confirmed by TEM and electron energy-loss spectroscopy, but most flakes were strongly rippled.

Via CVD of the decomposition gases of polymers under high-pressure Ar gas, aggregates of graphene flakes have been synthesized, called "Graphene Flowers" (trade name) for their appearance, as shown in Figure 3.13 [114,115]. "Petals" of the flower consist of a few graphene layers and their lateral size is 1-20 μm. Their growth rate was very high, so as to be possible to synthesize on a large scale, and they are available on the market as a dispersion in solvent.

3.4 Synthesis through the organic route

The organic synthesis route to graphene has been investigated, although there are various problems in obtaining large-sized graphene, owing to limited solubility and side reactions [116–123]. In Figure 3.14, one of the synthesis strategies toward graphene nanoribbons is outlined [123]. Polymer 5 was successfully synthesized from polymer 1 via the Suzuki-Miyaura coupling process. Polymer 5 in the figure is obtained as a black solid with a yield of 65% [116]. Drop-casting of its THF solution onto a silica surface gave nanoscrolls with a diameter of about 100 nm and a length of up to 5 μm, with well-ordered stacking of the layers [121]. Stacking of the molecules of these polymers could lead to nanotubes with a platelet nanotexture and interlayer spacing of 0.34 nm [124].

An idea for modifying the fragments of graphene via organic processes to construct a three-dimensional open-shell structure has been proposed [125]. These "open-shell graphene fragments" were shown to be of substantial interest from the point of view of fundamental science, as well as in their potential applications in materials science, in particular quantum electronic devices, by using phenalenyl derivatives (consisting of three π-conjugated benzene rings).

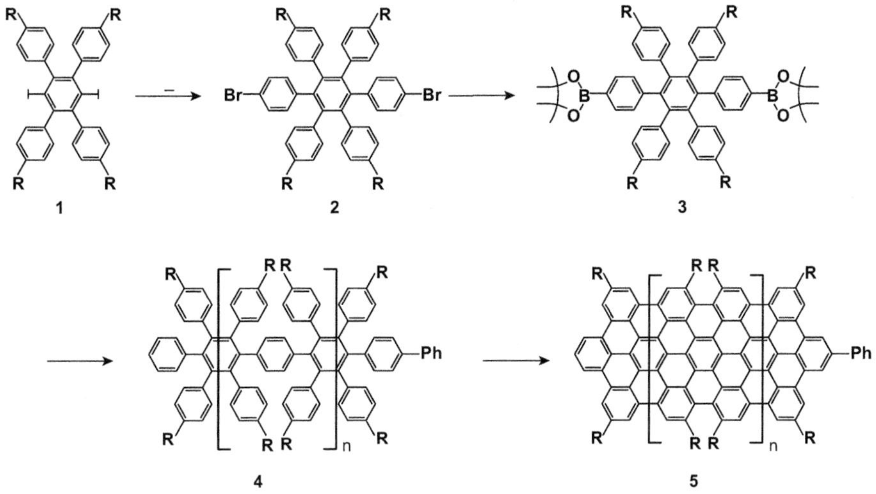

FIGURE 3.14 Scheme to Synthesize Graphene by the Organic Chemistry Route

From [123].

3.5 **Preparation through other processes**

Thin nanoribbons have been synthesized through the pyrolysis of an aerosol [126], as shown in Figure 3.15. The aerosol, prepared from ferrocene and thiophene in ethanol under ultrasonic agitation, was introduced with an Ar flow into a furnace heated to 1223 K. The ribbon-like products, having a width of 20–300 nm and a thickness less than 15 nm, consisted of flat regions and rippled areas, and lengths of several micrometers were obtained. The synthesized ribbon had high crystallinity, revealing three-dimensional electron diffraction patterns, and the presence of graphitic ABAB stacking was confirmed. High-temperature annealing of defects in these nanoribbons was performed by heat treatment at 2800 °C [127]. The formation of multiple loops of layers was observed preferentially at the edges of the ribbon above 1500 °C. Nanoribbons having a thickness of about 0.35 nm, a width of 8 nm, and a length of 1 μm have been prepared by heat treatment of nanoparticles of diamond deposited on HOPG at 1600 °C [128]. On the nanoribbons, a marked irreversibility in adsorption of CO_2 and H_2O was demonstrated, which may indicate a potential in the fabrication of novel types of catalysts and highly selective gas sensors [129].

An alternative route to the synthesis of graphene nanoribbons with straight edges is the unzipping of carbon nanotubes [130]. By plasma etching, graphene nanoribbons with smooth edges and a narrow width of 10–20 nm were successfully obtained [131]. The unzipping process is schematically shown in Figure 3.16. Carbon nanotubes (CNTs) were embedded in PMMA on a Si substrate and then the resultant PMMA/CNT film was peeled from the substrate, where a narrow strip of the CNT side wall was exposed to the atmosphere. This exposed strip was etched by Ar plasma. Mono- and bilayer ribbons were prepared from single-walled and double-walled CNTs (SWCNTs and DWCNTs), respectively. Starting from multi-walled

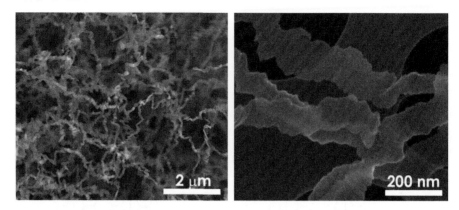

FIGURE 3.15 Nanoribbons Prepared From an Aerosol at 1223 K

From [126].

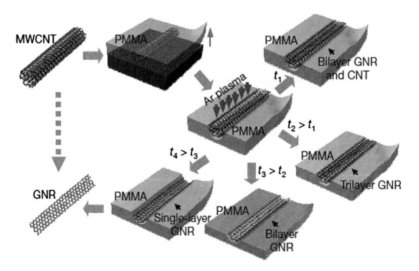

FIGURE 3.16 Unzipping of a Multi-walled Carbon Nanotube by Plasma Etching

Hydrogen arc-discharge.

From [131].

CNTs (MWCNTs) with a diameter of about 8 nm, nanoribbons consisting of a few layers and having a width of 10–20 nm were obtained in a yield of about 20%. The obtained ribbons were uniform in width along their length and very smooth on their edges. The intensity ratio I_D/I_G in the Raman spectrum increased from almost zero for the starting MWCNTs to 0.38–0.28 for the ribbons obtained, which can reasonably be supposed to be mainly due to their open edges. On these nanoribbons, electric-field dependence of conductivity, similar to that reported on thin flakes, was observed. Unzipping of MWCNTs by an abrupt N_2 gas expansion within their hollow core due to a thermal shock has also been performed [132].

Synthesis and characterization of nanoribbons have been reviewed in relation to graphene sheets and carbon nanotubes [4].

3.6 **Concluding remarks**

According to the experimental results published, thin flakes, including monolayer graphene, have different features depending on their preparation route, although this is not very reproducible at present.

Cleavage of graphite crystals is the best method in terms of the structural and electrical qualities of the flakes obtained, but it is challenging to bring this process to large-scale production [2]. The thickness of the thin flakes is difficult to tune through this process, and to get monolayer graphene is very laborious. The size and shape of the resultant flakes are very dependent on the graphite used, as demonstrated through pillaring of HOPG [10], and the crystallinity of the flakes can be high if highly crystalline graphite is used. However, it has to be pointed out that even the highest grade in commercially available HOPG is not a single crystal and consists of grains with different orientations of a-axes, even though c-axis orientation is extremely high [133]. The kish graphite selected for cleavage [8,9] has very high crystallinity, as shown by the high $\varrho_{300K}/\varrho_{4.2K}$ of 32.3, and the flakes after cleavage show a Shubunikov–de Haas oscillation (Figure 3.2B). By using micromechanical cleavage, the patterning of thin sheets can be done on pristine graphite [10,11,14], but special care concerning the crystal perfection of the prepared thin sheets seems to be needed, because the patterning of graphite has been carried out using oxygen plasma [12]. Scanning tunneling microscopy lithography has been successfully applied to control the size of graphene ribbons as well as the crystallographic orientation at their edges [134]. The nanopatterning of graphene flakes has been reviewed, comparing different techniques [135]. Also, the cleavage process has to be performed carefully, because some distortion may be induced in peeled flakes, particularly when flexible adhesive tape is used.

Preparation of thin flakes using graphite oxide (GO) has been reported in a number of papers. The process consists of many steps, including oxidation of the pristine graphite to GO, exfoliation to separate layers, reduction of GO, and annealing of the products. Each step involves various chemical reactions; either oxidation or reduction, associated with the attachment of oxygen-containing functional groups to carbon atoms and with the evolution of CO and CO_2 by removing these groups. Some of the carbon atoms in the pristine graphite are lost as CO and CO_2 during exfoliation and reduction. In addition, the bonding state of almost all carbon atoms has to be changed from sp^2 in pristine graphite to sp^3 in GO during the oxidation process, and vice versa during the exfoliation and reduction processes. These drastic changes in composition and bonding nature are reasonably supposed to give a number of structural imperfections in the carbon layers. For annealing of the layers, heat treatment at high temperatures is usually required. Many experiments have shown that high-temperature treatment is necessary to have less defective carbon layers, but it also improves stacking order of the layers and increases the number of stacked layers. GO flakes must be reduced and annealed at high temperatures, particularly anticipating high electrical conductivity, because of the large content of oxygen-containing functional groups in GO and also structural defects after its reduction. Study of the reduction process of GO flakes under TEM has demonstrated that the content of oxygen atoms detected by electron energy loss spectroscopy (EELS) decreased gradually with step-by-step Joule heating, and became negligibly small only after increasing bias up to about 20 V (thought to be about 2000 °C) [136]. Electrical conductance of the flake increased to about $1.5{\times}10^5$ S/m, a rise of six orders of magnitude, during this reduction process.

The process for the synthesis of monolayer graphene from GO has other problems to be solved. In order to get flakes consisting of single or few layers, the supernatant of the dispersed solution or sol of GO particles has to be used. Therefore, the yield of thin flakes from pristine graphite seems to be very low, less than few %. The thin flakes recovered from the supernatant are a mixture of monolayer to few-layer particles, and reproducibility in the thickness distribution in the product seems to be very poor. Most of the resultant thin flakes contain ripples and folds, and are partly scrolled. The process using a colloidal suspension of GO has been reviewed, focusing on the production of graphene and chemically modified graphene [138].

The process using first-stage graphite intercalation compounds with K (donor-type compound) is thought to give fewer structural defects, because the interaction between the graphite layer and intercalate K is only charge transfer, and the sp^2 bonding nature of carbon atoms in the layer is preserved; in other words, chemical oxidation and a change in the bonding nature in the layers, as occurs with GO, are not needed. From these K-GICs, the main product was nano-sized scrolls. From the first-stage graphite intercalation compound with nitric acid (acceptor-type compound), scrolls were also obtained. The formation of nanoscrolls through this process might open a new science and engineering of quasi-one-dimensional nanocarbons, even though the reproducibility of the nanoscroll formation and the possibility to create nanoscrolls consisting of a monolayer need to be studied, in addition to the reason why they are formed only in the process using intercalation compounds.

Chemical synthesis of graphene via organic processes may have the possibility to obtain graphene-like sheets with homogeneous size, but it might be very difficult to produce monolayer graphene. Unzipping of carbon nanotubes is a scientifically interesting process to control the number of stacked layers, but is difficult to establish as a practical fabrication process. The synthesis of nanoribbons via pyrolysis of organic precursors has the potential to expand the scale of synthesis, but most of the products consist of graphitic regular stacking [126].

The CVD process for the synthesis of graphene seems to be easier for patterning the graphene sheets [76,100] and also has better reproducibility than the other four routes. In the case of using a transition metal substrate, for example Ni, the precipitation of carbon atoms doped into the substrate has to be taken into account [76]. On thin sheets prepared using this route, however, the interaction with the surface of the substrate crystal is important, resulting in the epitaxial growth of graphene [10,97,100] and weakening of C-C bonds [94], although the results of micro-Raman scattering on few-layer sheets formed on SiC suggest a very weak electronic coupling with the substrate [139]. Marked weakening of the interaction with the substrate was reported with increasing number of the layers stacked in the flake [106]. On polycrystalline Cu and Pt substrates, however, single-grain (single-crystal) graphene was successfully grown, suggesting no epitaxial relationship between graphene and the substrate [87,73]. Grain boundary in graphene was reported to have significant effects on the electronic properties of graphene sheets [87], but also reported to give no drastic change, although they severely weaken the mechanical strength of graphene sheets.

Microscopic corrugation, rippling, and partial scrolling at the edges were frequently observed on thin sheets prepared by each of these routes. The formation of ultra-flat graphene was reported through micromechanical peeling by selecting cleaved mica as an appropriate substrate [17]. However, thin sheets prepared through this method were shown to scroll at their edges and also fold when the sheet was detached from the scaffold [14]. Monolayer graphene is susceptible to structural distortions and the suspended graphene flakes show spontaneous rippling of about 1 nm. Therefore, it is very difficult to prepare monolayer graphene without any ripples. In order to avoid any structural distortion and scrolling at the edges, the selection of an appropriate scaffold or substrate seems to be the key factor. This rippling was invoked to explain the thermodynamic stability and distinctive electronic properties [140–144], because monolayer graphene does not have π-stacking, which favors flat structure. Point and line defects in the graphene layer, introduced either intrinsically or extrinsically, have been reviewed in relation to the electrical, magnetic, and mechanical properties [145].

The chemical properties of monolayer graphene are expected to be largely different from those of bulk graphite, although graphite is known as a chemically inert material, and its chemical stability in specific atmospheres is particularly important for its applications in nanodevices. Oxidation of thin flakes was reported to occur in O_2/Ar gas flow at a temperature of 200–600 °C [146]. Monolayer regions in a flake were oxidized faster than three-layer regions, of which the oxidation rate was comparable with bulk natural graphite, and resulted in the random formation of etch pits on the surface. Reversible hydrogenation was observed on thin flakes [146,147]. Monolayer graphene gave a prominent D band in its Raman spectrum after hydrogenation by electron beam irradiation in the presence of hydrogen silsesquioxane (HSQ), although both the pristine graphene and that irradiated by an electron beam without HSQ showed negligibly small D-band intensity. A bilayer flake did not give a D band even after hydrogenation [146]. The hydrogenation rate of monolayer graphene was faster than that of bilayer flakes, as is also the case for their oxidation. Hydrogenation of the monolayer graphene made its lattice spacing shorter, while retaining its hexagonal lattice, and changed its electronic state from zero-overlap semi-metallic to insulating [147]. Site-specific single-atom spectroscopy at a graphene boundary was performed in a low-voltage scanning transmission electron microscope and energy-loss near-edge fine structure of carbon at the edge of the graphene was determined [148], which may open a new insight into the electronic and bonding structures at the edges of graphene and other carbon materials. Edge structure of graphene has been reviewed in relation to its preparation process [149].

The thin flakes may be used either as a nano-component, using their different functions, in various devices or as a transparent conductive thin film with relatively large area. It is also known that the electrical, thermal, and optical properties of flakes depend strongly on their thickness, as shown by a number of papers and also in Figures 3.2 and 3.3. Therefore, the preparation of thin flakes with a homogeneous thickness is also required for device applications. To satisfy this requirement, further improvements are needed in all processes for either preparation or synthesis of graphene. CVD may have the highest possibility to realize homogeneous thickness of deposited film, and the roll-to-roll technique developed [85], which can transfer the large-area thin films to the target substrate, may reinforce the possibility for CVD. For use as a transparent conductive thin film, on the other hand, homogeneity in each flake is not essential, but all flakes are required to be thin enough to prepare a transparent film. In order to prepare transparent films, the supernatant of the suspension of the flakes has to be used, which results in a very low yield, as described above. Exfoliation of graphite in chlorosulfonic acid was reported to be an efficient method to get a supernatant containing monolayer graphene, much more efficient than the dispersion of GO flakes, but the suspension concentration of thin flakes was up to 2 mg/cm^3 [58]. Some papers have claimed that a large amount of thin flakes could be synthesized [150,151], but the products are black powder and not transparent after forming into a film. The opaque black films reported should be compared with thin, flexible graphite sheets, which are industrially prepared from exfoliated graphite and widely used as gaskets, seals, and packings [26]. More precise information is required to be reported on the homogeneity of thickness within flakes and among flakes prepared, and also on the yield of thin flakes.

References

[1] Boehm HP, Setton R, Stumpp E. Carbon 1986;24:241–5.
[2] Soldano C, Mahmood A, Dujardin E. Carbon 2010;48:2127–50.
[3] Allen MJ, Tung VC, Kaner RB. Chem Rev 2010;110:132–45.
[4] Terronesa M, Botello-Méndez AR, Campos-Delgado J, et al. Nano Today 2010;5: 351–72.
[5] Inagaki M, Kim YA, Endo M. J Mater Chem 2011;21:3280–94.
[6] Jia X, Campos-Delgado J, Terrones M, et al. Nanoscale 2011;3:86–95.
[7] Nobelprize.org. The 2010 Nobel Prize in Physics: Press Release; 19 Oct 2010. http://nobel prize.org/nobel_prizes/physics/laureates/2010/press.htm.
[8] Ohashi Y, Koizumi T, Yoshikawa T, et al. TANSO 1997; No.180: 235–8.
[9] Ohashi Y, Hironaka T, Kubo T, et al. TANSO 2000; No. 195: 410–3.
[10] Novoselov KS, Geim AK, Morozov SV, et al. Science 2004;306:666–9.
[11] Zhang Y, Small JP, Pontius WV, et al. Appl Phys Lett 2005;86:073104.

[12] Lu X, Huang H, Nemchuk N, et al. Appl Phys Lett 1999;75:193–5.

[13] Blake P, Hill EW, Castro Neto AH, et al. Appl Phys Lett 2007;91:063124.

[14] Meyer JC, Geim AK, Katsnelson MI, et al. Nature 2007;446:60–3.

[15] Meyer JC, Geim AK, Katsnelson MI, et al. Solid State Commun 2007;143:101–9.

[16] Neubeck S, You YM, Ni ZH, et al. Appl Phys Lett 2010;97:053110.

[17] Lui CH, Liu L, Mak KF, et al. Nature 2009;462:339–41.

[18] Shukia A, Kumar R, Mazher J, et al. Solid State Commun 2009;149:718–21.

[19] Chen G, Weng W, Wu D, Wu C, Lu J, Wang P, Chen X. Carbon 2004;42:753–9.

[20] Savath Z, Darabout Al, Nemes-Incze P, et al. Carbon 2007;45:3022–6.

[21] Hernandez Y, Nicolosi V, Lotya M, et al. Nature Nanotech 2008;3:563–8.

[22] Lotya M, Hernandez Y, King PJ, et al. J Am Chem Soc 2009;131:3611–20.

[23] Economopoulos SP, Rotas G, Miyata Y, et al. ACS Nano 2010;4:7499–507.

[24] Vadukumpully S, Paul J, Valiyaveettil S. Carbon 2009;47:3288–94.

[25] Vadukumpully S, Paul J, Mahanta N, et al. Carbon 2011;49:198–205.

[26] Inagaki M, Kang FY, Toyoda M. Chem Phys Carbon 2004;29:1–69.

[27] Hummers W, Offeman R. J Am Chem Soc 1958;80:1339.

[28] Staudenmaier L. Ber Dtsch Chem Ges 1898;31:1481–7.

[29] Brodie BC. Ann Chim Phys 1860;59:466–72.

[30] Kang F, Leng Y, Zhang TY. Carbon 1997;35:1089–96.

[31] Peckett JW, Trens P, Gougeon RD, et al. Carbon 2000;38:345–53.

[32] Toyoda M, Shimizu A, Iwata H, et al. Carbon 2001;39:1697–707.

[33] Toyoda M, Katoh H, Inagaki M. Carbon 2001;39:2231–7.

[34] Weng W-G, Chen G-H, Wu D-J, et al. Synth Met 2003;139:221–5.

[35] Schloz W, Boehm HP. Z Anorg Allg Chem 1969;369:327–40.

[36] Yazami R, Ph Touzain, Chabre Y, et al. Rev Chim Miner 1985;22:398–411.

[37] Nakajima T, Matsuo Y. Carbon 1994;32:469–75.

[38] Schniepp HC, Li J-L, McAllister MJ, et al. J Phys Chem Lett 2006;110:8535–9.

[39] MacAllister MJ, Li J-L, Adamson DH, et al. Chem Mater 2007;19:4396–404.

[40] Kotov NA, Dekany I, Fendler JH. Adv Mater 1996;8:637–41.

[41] Akhavan O. Carbon 2010;48:509–19.

[42] Gilje S, Han S, Wang M, et al. Nano Lett 2007;7:3394–8.

[43] Tung VC, Allen MJ, Yang Y, et al. Nat Nanotech 2009;4:25–9.

[45] Stankovich S, Piner RD, Chen X, et al. J Mater Chem 2006;16:155–8.

[47] Stankovich S, Dikin DA, Dommett GHB, et al. Nature 2006;442:282–6.

[48] Yan J, Wei T, Shao B, et al. Carbon 2010;48:487–93.

[49] Yu D, Dai L. J Phys Chem Lett 2010;1:467–70.

[50] Bourlinos AB, Gournis D, Petridis D, et al. Langmuir 2003;19:6050–5.

[51] Chen W, Yan L, Bangal PR. Carbon 2010;48:1146–52.

[52] Williams G, Seger B, Kamat PV. ACS Nano 2008;2:1487–91.

[53] Wu Z-S, Ren W, Gao L, et al. Carbon 2009;47:493–9.

[54] Wang SJ, Geng Y, Zheng Q, et al. Carbon 2010;48:1815–23.

[55] Wu Z-S, Ren W, Gao L, et al. ACS Nano 2009;3:411–7.

[56] Chen C, Yang Q-H, Yang Y, et al. Adv Mater 2009;21:3007–11.

[57] Wei Z, Barlow DE, Sheehan PE. Nano Lett 2008;8:3141–5.

[58] Behabtu N, Lomeda JR, Green MJ, et al. Nat Nanotechnol 2010;5:406–11.

[59] Li X, Zhang G, Bai X, et al. Nat Nanotechnol 2008;3:538–42.

[60] Su C-Y, Lu A-Y, Xu Y, et al. ACS Nano 2011;5:2332–9.

[61] Wang J, Manga KK, Bao Q, et al. J Am Chem Soc 2011;133:8888–91.

[62] Shioyama H. J Mater Sci Lett 2001;20:499–500.

[63] Shioyama H. J Mater Chem 2001;11:3307–9.

[64] Shioyama H, Akita T. Carbon 2003;41:179–81.

[65] Viculis LM, Mack JJ, Kaner RB. Science 2003;299:1361.

[66] Valles C, Drummond C, Saadaoui H, et al. J Am Chem Soc 2008;130:15802–4.

[67] Savoskin MV, Mochalin VN, Yaroshenko AP, et al. Carbon 2007;45:2797–800.

[68] Oshima C, Nagashima A. J Phys Condens Matter 1997;9:1–20.

[69] Gall NR, Rut'kov EV, Tontegode AV. Int J Mod Phys B 1997;11:1865–911.

[70] Wintterlin J, Boequet M-L. Surf Sci 2009;603:1841–52.

[71] Land TA, Michely T, Behm RJ, et al. Surf Sci 1992;264:261–70.

[72] Fujita T, Kobayashi W, Oshima C. Surf Interface Anal 2005;37:120–3.

[73] Gao L, Ren W, Xu H, et al. Nat Commun 2012;3:699.

[74] Ueta H, Saida M, Nakai C, et al. Surf Sci 2004;560:183–90.

[75] Park HJ, Meyer J, Roth S, et al. Carbon 2010;48:1088–94.

[76] Kim KS, Zhao Y, Jang H, et al. Nature 2009;457:706–10.

[77] Papagno L, Caputi L. Phys Rev B 1984;29:1483–6.

[78] Rokuta E, Hasegawa Y, Itoh A, et al. Surf Sci 1999;427-428:97–101.

[79] Reins A, Jia X, Ho J, et al. Nano Lett 2009;9:30–5.

[80] Nagashima A, Gamou Y, Terai M, et al. Phys Rev B 1996;54:13491–4.

[81] Oshima C, Itoh A, Rokuta E, et al. Solid State Commun 2000;116:37–40.

[82] Tanaka T, Itoh A, Yamashita K, et al. Surf Rev Lett 2003;10:697–703.

[83] Li X, Cai W, An J, et al. Science 2009;324:1312–4.

[84] Li X, Zhu Y, Cai W, et al. Nano Lett 2009;9:4359–63.

[85] Bae S, Kim H, Lee Y, et al. Nat Nanotechnol 2010;5:574–8.

[86] Juang Z-Y, Wu C-Y, Lu A-Y, et al. Carbon 2010;48:3169–74.

[87] Rasool HI, Song EB, Mecklenburg M, et al. J. Am Chem Soc 2011;133:12536–43.

[88] Yu Q, Jauregui LA, Wu W, et al. Nat Mater 2011;10:443–9.

[89] Kim J, Ishihara M, Koga Y, et al. Appl Phys Lett 2011;98:091502.

[90] Coraux J, N'Diaye AT, Busse C, et al. Nano Lett 2008;8:565–70.

[91] Sutter PW, Flege J-I, Sutter EA. Nat Mater 2008;7:406–11.

[92] Marchini S, Guenther S, Wintterlin J. Phys Rev B 2007;76:075429.

[93] Vaari J, Lahtinen J, Hautojarvi P. Catal Lett 1997;44:43–9.

[94] Nagashima A, Nuka K, Itoh H, et al. Surf Sci 1993;291:93–8.

[95] Hasegawa N, Gamo Y, Terai M, et al. TANSO 1997; No. 180: 229–234 [in Japanese].

[96] Terai M, Hasegawa N, Okusawa M, et al. Appl Surf Sci 1998;130-132:876–82.

[97] Itoh H, Ichinose T, Oshima C, et al. Surf Sci Lett 1991;254:L437–42.

[98] Berger C, Song Z, Li T, et al. J Phys Chem B 2004;108:19912–6.

[99] Park JH, Mitchel WC, Smith HE, et al. Carbon 2010;48:1670–92.

[100] Berger C, Song Z, Li X, et al. Science 2006;312:1191–6.

[101] Van Bommel AJ, Crombeen JE, Van ToTooren A. Surf Sci 1975;48:463–72.

[102] Faubeaux I, Themlin J-M, Charrier A, et al. Appl Surf Sci 2000;162/163:406–12.

[103] Charrier A, Coati A, Agunova T, et al. J Appl Phys 2002;92:2479–84.

[104] Hass J, Feng R, Li T, et al. Appl Phys Lett 2006;89:143106.

[105] Kusunoki M, Suzuki T, Hirayama T, et al. Appl Phys Lett 2000;77:531–4.

[106] Nagashima A, Itoh H, Ichinokawa T, et al. Phys Rev B 1994;50:4756–63.

[107] Kholmanov I, Cavaliere E, Fanetti M, et al. Phys Rev B 2009;79:233403.

[108] Kholmanov IN, Cavaliere E, Cepek C, et al. Carbon 2010;48:1619–25.

[109] Kholmanov IN, Edgeworth J, Cavaliere E, et al. Adv Mater 2011;23:1675–8.

[110] Affoune AM, Prasad BLV, Sato H, et al. Chem Phys Lett 2001;348:17–20.
[111] Enoki T. Phys Solid State 2004;46:635–40.
[112] Wei D, Liu Y, Zhang H, et al. J Am Chem Soc 2009;131:11147–54.
[113] Dato A, Radmilorie V, Lee Z, et al. Nano Lett 2008;8:2012–6.
[114] Muramatu K. TANSO 2013;No. 256:60–66 [in Japanese].
[115] Muramatu K, Sutani K, Toyoda M. Converter 2012;Nov/Dec:2–7.
[116] Simpson CD, Brand JD, Berresheim AJ, et al. Chem Eur J 2002;8:1424–9.
[117] Wu J, Gherghel L, Watson MD, et al. Macromolecules 2003;36:7082–9.
[118] Tahara K, Tobe Y. Chem Rev 2006;106:5274–90.
[119] Wu J, Pisula W, Muellen K. Chem Rev 2007;107:718–47.
[120] Dou X, Yang X, Bodwell GJ, et al. Org Lett 2007;9:2485–8.
[121] Hamaoui BE, Zhi L, Pisula W, et al. Chem Commun 2007:2384–6.
[122] Wu D, Zhi L, Bodwell GJ, et al. Angew Chem Int Ed 2007;46:5417–20.
[123] Yang X, Douy X, Rouhanipour A, et al. J Am Chem Soc 2008;130:4216–7.
[124] Zhi L, Wu J, Li J, et al. Angew Chem Int Ed 2005;44:2120–3.
[125] Morita Y, Suzuki S, Sato K, et al. Nat Chem 2011;3:197–204.
[126] Campos-Delgado J, Romo-Herrera JM, Jia X, et al. Nano Lett 2008;8:2773–8.
[127] Campos-Delgado J, Kim YA, Hayashi T, et al. Chem Phys Lett 2009;469:177–82.
[128] Cancado LG, Pimenta MA, Medeiros-Ribeiro G, et al. Phys Rev Lett 2004;93;047403.
[129] Asai M, Ohba T, Iwanaga T, et al. J Am Chem Soc 2011;133:14880–3.
[130] Kosynkin DV, Higginbotham AL, Sinitskii A, et al. Nature 2009;458:872–6.
[131] Jiao L, Zhang L, Wang X, et al. Nature 2009;458:877–80.
[132] Morelos-Gomez A, Vega-Dıaz SM, Gonzalez VJ, et al. ACS Nano 2012;6:2261–72.
[133] Yoshida A, Hishiyama Y. J Mater Res 1992;7:1400–5.
[134] Tapaszto L, Dobrik G, Lambin P, Biro LP. Nat Nanotechnol 2008;3:397–401.
[135] Biro LP, Lambin P. Carbon 2010;48:2677–89.
[136] Xu Z, Bando Y, Liu L, et al. ACS Nano 2011;5:4401–6.
[138] Faugeras C, Nerriere A, Potemski M, et al. Appl Phys Lett 2008;92; 011914.
[139] Fasolino A, Los JH, Katsnelson MI. Nat Mater 2007;6:858–61.
[140] Huang PY, Ruiz-Vargas CS, Van der Zande AM, et al. Nature 2011;469:389–93.
[141] Morozov SV, Novoselov KS, Katsnelson MI, et al. Phys Rev Lett 2006;97;016801.
[142] Martin J, Akerman N, Ulbricht G, et al. Nat Phys 2008;4:144–8.
[143] Deshpande A, Bao W, Miao F, et al. Phys Rev B 2009;79:205411.
[144] Banhart F, Kotaboski J, Krasheninnikov AV. ACS Nano 2011;5:26–41.
[145] Liu L, Ryu S, Tomasik MR, et al. Nano Lett 2008;8:1965–70.
[146] Ryu S, Han MY, Maultzsch J, et al. Nano Lett 2008;8:4597–602.
[147] Elias DC, Nair RR, Mohiuddin MG, et al. Science 2009;323:610–3.
[148] Suenaga K, Koshino M. Nature 2010;468:1088–90.
[149] Jia X, Campos-Delgado J, Terrones M, et al. Nanoscale 2011;3:86–95.
[150] Li D, Mueller MB, Gilje S, et al. Nat Nanotechnol 2008;3:101–5.
[151] Choucair M, Thordarson P, Stride JA. Nat Nanotechnol 2009;4:30–3.

Further Reading

[44] Fan X, Peng W, Li Y, et al. Adv Mater 2008;20:4490–3.
[46] Kotov NA. Nature 2006;442:254–5.
[137] Park S, Ruoff RS. Nat Nanotechnol 2009;4:217–24.

Carbonization Under Pressure

Most of the reactions causing carbonization of precursors are accompanied by the evolution of various gases—different hydrocarbons, carbon oxides, and hydrogen, as shown in Figure 4.1—so that pressurization modifies the carbonization reactions. As a consequence, the carbonization behavior of the carbon precursor changes under pressure and the resultant carbon is different in structure, properties, and even particle morphology from that obtained without pressure. The departure of various hydrocarbons and carbon oxides results in the loss of carbon atoms from the precursor; in other words, lowers the carbonization yield. Under pressure, therefore, departure of these gases, particularly hydrocarbon gases, is suppressed and the carbonization yield may increase. It may be possible to get carbon residues from some organic polymers that depolymerize at high temperatures under normal pressure and do not give any carbon residues. In addition, pressure may influence the solubility, viscosity, density, and phase separation of carbonaceous intermediates and consequently the graphitization behavior of the resultant carbons. Carbonization under pressure is expected to give the novel possibility of controlling the structure and texture of resultant carbons. The principal purpose for carbonization under pressure is the modification of carbonization behavior to improve the carbon yield, to densify the resultant carbon, to change graphitization behavior, and also to obtain specific particle morphology of the resultant carbons.

There has been much experimental work performed on carbonization under pressure, which may be classified into three routes: (1) carbonization under pressure built up by the decomposition gases of the precursor, (2) carbonization under hydrothermal conditions, and (3) reduction of CO_2 under pressure. Here, the experimental work on carbonization under pressure is reviewed, mainly focusing on the yield and morphology of resultant carbons in relation to the carbonization conditions. The conditions to obtain spherical carbon materials are discussed in relation to the temperature-pressure conditions during carbonization and also the chemical composition of precursors. Carbonization under built-up pressure has been reviewed, focusing on the formation of carbon spherules [1], and also including hydrothermal and supercritical conditions [2].

Prerequisite for readers: Chapter 2.4 (Novel techniques for carbonization) in *Carbon Materials Science and Engineering: From Fundamentals to Applications*, Tsinghua University Press.

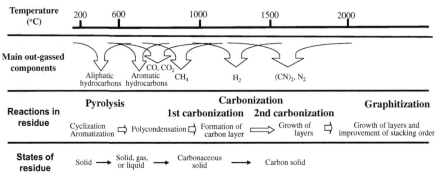

FIGURE 4.1 The Carbonization Process

4.1 Carbonization under built-up pressure

4.1.1 Setup for carbonization under pressure

Carbonization under pressure can be done in the atmosphere of the gases produced by the carbon precursor during its decomposition (pyrolysis and carbonization) through two procedures: (1) heating the carbon precursor from room temperature to carbonization temperature under a constant pressure and (2) building up the pressure gradually with increasing temperature by the decomposition gases in a closed vessel.

By the first procedure, pressure can be kept constant but special equipment, an autoclave, has to be used for heating and compressing the samples up to carbonization temperature. The precursor is sealed into a thin-walled gold capsule and heated up to a programmed temperature in a pressure bomb, where pressure built up inside the capsule owing to the evolution of decomposition gases from the carbon precursor is balanced with the pressure in the bomb due to the expansion/shrinkage of the gold capsule. The carbonization is performed under homogeneity in temperature and pressure. When the gold capsule is broken during carbonization treatment, the obtained carbon is excluded from the samples for further investigation, except in the case where the capsules are intentionally not sealed. By the second procedure, pressure is gradually built up by increasing the temperature, but it is difficult to keep constant, and depends strongly on the temperature and also the amount of sample used.

4.1.2 Optical texture and carbonization yield

The powder mixtures of a pitch with phenol resin, novolac- or resol-type, have been carbonized at 600 °C under 30 MPa [3,4]. The heat treatment of the precursor mixtures was performed both in a sealed gold capsule (closed system) and in an open capsule (open system), and the results were compared with those obtained under atmospheric pressure of an inert gas. Pressure carbonization modified the optical texture of the resultant carbons from coarse mosaic to isotropic by gradual decrease in the size of mosaic units with increasing mixing ratio of phenolic resin, as shown in Figure 4.2. Under atmospheric pressure, these two precursors had to be mixed by

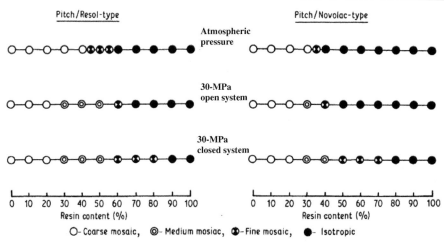

FIGURE 4.2 Optical Texture of Carbon Prepared From the Mixture of a Pitch with Phenol Resin

Pitch mixed with resol- and novolac-type resins, at 600 °C and 30 MPa.

From [4]

using pyridine as a common solvent, in order to get a homogeneous optical texture. Even so, the optical texture of the resultant carbons changed suddenly from coarse mosaic to isotropic at a phenol resin content of around 50 mass%, with a narrow range of fine mosaic texture. Under pressure, on the other hand, the change in optical texture occurred gradually with increasing mixing ratio of phenol resin, in particular in the closed system.

As a consequence of the modification of optical texture, graphitizability of the resultant carbons changed from graphitizing to non-graphitizing gradually, with the increase in mixing ratio of phenol resin. A high carbonization yield, more than 80 mass%, was obtained in a sealed tube under 30 MPa (in the closed system), although it was about 60 mass% under atmospheric pressure and also in the open system even under pressure. Similar changes in carbonization yield and optical texture were observed in the mixture of the benzene-soluble fraction of the pitch with phenol resin under pressure [5].

The improvement in carbonization yield was found to depend strongly on the precursor. A high carbonization yield at 650 °C was obtained under 30 MPa for the pitches fractionated by using organic solvents [6]. As shown in Figure 4.3, the fractions with lower molecular weight, BS-1 to BS-5 (fractionated by using the mixture of benzene and n-hexane in different ratios), gave marked improvement in carbonization yield under different pressure conditions, particularly in the closed system, in comparison with the benzene-insoluble but pyridine-soluble fraction (SI-PS) and pyridine-insoluble fraction (PI). Even from volatile organics, such as polyethylene (PE) that is vaporized out under atmospheric pressure by depolymerization, carbon residues were obtained under pressure in the closed system [7].

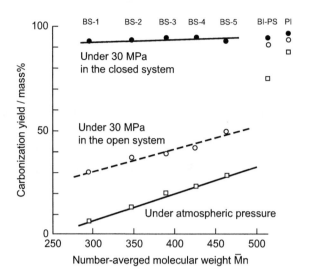

FIGURE 4.3 Carbonization Yield as a Function of the Molecular Weight of the Fractionated Pitches and of the Pressure During Carbonization

From [6]

4.1.3 Particle morphology

Carbons with botryoidal particles consisting of spherical primary particles have been obtained from anthracene, biphenyl, and their mixtures under a pressure above 200 MPa at a temperature around 550 °C [8]. The primary particles were thought to be mesophase spheres with anisotropic nanotexture and the pressure was thought to prevent their complete coalescence to bulk mesophase. The formation conditions of mesophase and the graphitizability of the resultant cokes have been studied [9]. On the mixtures of a pitch with poly(vinyl chloride) (PVC), the morphology of particles changed from lump-type to botryoidal with gradual diminution in the size of the primary particles under 30 MPa [10].

The carbonization of a naphtha-tar pitch and aromatic hydrocarbons, anthracene and phenanthrene, was carried out at 360–700 °C under 60 MPa [11]. Some acceleration of the formation of mesophase spheres and also an increase in carbonization yield were observed. A vertical distribution of texture along the carbonized specimen is observed under polarized-light optical microscopy, as shown in Figure 4.4. In the upper part of the specimen (part I), mesophase spheres with different sizes are scattered in the isotropic matrix. In the middle part (part II), mesophase spheres in high concentration, appearing separately and having a homogeneous size, are observed. Mosaic texture due to the complete coalescence of mesophase spheres is seen in the lower part (part III). At the bottom, there are deformed mesophase spheres and incompletely coalesced texture, probably due to the presence of free carbon particles (not shown in the figure). This vertical distribution in the texture was thought to be formed by the precipitation of textured particles (mesophase spheres) with different

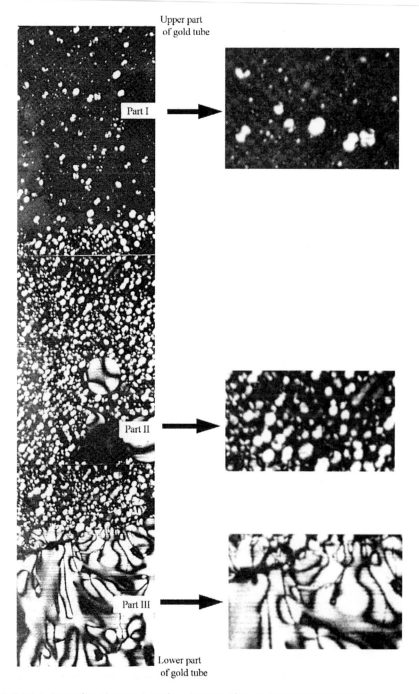

FIGURE 4.4 Formation of Mesophase in a Coal-tar Pitch at 700 °C Under 60 MPa

specific gravities in the molten specimen. On a naphtha-tar pitch, the anisotropic area in bulk mesophase became larger with increasing the pressure from 5 to 11 MPa [12].

Powder mixtures of PE with 10 and 30 mass% PVC have been carbonized in a gold tube under a pressure of 10–30 MPa at a temperature of 300–650 °C [7,13]. Above 600 °C under 30 MPa, the solid carbon was obtained in a yield of about 40 mass%. By the reduction of pressure from 30 to 10 MPa, the solid carbon was obtained above 550 °C. The resultant carbon found at the bottom of the gold tube consisted of carbon spheres with homogeneous size of 1–2.5 μm, as shown by the scanning electron microscopy (SEM) image in Figure 4.5A. Nanotexture of these spheres was found to be radial point orientation by high-resolution transmission electron microscopy [14], as shown schematically in Figure 4.5B. These spheres have a nanotexture a little different from mesophase spheres, which have often been observed in pitches: radial orientation even at the center of the carbon sphere is produced from PE/PVC mixtures under pressure, but radial orientation is not seen at the center of mesophase spheres.

The addition of 5–20 mass% PVC produced carbon spheres at a yield of about 45 mass%. Above 20 mass% PVC, the yield of carbon increased but aggregated botryoidal particles existed in the resultant carbon. Carbon obtained above 75 mass% PVC had coke-like morphology. The addition of poly(vinylidene chloride) (PVDC) was found to have the same effect on pressure carbonization of PE [15]. Polypropylene (PP) was able to replace PE and produce carbon spheres through pressure carbonization. On the other hand, polystyrene (PSt) was found not to be carbonized up to 650 °C, suggesting that a much higher temperature would be needed to get carbon spheres. In Figure 4.6, the carbon spheres obtained are shown.

In order to study what molecular structure of organic precursors gave carbon spheres under pressure, various dioctyl and dicetyl esters and their mixtures with either PE or PVC were carbonized at 650 °C under 30 MPa [16,17]. The

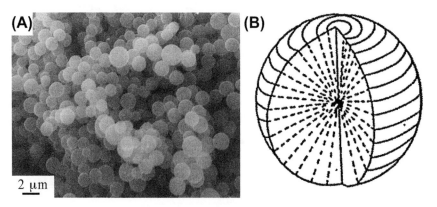

FIGURE 4.5 Carbon Spherules Formed From the Mixture of PE with PVC Under Pressure

(A) SEM image, and (B) scheme of the nanotexture.

From [14]

precursors used are listed in Figure 4.7 with their abbreviations, along with SEM images of the carbons obtained from these esters. Single spheres with uniform size of 2–3 µm were obtained from a dioctyl ester, DOP. From other dioctyl esters, DOS, DOMO, and DOME, spherical particles of carbon were also obtained, but some of them were ellipsoidal rather than spherical. The mixtures of DOME with 20 and 30 mass% PE also gave well-separated single carbon spheres, of 4-6 µm with a small amount of ellipsoidal particles. Although coalesced particles were produced from the dicetyl esters DCP and DCS, the addition of 5 and 10 mass% PVC to DCP resulted in single spherules of 4 and 2.5 µm size, respectively, and the addition of 10 mass% PVC into DCS was effective to get single spheres of 2 µm size.

From a natural polymer (outer walls of fresh alga after elimination of the cells) consisting of long-chain aliphatic groups (up to 31 carbon atoms) bonded to each other by oxygen (ether and ester functional groups), single carbon spheres were obtained under 30 MPa at 500–650 °C [18,19].

PE + 5 mass% PVC   PP + 20 mass% PVC

PE + 10 mass% PVDC   PP + 10 mass% PVDC

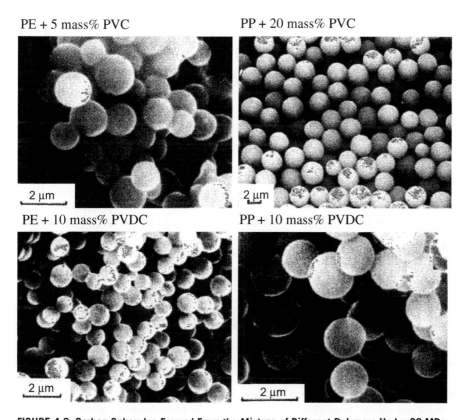

FIGURE 4.6 Carbon Spherules Formed From the Mixture of Different Polymers Under 30 MPa

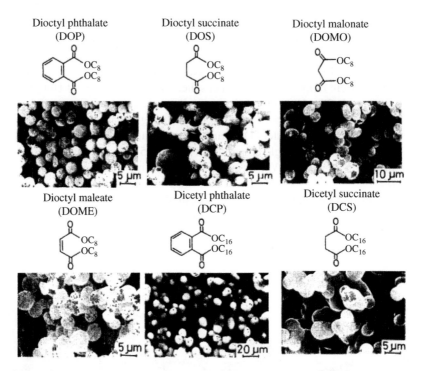

FIGURE 4.7 **Formation of Carbon Spheres From Different Organic Compounds at 650 °C Under 30 MPa**

From [16].

From divinylbenzene, carbon spheres with isotropic glass-like carbon nanotexture have been synthesized at 700 °C under 100–125 MPa [20,21]. In an autoclave, carbon spheres with a size of 1–2 µm were obtained from absolute ethanol with nickel acetate at 600 °C [22]. Ethanol was able to give carbon microbeads in an autoclave at 500–600 °C for 48 h in coexistence with $LaNi_5$ as a catalyst [23]. The carbon microbeads were composed of spherical and spindle-like particles; the former had a diameter of 2–5 µm and the latter had a diameter of 2–4 µm and a length of 5–10 µm.

4.2 Carbonization under hydrothermal conditions

Carbonization under hydrothermal conditions (hydrothermal carbonization) can be divided into two categories, depending on the temperature applied: high temperatures above 400 °C and low temperatures below 250 °C. Hydrothermal carbonization at high temperature gives various carbon materials, multi-walled carbon nanotubes, fullerenes, and carbon spheres with different nanotextures. In hydrothermal carbonization at a low temperature, the pressure is usually built up by water vapor and so depends on the volume of the autoclave and also on the amount of water. Therefore, the higher temperature causes higher pressure, and consequently the pressure is not presented in

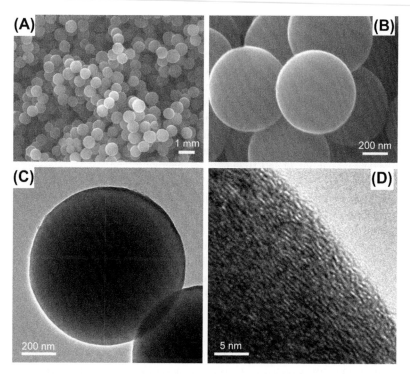

FIGURE 4.8 Carbon Spheres Formed From Benzene in Supercritical Water at 400 °C Under 48 MPa

(A and B) SEM images, (C and D) TEM images.

From [28]

most reports in the literature. Hydrothermal carbonization at low temperature has been applied to various sugars, where carbon spheres have been obtained as main products.

Hydrothermal processing of PE at 700–800 °C under 60–100 MPa in the presence of Ni gave multi-walled carbon nanotubes (MWCNTs) [24]. The PE specimens were placed in a gold capsule together with Ni metal powder and water. The characteristics of the MWCNTs synthesized are their thin walls (about 10% of the inner diameters of 20–800 nm) and their crystallinity, which was reported to be comparable to that of MWCNTs prepared by arc-discharging. The same hydrothermal processing on amorphous carbons, which were prepared from natural woods with phenol resin, without intensive addition of nickel resulted in multi-walled carbon nanocells of less than 100 nm, together with MWCNTs [25,26]. MWCNTs synthesized from amorphous carbon under hydrothermal conditions at 800 °C and 100 MPa showed a low Raman intensity ratio, I_D/I_G, of 0.03, suggesting high structural perfection [27].

In supercritical water (above 374 °C and 22.1 MPa), single carbon spheres with submicron size and a nanotexture of concentric point orientation were obtained from benzene by adding 53 mass% hydrogen peroxide (H_2O_2) at 400 °C and 48 MPa [28]. SEM and transmission electron microscopy (TEM) images are shown in Figure 4.8. The size of the spheres was homogeneous, the mean diameter being 720 ± 112 nm. From *n*-hexane, on the other hand, botryoidal particles (interlinked spheres) were obtained in the same temperature conditions but under 71 MPa by the addition of 48 mass% H_2O_2. The so-called supercritical water oxidation is widely applied for the treatment of organic waste [29,30], where a stoichiometric amount of H_2O_2 is used to complete the oxidation of organic compounds. However, the addition of only 6–8 mol% of stoichiometric H_2O_2 was found to result in the formation of carbon spheres in supercritical water. H_2O_2 was thought to act as an initiator of radical reactions of the precursors. The yield of carbon from benzene and *n*-hexane was about 34 and 10 mass%, respectively. The same processing using ferrocene gave magnetic particles of nano- to micrometer sizes, which were composed of an iron-oxide core, i.e. magnetite Fe_3O_4 and maghemite γ-Fe_2O_3, and a carbon shell [31].

Supercritical water has been used as an activating agent for glass-like carbon spheres prepared from novolac-type phenol resin [32]. Mesoporous carbon spheres with surface area determined by Brunauer-Emmett-Teller method (BET surface area) of 920 m^2/g and total pore volume of 0.45 cm^3/g were obtained in supercritical water at 650 °C under 36 MPa, the development of mesopores being a little more marked in comparison with conventional steam activation.

Hydrothermal carbonization has been applied to various biomasses, including sucrose, glucose, fructose, cyclodextrin, and starch, and biomass derivates, such as 5-hydroxymethyl-furfural-1-aldehyde (HMF) and furfural, to synthesize carbon spheres below 240 °C [33–40]. A review on hydrothermal carbonization applied to biomass has been published [41]. The size of carbon spheres depends strongly on the carbonization conditions, the precursor saccharide used and its concentration, the hydrothermal temperature, and holding time, etc. Under hydrothermal conditions at 190 °C, the sphere size was about 0.25 μm from a saccharide solution of 0.15 mol/L and about 5 μm from 1.5–3.0 mol/L [33]. Carbon spheres prepared from α-, β-, and γ-cyclodextrins at 160 °C under about 0.4 MPa had sizes of 70–150 nm, and those from the latter two cyclodextrins were composed of a hydrophobic core with a hydrophilic shell, the shell having more functional groups containing -OH [36]. The elemental composition of these spheres was expressed by the atomic ratios of O/C = 0.33–0.37 and H/C = 0.73–0.78 [40]. The spheres synthesized below 240 °C were able to keep their spherical morphology after heat treatment at 1000 °C in argon flow [33]. By mixing either $HAuCl_4$ or colloidal Ag particles into glucose solution, carbon spheres with either Pt or Ag nanoparticles at the core have been synthesized [34]. Loading of nanoparticles of different noble metals, such as Au, Ag, and Pd, has been performed [40,42,43]. These carbon spheres were used as a template for the preparation of hollow spheres of inorganic compounds, such as SiO_2, Ga_2O_3, GaN, and WO_3 [35,44–46]. From the mixture of monosaccharides (xylose and fructose) with phenolic compounds (phenol, resorcinol, and phloroglucinol), carbon spheres were synthesized under hydrothermal conditions at 130–170 °C, having a size of 170 nm to 10 μm and the elemental composition of O/C = 0.37–0.56 and H/C = 0.78–0.99 [47]. From cellulose dispersed in water, carbon spheres with a size of 2–5 μm have also been obtained at 220–250 °C [48].

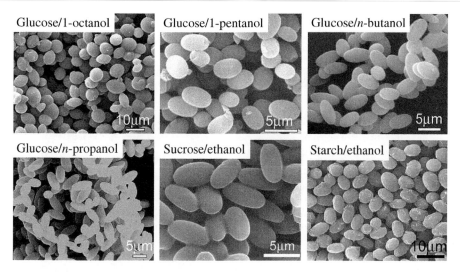

FIGURE 4.9 Morphology of Carbon Particles Obtained From Different Carbohydrate/Alcohol Solutions Under Hydrothermal Conditions at 550 °C for 12 h

From [52].

Single carbon spheres of 1–2 μm have been synthesized from glucose at 500 °C for 12 h, but at 600 °C the spheres were aggregated [49].

Synthesis of carbon materials has been performed in a Teflon-lined autoclave by using different solvents with a carbon precursor (solvothermal synthesis). Carbon materials with different morphologies—hollow spheres, nanosheets, nanoparticles, flakes, etc.—were obtained from the mixture of various carbon precursors (tribromo-compounds of phenol, benzene, toluene, aniline, and benzoic acid) with ferrocene in the mixed solvents of toluene and ethanol at 250 °C [50]. Hollow carbon microspheres with a diameter of 300 nm to 5 μm were obtained from 2,4,6-tribromophenol mixed with ferrocene by the addition of ammonia in various fractions. The thickness of the walls of the microspheres was 50–70 nm. The heat-treatment of a mixture of metallic sodium and ethanol in equal molar ratio at 220 °C gave carbon nanosheets, after pyrolysis and sonication [51]. The nanosheets had an elemental composition of 78.5% C, 2.7% H, and 18.8% O. Mono-dispersed hollow spheres of carbon were obtained by solvothermal treatment of a toluene/ethanol solution of 2,4,6-tribromophenol and ferrocene with ammonia at 250 °C; the surface of these spheres was modified by amino and hydroxyl groups [47]. Spheroidal particles of carbon have been synthesized by carbonization of the alcohol solutions of carbohydrates at 550 °C in an autoclave (supposedly under 22.8 MPa) [52]. Morphology of the carbon particles depends on the carbohydrate and alcohol used, as shown in Figure 4.9. When an ethanol solution of glucose was used, colloidal spheres with a size of 200–300 nm were formed after heating to 200 °C, and their diameters increased to 500–600 nm without appreciable aggregation at 300 °C. Above 350 °C, the spheres tended to be fused and aggregated into larger sizes above 350 °C, and changed to spheroidal particles above 550 °C.

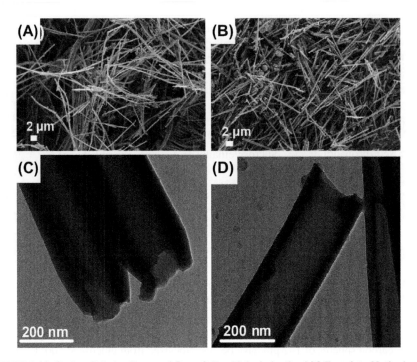

FIGURE 4.10 Carbon Tubules Prepared From 2-Furaldehyde by the AAO Template Method Under Hydrothermal Conditions

(A and C) as-prepared, (B and D) after treatment at 750 °C; A and B are SEM images and C and D are TEM.

Courtesy of Dr S. Kubo of AIST, Japan

Hydrothermal carbonization of carbohydrates has been performed coupled with template carbonization to control the morphology and pore structure in the resultant carbon [53]. Carbon tubules were obtained by using anodic aluminum oxide membranes with pores of c. 200 nm diameter [53,54]. Uniform diameter and length of tubular carbon can be obtained after heat treatment to 750 °C, as shown the SEM and TEM images in Figure 4.10.

4.3 Carbonization under supercritical conditions

The reduction of CO_2 under its supercritical condition (critical point: 31 °C and 7.4 MPa) using metals as reductants was reported to give various carbon materials: carbon nanotubes, diamond, graphite, amorphous carbon, and carbon spheres [55–60]. Most experiments were performed by sealing high-purity CO_2 gas together with a metal, either an alkali or an alkaline-earth metal, in an autoclave. The recovered products, after washing out the metal carbonate, are usually mixtures of different carbon materials.

Crystalline diamond particles, accompanied by some graphite particles, have been synthesized at a temperature of 440–600 °C under about 80 MPa by using alkali metals (Li, Na, and K) as the reductant [56,59]. The size and morphology of diamond crystals depended strongly on the reductant alkali metal: the size of the crystals synthesized using Li and Na reached up to about 4.3 and 250 μm, respectively. For small crystals, octahedral morphology was observed, but the use of K resulted in large-sized hexagonal crystals, up to 450 μm. Alkali metals with lower melting point gave a higher yield of diamond. At a low temperature of 400 °C, no diamond was detected.

The formation of carbon nanotubes (CNTs) occurred when alkali metal (Li) and alkaline-earth metals (Mg and Ca) were used as reductant [55,57,58,60]. The reduction of supercritical CO_2 with metallic Mg at 1000 °C under 1000 MPa gave CNTs together with nested fullerenes [55]. By using Li, CNTs were obtained at a temperature above 550 °C under 71 MPa [57]. They had relatively large diameters and lengths, up to about 50 nm in diameter and larger than 1.5 μm in length. The CNTs obtained at 700 °C had relatively high crystallinity and significantly increased lengths, although the yield of CNTs decreased. Also, carbon spheres have been obtained together with CNTs under a high pressure at 101 MPa and 650 °C in the coexistence of Li [58]. The carbon spheres obtained had a size of 70–900 nm and a core-shell structure, consisting of an amorphous core and a crystalline shell. Carbon spheres with a size of 500–1500 nm were the main product at 550 °C in the coexistence of Ca [60]. At 450 °C no reduction was detected and at 500 °C carbon nano-rods and graphite were formed. With an increase in temperature above 550 °C, the particles deformed to elliptic with larger sizes. No formation of CNTs was observed in the coexistence of Ca.

By heating a mixture of metallic Mg and $Co(CO)_3NO$ to 900 °C in an autoclave, carbon nanoflasks and CNTs were obtained, the CNTs being filled with cobalt metals of either continuous rods or nanoparticles [61,62]. From the mixture of metallic Mg, Na_2CO_3, and CCl_4 in benzene, hollow carbon spheres were obtained together with some carbon particles having filamentous and sheet-like morphologies at 450 °C in an autoclave [63]. The outer diameter of the hollow spheres was 150–600 nm and their wall thickness was about 6 nm. The proposed mechanism is that Na_2CO_3 is reduced by Mg to form carbon with metallic Na and MgO, and then metallic Na reduces CCl_4 to give carbon and NaCl. Carbon sources were thought to be Na_2CO_3 and CCl_4, but the contribution of benzene to the formation of carbon was not discussed.

4.4 Concluding remarks

4.4.1 Temperature and pressure conditions for carbonization

Carbonization under pressure has been reviewed and discussed, classifying it into three routes: carbonization under built-up pressure by decomposition gases from carbon precursors, that under hydrothermal conditions, and the reduction of CO_2 under pressure. Under pressure built up by the decomposition gases from carbon precursors, such as pitches, the acceleration of the formation of mesophase spheres was confirmed and their coalescence was suppressed at a temperature a little higher than under atmospheric pressure, even though their coalescence could not be inhibited. A marked increase in carbonization yield was observed on all precursors, suggesting that the evolution of hydrocarbon gases from the organic precursors was strongly suppressed during pyrolysis and carbonization. However, it has to be pointed out that an efficient increase in carbonization yield was possible only in the sealed capsule (closed system), where all decomposition gases were included in the capsule. When the capsule was open, even in the autoclave, a marked improvement in carbonization yield was not observed in most cases, because the decomposition gases were deposited on low-temperature parts of the autoclave. In order to get carbon materials from polyethylene, which does not give any carbon residues under atmospheric pressure, the whole of the autoclave itself had to be heated to a high temperature [64].

Different carbon materials—carbon nanotubes, fullerenes, diamond, and carbons with various morphologies—were often formed as a mixture. However, carbon materials with spherical morphology (carbon spheres) were synthesized without appreciable amounts of other forms of carbon by selecting the conditions of pressure carbonization. On the basis of the results published, the formation conditions of single carbon spheres with different nanotextures are discussed here with relation to the temperature-pressure conditions and also the chemical composition of the precursors. The ranges of pressure and temperature for the formation of single carbon spheres are summarized in Figure 4.11, with a brief note on the carbonization process, the principal precursor used, and the size of carbon spheres obtained. The spherical morphology of carbon materials has been reviewed, focusing on their nanotexture

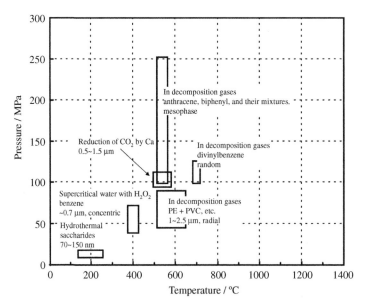

FIGURE 4.11 Pressure-temperature Conditions for the Formation of Carbon Spheres From Different Precursors.

and preparation processes [65]. Carbonization of different organic precursors in the atmosphere of their decomposition gases has been carried out in relatively limited ranges of temperature (500–700 °C) and pressure (50–250 MPa). In supercritical water, carbon spheres were obtained from benzene at a slightly lower temperature of 400 °C by adding a small amount of H_2O_2 as an initiator of radical reactions. Under hydrothermal conditions, pressure carbonization of saccharides occurred at the even lower temperature of 200 °C, to give carbon spheres that still contained hydrogen and oxygen, as shown by the H/C of around 0.7–0.8 and the O/C of about 0.3–0.4.

4.4.2 Composition of precursors for the formation of carbon spheres

The pyrolysis and carbonization of natural carbon precursors, including coals, have been discussed by using the so-called van Krevelen diagram, which was represented by plotting the atomic ratio H/C against O/C of the precursors [66]. The van Krevelen diagram has also been used on the formation of carbon spheres from saccharides under hydrothermal conditions [42,49]. Here, the formation of carbon spheres was discussed by applying a small modification to the van Krevelen diagram based on the experimental results of the pressure carbonization carried out using various combinations of organic compounds [7,12–19]. In Figure 4.12, the

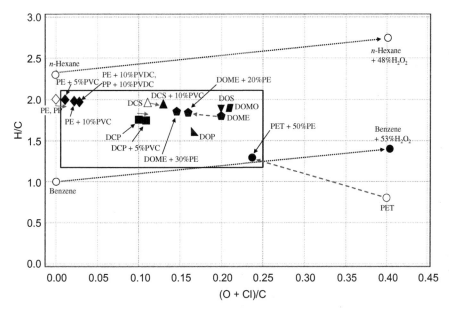

FIGURE 4.12 Plot of H/C vs (O+Cl)/C (Modified Van Krevelen Diagram) for the Formation of Single Carbon Spheres Under Pressure

Solid marks indicate the formation of single carbon spheres, open marks show the formation of botryoidal particles.

experimental points that gave single carbon spheres under pressure are plotted on the H/C vs (O+Cl)/C diagram, together with the precursors.

From DOS, which has octyl chains consisting of 8 carbon atoms (H/C = 1.9 and O/C = 0.2), single spheres were obtained at 650 °C under 30 MPa without any additives (Figure 4.7). From DCS, having longer cetyl chains of 16 carbon atoms (H/C = 1.94 and O/C = 0.1), however, not individual spheres, but botryoidal particles were obtained (Figure 4.7). The addition of 10 mass% PVC to DCS resulted in the formation of well-separated spheres, and shifted the experimental point to H/C = 1.92 and (O+Cl)/C = 0.13 in Figure 4.12. DCP with cetyl chains (H/C = 1.75 and O/C = 0.10) gave well-separated spheres (Figure 4.7), and even the addition of 5 mass% PVC to DCP (H/C = 1.74 and (O+Cl)/C = 0.11) resulted in the formation of separated spheres. In the case of PE and PP, a small amount of PVC or PVDC was necessary to get well-separated spheres. On the other hand, poly (ethylene terephthalate) (PET) with H/C = 0.8 and O/C = 0.4 did not give spheres, and the addition of 50 mass% PE was necessary to get single carbon spheres under pressure, the point shifting to H/C = 1.29 and O/C = 0.24. These experimental results suggest that there might be an optimal chemical composition of the starting precursors, H/C and (O+Cl)/C, for the formation of single carbon spheres under pressure built up by decomposition gases: H/C is the fraction of chain components to give residual carbons and (O+Cl)/C is the fraction of functional groups. The latter was thought to initiate the dehydrogenation of the former to give carbons. The van Krevelen plot (Figure 4.12) shows that carbon spheres are formed from precursors in the compositional range of H/C = 1.3-2.0 and the wide range of (O+Cl)/C = 0.01–0.25, although there is the one exception of DCS.

In supercritical water, single carbon spheres were obtained at the low temperature of 400 °C from the mixture of benzene (H/C = 1.00) with 53 mass% H_2O_2 [28], as plotted in Figure 4.12. The given amount of H_2O_2 corresponds to no more than 8.1 mol%, compared with the stoichiometric amount required for the complete oxidation of benzene. The additive H_2O_2 was expected to change to hydroxyl radicals (•OH) by homolysis and to form free radical species of hydrocarbons by hydrogen abstraction reactions with •OH, owing to its quantitative insufficiency. In the van Krevelen diagram, however, even the addition of insufficient H_2O_2 makes the O/C ratio increase to the same value as that for PET. The successful formation of single carbon spheres from benzene seems to support the •OH-induced carbon formation mechanism, different from that based on the function of the oxygen-containing functional groups in precursors. From n-hexane, no single carbon spheres, only interconnected small spheres like botryoidal particles, were obtained even after the addition of 48 mass% H_2O_2, of which the O/C is the same as for PET and benzene with 53 mass% H_2O_2[28]. The result seems to suggest that the •OH-induced hydrogen abstraction in the given amount of H_2O_2 is not sufficient for n-hexane with its high H/C (=2.33), which is about twice as high as that for benzene.

As mentioned above, the addition of PVC achieved the formation of single spheres, despite the slight shift of (O+Cl)/C. The strong effect of PVC suggests its ability as an initiator of radical reactions. At temperatures above 200 °C, the cleavage of the C–Cl bond of PVC produces chlorine radicals (•Cl) and dechlorinated radicals (denoted as P–(CH$_2$•CH)–P) [67]. The •Cl radicals react with other P–(CH$_2$CHCl)–P to form HCl and P–(•CHCHCl)–P (i.e. a dehydrochlorination step). The latter radicals can also be formed by the reaction with different radicals, •R, present in the system. The successive β-scission of P–(•CHCHCl)–P forms new propagating •Cl and molecular species with a double bond in the chain structure, P–(CH = CH)–P. The dehydrochlorination followed by β-scission leads to the formation of polyene molecules, –(CH = CH)$_n$–. The polyene molecules can undergo cyclization reactions (inter- and intramolecular cyclization) to form aromatic species via complex molecular and radical reactions, although reticulation and cross-linking reactions of polymer structures are included. Then, the aromatic species form polyaromatic hydrocarbons (PAH) via condensation reactions and would finally form carbonaceous residues. When PVC coexists in the system, organic precursors might effectively be integrated into the series of radical and molecular reactions initiated by •Cl radicals. Figure 4.12 shows clearly that the •Cl-induced reactions are more effective for the formation of single spheres than the function of the oxygen-containing functional groups in precursors.

From anthracene (H/C = 0.71) and its mixture with biphenyl (H/C = 0.83), botryoidal particles were obtained under pressure [8]. Based on the van Krevelen diagram (Figure 4.12), it might be possible to give separated carbon spheres if a small amount of PVC were added. From divinylbenzene (H/C = 1.00), carbon spheres with random nanotexture, different from the oriented nanotexture of the above-mentioned spheres, were obtained at 700 °C and 100–125 MPa [9], similar to the previously mentioned conditions. On the van Krevelen diagram, divinylbenzene locates at the same position as benzene. It might be interesting to explore the effect of an additive, either PVC or H$_2$O$_2$, on pressure carbonization of these aromatics to produce single spheres with oriented nanotexture.

In the modified van Krevelen diagram, Figure 4.12, it should be noted that the range of (O+Cl)/C for the formation of single carbon spheres can be divided into two regions, i.e. the reaction systems containing oxygen (O/C = 0.10–0.25) and those containing chlorine (Cl/C = 0.01–0.028). This implies that their efficiency in contributing to the formation of single spheres is different, as pointed out above. By the addition of a small amount of PVC, inducing a slight shift in (O+Cl)/C, the ability to form single carbon spheres is markedly improved, as seen on PE and DCS. In addition, oxygen atoms in different radicals, such as etheric, esteric, or hydroxylic, are not differentiated. Also, chlorinated compounds other than PVC and PVDC have not yet been examined. The role of components containing Cl and/or O has to be studied in more detail, taking into account the differences in their chemical state.

From various saccharides (carbohydrates C$_m$(H$_2$O)$_n$), carbon spheres were obtained under mild hydrothermal conditions of temperature and pressure [33–42,49], as shown in Figure 4.11, and their formation was discussed mainly based on the dehydration of saccharides [33,36,47]. Under hydrothermal conditions, for example, glucose and fructose (C$_6$H$_{12}$O$_6$) were converted firstly into 5-hydroxymethyl-2-furaldehyde (HMF, C$_6$H$_6$O$_3$) above 160 °C [36,40] and at 120–140 °C [36], respectively, and their polymerization followed by carbonization formed carbon spheres. Most of the saccharides and their dehydrated intermediate structure (HMF) have high O/C ratios, higher than those used for the formation of carbon spheres under pressure (Figure 4.12). Carbonization of saccharides is considered to be achieved via two reaction routes: intermolecular dehydration of HMFs with hydroxyl groups and condensation of aldehydes formed via fragmentation reactions of monosaccharides [48]. The intermolecular dehydration significantly decreases O/C ratios. This is a typical example of the role of oxygen-containing functional groups in the formation of single spheres. The spheres derived from saccharides were confirmed to keep their spherical morphology even after heat treatment up to 1000 °C in an inert atmosphere [33], but their nanotexture and the structural evolution in these spheres during high-temperature treatment, higher than 1000 °C, have not yet been studied in detail.

References

[1] Inagaki M, Sakai M. TANSO 1988; No.134: 175–187 [in Japanese].
[2] Inagaki M, Park KC, Endo M. New Carbon Mater 2010;25:409–20.
[3] Ogawa I, Yoshida H, Kobayashi K, et al. J Mater Sci 1985;20:414–20.
[4] Ogawa I, Sakai M, Inagaki M. J Mater Sci 1985;20:17–22.
[5] Inagaki M, Ibuki T, Kobayashi K, et al. Carbon 1990;28:559–64.
[6] Inagaki M, Kuroda K, Sakai M, et al. Carbon 1984;22:335–9.
[7] Inagaki M, Kuroda K, Sakai M. High Temp High Press 1981;13:207–13.
[8] Marsh H, Dachille F, Melvin J, et al. Carbon 1971;9:159–77.
[9] Hirano S, Ohta I, Naka S. Nippon Kagaku Kaishi 1981;1356; [in Japanese].
[10] Inagaki M, Urata M, Sakai M. J Mater Sci 1989;24:2781–6.
[11] Inagaki M, Ishihara M, Naka S. High Temp High Press 1976;8:279–91.
[12] Sanada Y, Furuta T, Kumai J, et al. Sekiyu Gakkai Shi 1975;18:113–8; [in Japanese].
[13] Inagaki M, Kuroda K, Sakai M. Carbon 1983;21:231–5.
[14] Hishiyama Y, Yoshida A, Inagaki M. Carbon 1982;20:79–84.
[15] Inagaki M, Kuroda K, Inoue N, et al. Carbon 1984;22:617–9.
[16] Washiyama M, Sakai M, Inagaki M. Carbon 1988;26:303–7.
[17] Inagaki M, Washiyama M, Sakai M. Carbon 1988;26:169–72.
[18] Ayache J, Oberlin A, Inagaki M. Carbon 1990;28:337–51.
[19] Ayache J, Oberlin A, Inagaki M. Carbon 1990;28:353–62.
[20] Hirano S, Dachille F, Walker Jr PL. High Temp High Press 1973;5:207–20.
[21] Hirano S, Ozawa M, Naka S. J Mater Sci 1981;16:1989–93.
[22] Mi Y-Z, Liu Y- L. New Carbon Mater 2009;24:375–8.
[23] Mi Y, Liu Y, Yuan D, Zhang J. Chem Lett 2005;34:846–7.
[24] Gogotsi Y, Libera JA, Yoshimura M. J Mater Res 2000;15:2591–4.
[25] Calderon Moreno JM, Swamy SS, Fujino T, et al. Chem Phys Lett 2000;329:317–22.
[26] Calderon Moreno JM, Fujino T, Yoshimura M. Carbon 2001;39:618–21.
[27] Calderon Moreno JM, Yoshimura M. J Am Chem Soc 2001;123:741–2.
[28] Park KC, Tomiyasu H, Morimoto S, et al. Carbon 2008;46:1804–8.
[29] Gloyna EF, Li L, McBrayer RN. Water Sci Technol 1994;30:1–0.
[30] Gloyna EF, Li L. Environ Prog 1995;14:182–92.
[31] Park KC, Wang F, Morimoto S, et al. Mater Res Bull 2009;44:1443–50.
[32] Cai Q, Huang Z-H, Kang F, et al. Carbon 2004;42:775–83.

[33] Wang Q, Li H, Chen L, et al. Carbon 2001;39:2211–4.

[34] Sun X, Li Y. Angew Chem Int Ed 2004;43:597–601.

[35] Zheng M, Cao J, Chang X, et al. Mater Lett 2006;60:2991–3.

[36] Yao C, Shin Y, Wang L-Q, et al. J Phys Chem C 2007;111:15141–5.

[37] Titirici M-M, Antonietti M, Baccile N. Green Chem 2008;10:1204–12.

[38] Shin Y, Wang L-Q, Bae I-T, et al. J Phys Chem C 2008;112:14236–40.

[39] Mi YZ, Hu WB, Dan YM, et al. Mater Lett 2007;62:1194–6.

[40] Sevilla M, Fuertes AB. Chem Eur J 2009;15:4195–203.

[41] Hu B, Wang K, Wu L, et al. Adv Mater 2010;22:813–28.

[42] Yu S-H, Cui X, Li L, et al. Adv Mater 2004;16:1636–40.

[43] Sun X, Li Y. Langmuir 2005;21:6019–24.

[44] Sun X, Li Y. Angew Chem Int Ed 2004;43:3827–31.

[45] Li X-L, Lou T-J, Sun X-M, et al. Inorg Chem 2004;43:5442–9.

[46] Sun X, Liu J, Li Y. Chem Eur J 2006;12:2039–47.

[47] Ryu J, Suh Y-W, Suh DJ, Ahn DJ. Carbon 2010;48:1990–8.

[48] Sevilla M, Fuertes AB. Carbon 2009;47:2281–9.

[49] Mi Y, Hu W, Dan Y, Liu Y. Mater Lett 2008;62:1194–6.

[50] Lai L, Huang G, Wang X, et al. Carbon 2010;48:3145–56.

[51] Choucair M, Thordarson P, Stride JA. Nat Nanotechnol 2008;4:30–3.

[52] Zheng M, Liu Y, Jiang K, et al. Carbon 2010;48:1224–33.

[53] Kubo S, Demir-Cakan R, Zhao L, et al. Chem Sus Chem 2010;3:188–94.

[54] Kubo S, Tan I, White R, et al. Chem Mater 2010;22:6590–7.

[55] Motiei M, Hacohen YR, Calderon-Moreno J, et al. J Am Chem Soc 2001;123:8624–5.

[56] Lou Z, Chen Q, Zhang Y, et al. J Am Chem Soc 2003;125:9302–3.

[57] Lou Z, Chen Q, Wang W, et al. Carbon 2003;41:3063–74.

[58] Lou Z, Chen Q, Gao J, et al. Carbon 2004;42:229–32.

[59] Lou Z, Chen Q, Zhang Y, et al. J Phys Chem B 2004;108:4239–41.

[60] Lou Z, Chen C, Zhao D, et al. Chem Phys Lett 2006;421:584–8.

[61] Liu S, Tang X, Yin L, et al. J Mater Chem 2000;10:1271–2.

[62] Liu S, Zhu J, Mestai Y, et al. Chem Mater 2000;12:2205–11.

[63] Liu J, Shao M, Tan Q, et al. Carbon 2003;41:1682–5.

[64] Inagaki M, Sakai M. TANSO 1988; No. 134: 175–187 [in Japanese].

[65] Inagaki M, Park C-R, Skowronski JM, et al. Adv Sci Technol 2008;26:735–87.

[66] Van Krevelen DW. Fuel 1950;29:269–84.

[67] Marongiu A, Faravelli T, Bozzano G, et al. J Anal Appl Pyrolysis 2003;70:519–53.

Stress Graphitization

It was proposed in 1951 that the stress caused by anisotropic thermal expansion of carbon layers could promote graphitization of carbon [1]. In 1965, marked acceleration of graphitization under a pressure of 1 GPa was reported [2], and experiments on various carbon materials and precursors followed. The application of pressure in a particular temperature range during the early stages of pyrolysis of organic precursors was also found to be important for morphology control and graphitizability improvement of the resultant carbons, as discussed in Chapter 4. Stress accumulation at the interface between two carbon materials with quite different shrinkages during high-temperature treatment was first found experimentally to promote local graphitization in a composite of carbon fiber with glass-like carbon matrix in 1974 [3]. However, an experimental verification of the occurrence of stress high enough for the acceleration of graphitization of carbon by thermal expansion anisotropy has not yet been reported.

It has been observed that graphite crystals occurring in nature have a very high degree of crystalline perfection in the presence of other minerals. The thermal history of natural graphite has been pointed out to be very mild, at temperatures lower than 1000 °C and pressure of 0.5 GPa, which was geologically estimated from the surrounding minerals [4,5]. These conditions for the formation of natural graphite are much lower in temperature than the conditions experimentally determined for the graphitization of carbon materials. Consequently, experimental studies have been conducted on the effect of the coexistence of minerals on graphitization in combination with pressure. Significant acceleration of graphitization has been found to take place in the presence of calcium compounds.

In the present chapter, graphitization of various carbon materials under pressure and the acceleration of graphitization with the combination of pressure and coexistence of some minerals are reviewed. The structural changes due to stress, i.e. stress graphitization, have been reviewed, including graphitization in carbon/carbon composites [6], and the occurrence of graphite in nature has also been reviewed [7].

Prerequisite for readers: Chapters 2.5 (Structural development in carbon materials) and 2.6 (Acceleration of graphitization) in *Carbon Materials Science and Engineering: From Fundamentals to Applications*, Tsinghua University Press.

Advanced Materials Science and Engineering of Carbon.

5.1 Graphitization under pressure

The graphitization of carbons under pressure was first shown, using a carbon derived from pitch coke, to occur at 1600 °C under 1 GPa [2,8]. This temperature was notably low in light to the fact that it was necessary to use heat-treatment above 2500 °C under atmospheric pressure. Subsequent studies have shown that the pressure can be reduced to about 0.3 GPa [9], and also that even non-graphitizing carbons, such as a carbon derived from phenol resin, can be converted to graphite at around 1600 °C under pressure [10]. These experimental results have been reviewed [6,8,11].

Graphitization under pressure has been performed by using a piston-cylinder type of apparatus, where the sample carbon is placed in glass-like carbon tubes with graphite lids to avoid its contact with the boron nitride and pyrophyllite used as pressure-transmitting media. For details of the experiments, the reader is referred to the review [8]. The carbon samples used for these studies had a wide range of nanotextures from highly ordered, with plane, axial, and point orientations, to random or disordered. Most of the samples were in powder form with a grain size in a range of 40 to 70 μm.

5.1.1 Structural change in carbons

In Figure 5.1, changes in 004 diffraction profile with heat-treatment temperature (HTT) under atmospheric pressure and a pressure of 0.5 GPa are compared on a typical graphitizing carbon, which was prepared from poly(vinyl chloride) by carbonization at 680 °C (PV-7) [12].

Under atmospheric pressure, the 004 diffraction line of the coke shifts gradually towards the high-angle side and its profile sharpens in a symmetrical manner with increasing HTT, as is seen with various graphitizing carbons (Figure 5.1A). This behavior suggests that graphitic three-dimensional stacking occurs randomly in the crystallites at high temperatures. Consequently, the average interlayer spacing along the c-axis, d_{002}, gradually approaches the spacing of graphite, 0.3354 nm (54.6° in 2θ with Cu Kα), with increase in HTT.

Under 0.5 GPa pressure, however, the profile of the 004 diffraction line appears to be a composite after heat treatment at 1460 and 1520 °C (Figure 5.1B), consisting of two peaks corresponding to d_{002} of about 0.336 and 0.343 nm (54.5 and 53.5° in 2θ). With the increase in HTT under pressure, the peak located on the high-angle side grows, but the intensity of the peak on the low-angle side decreases. The profile of the 002 diffraction line changes with HTT in exactly the same manner. The peak on the high-angle side is reasonably supposed to have graphitic structure, and that on the low-angle side to have turbostratic characteristics. This suggests that the graphitization process under pressure proceeds as two-phase graphitization; some of the crystallites change to graphitic form, but the others remain turbostratic, and the relative number of crystallites with graphitic structure increases with higher HTT and longer residence time [12,13].

It should be noted that the structural change, i.e. graphitization, observed from 00*l* diffraction profiles for PV-7 starts at about 1600 °C under atmospheric pressure. Under 0.5 GPa, however, at the same temperature the transformation is almost

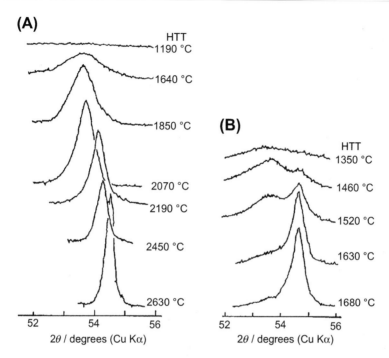

FIGURE 5.1 Changes in 004 Diffraction Profile of Poly(Vinyl Chloride) Coke PV-7 with HTT

(A) Under atmospheric pressure, and (B) under 0.5 GPa.

From [12]

finished. Therefore, it appears that pressure greatly accelerates the graphitization process of carbons. This acceleration has also been observed at pressures of about 0.3 GPa [9]. Under 1 GPa, the graphitization of PV-7 was so fast that the appearance of a composite profile of the 00l diffraction line was not clearly observed [2].

Examination with a transmission electron microscope showed the coexistence of two kinds of flaky particles; one having diffraction spots with hexagonal symmetry, exactly the same as graphite, and the other consisting of diffuse diffraction rings [12]. This observation corresponds exactly to the presence of two component peaks with d_{002} of about 0.336 and 0.343 nm in X-ray diffraction profiles. Therefore, the components corresponding to the peaks on the high- and low-angle sides have been called the graphitic and turbostratic components (G- and T-components), respectively.

The composite profiles of 00l diffraction lines have been graphically separated into two component peaks, and d_{002} was measured for each peak. The area of the G-component relative to the total area of the observed 00l line was used as a measure of the content of the G-component. In Figure 5.2, these two parameters are plotted against HTT for different residence times under 0.5 GPa [12].

The process under pressure appears to be divided into two steps, before and after the d_{002} reaches about 0.343 nm (Figure 5.2A). In the first step, d_{002} decreases to

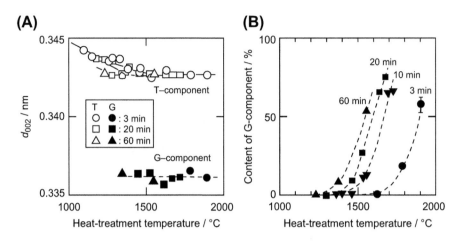

FIGURE 5.2 Changes in d_{002} and the Content of G-component with HTT and Residence Time Under 0.5 GPa

(A) d_{002} of the G- and T-components, and (B) the content of the G-component.

From [12]

0.343 nm with the increase of HTT. In this step, the 00*l* profile is broad and seems to consist of a single component. In the second step, however, a new G-component appears with a d_{002} of about 0.336 nm and its content increases with increasing HTT and residence time (Figure 5.2B).

In the second step of structural change, a big difference is observed, depending on whether the heat treatment was under atmospheric or high pressure. The high pressure process, above 0.3 GPa, may be called heterogeneous graphitization because of the appearance of composite X-ray diffraction profiles (Figure 5.1B), in contrast to the homogeneous process that occurs under atmospheric pressure where the symmetrical single profile shifts gradually to the high-angle side (Figure 5.1A). These two processes, heterogeneous and homogeneous, appear to depend on whether the graphitic and turbostratic structures occur in each crystallite or randomly in a single crystallite. The activation energy and activation volume for the heterogeneous process under pressure were estimated to be 340–500 kJ/mole and about −8 cc/mole, respectively, based on kinetic studies of the formation of the G-component under 0.5 GPa [12]. In contrast, the activation energy for the homogeneous graphitization process under atmospheric pressure was determined to be about 1000 kJ/mol [14–16].

As seen in the heterogeneous process under pressure, the graphitic component appeared in the second step, i.e. after the interlayer spacing reached 0.343 nm. This indicates that graphitic regular stacking of the layers (graphitization in its strict meaning) occurs in the second step of structural change. On the basis of the changes of structural parameters and properties, there is a good correlation with the previous discussion, i.e. graphitization under atmospheric pressure initiates for graphitizing carbons at a d_{002} of about 0.343–0.342 nm [17].

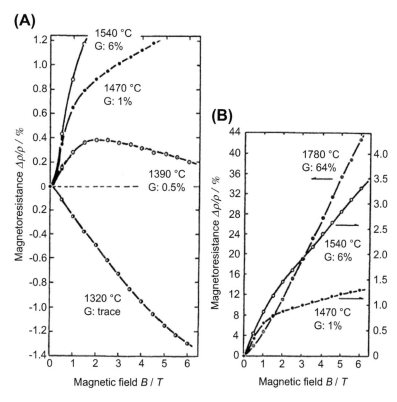

FIGURE 5.3 Magnetic Field Dependence of Magnetoresistance, Δρ/ρ, on Carbon Blocks Heat-treated at Different Temperatures Under 0.5 GPa and with Different Contents of G-component. (A) Change the sign of Dr/r with increasing HTT, (B) rapid increase in Dr/r and G-component with a small increase in HTT.

From [18]

Carbon blocks sintered under high pressure showed a strong dependence of transverse magnetoresistance, Δρ/ρ, on magnetic field, B [18]. Typical field dependence is shown in Figure 5.3 for sintered blocks obtained from PV-7 at different HTTs under 0.5 GPa and containing different amounts of the G-component.

The sample heat-treated at 1320 °C, containing only a trace of the G-component, shows negative values of Δρ/ρ at all magnetic fields (Figure 5.3A), which is characteristic for turbostratic carbons. But the one heat-treated at 1780 °C and containing more than 60% of the G-component, shows positive values (Figure 5.3B), which is characteristic of graphitic samples. For the samples heat-treated between 1390 and 1540 °C, however, non-linear dependence of Δρ/ρ on B is observed. This magnetic field dependence of Δρ/ρ was found to be different from that observed on carbons consisting of a single component, such as pyrolytic carbons and cokes.

This dependence of Δρ/ρ on B was experimentally reproduced by using a model sample, that consisted of a 2900 °C-treated pyrolytic graphite with crystalline structure (corresponding to the G-component) and a 1900 °C-treated coke with a turbostratic structure (corresponding to the T-component); they were electrically connected in series [18]. With this model sample, non-linear dependence of $\Delta\rho/\rho$ on B was observed, being very similar to that shown in Figure 5.3. The result supports the coexistence of two structural components, graphitic and turbostratic, in the samples heat-treated under pressure.

Pre–heat treatment at high temperatures under atmospheric pressure was observed to have a significant influence on graphitization under pressure using pre-heat-treated PV-7 cokes [19] and mesophase spheres (mesocarbon microbeads, MCMBs) [20]. Under 0.5 GPa, graphitization of the as-prepared MCMBs appears to initiate above 1800 °C, although a small amount of the turbostratic component does remain even after 2000 °C treatment under pressure. But almost no graphitization took place for the spheres pre-heat-treated to 2000 °C [20]. Similar retardation of the graphitization process under pressure was observed for PV-7 cokes, which were pre-heat-treated to 1500, 1700, and 2000 °C under atmospheric pressure [19]. Texture of the starting carbon material also appeared to have an influence on the graphitizability under pressure. For example, commercial Gilsonite™ and fluid cokes, which had concentric orientation of carbon layers in the spherical particles, showed slight acceleration of graphitization even with heat treatment above 1500 °C under 0.5 GPa, in comparison to PV-7 [21,22].

The acceleration of graphitization under 0.5 GPa was observed to be much more pronounced for carbons with random nanotexture than for those with oriented nanotexture [23]. In Figure 5.4A, the change in 002 diffraction profile with HTT under 0.5 GPa is shown for a carbon prepared from a phenol resin at 700 °C (named PH-7).

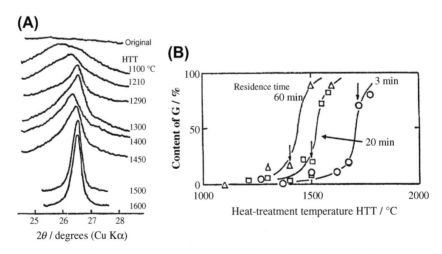

FIGURE 5.4 Heat Treatment of Phenol-resin-derived Carbon (PH-7) Under 0.5 GPa

(A) 002 diffraction profile, and (B) content of the G-component with HTT.

From [23]

Composite profiles of the 002 line, consisting of a sharp peak with d_{002} of 0.336 nm and a broad peak of 0.343 nm, i.e. G- and T-components, are observed around 1300 °C (Figure 5.4A). In the case of carbons with random nanotexture, the peak for the T-component was so broad that the separation of the two component peaks was sometimes not clear. A single sharp peak at 0.336 nm has been obtained after heat treatment above 1500 °C under 0.5 GPa [23–25]. The existence of two components was more clearly observed under a transmission electron microscope, showing two kinds of particles with diffraction patterns of diffuse rings and of sharp spots with hexagonal symmetry. The presence of the G-component has also been proved by electron spin resonance measurements [23]. In Figure 5.4B, the content of G-component formed from PH-7 is plotted as a function of HTT and residence time under pressure [26]. The increase in the content of G is very sudden and occurs at lower temperatures by employing longer residence times. Glass-like carbon beads with a random nanotexture were found to graphitize very abruptly at around 1600 °C under 0.5 GPa [24,27], as did PH-7 (Figure 5.4A). Very little effect was observed on this graphitization process, when the same beads received a pre–heat treatment to 2100 °C under atmospheric pressure [27].

5.1.2 Mechanism

The graphitization process under high pressure is shown in Figure 5.5, which summarizes the experimental results obtained through different techniques for various carbon materials at high temperatures under 0.5 GPa.

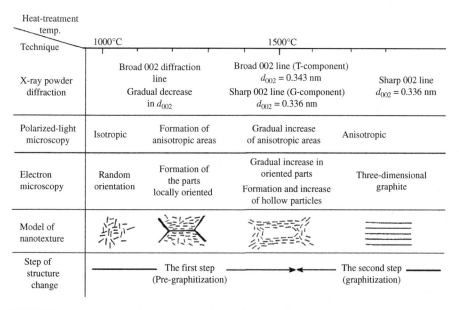

FIGURE 5.5 The Graphitization Process of Carbon Under Pressure

From [6]

In the first step, the profiles of the 002 and 004 X-ray diffraction lines are broad, shifting gradually to the high-angle side with increasing HTT. This step may be called pre-graphitization, since it is a preliminary step for the subsequent graphitization. In order to complete this first step, it is necessary to heat-treat most carbons to about 1500 °C under 0.3–1 GPa, which results in a d_{002} spacing of 0.343 nm. In the second step, above 1100 °C, the regions where the stress is concentrated, primarily at the contact points between carbon particles, tend to change to an oriented texture. Through detailed analysis using high-resolution transmission electron microscopy (TEM), these oriented regions usually consist of small hollow particles with walls, which are formed by small hexagonal carbon layers with a rather high degree of preferred orientation but still turbostratic stacking [25]. In Figure 5.6, TEM images (bright-image, 002 dark-field, and 10 dark-field images) are shown for a particle. At position **3**, a bright area on the 10 dark-field image (Figure 5.6C) reveals that the carbon layers are perpendicular to the incident electron beam. At positions **1** and **2**, shown in different orientations, bright areas on the 002 dark-field images (Figures 5.6B and D) reveal carbon layers parallel to the incident beam.

Above 1500 °C, the structure of carbons changes from turbostratic to graphitic through collapsing of the hollow particles to a lamellar form, in which carbon layers can readily grow and their stacking becomes graphitic. This collapsing of hollow particles followed by graphitization is assisted by the stress concentrated at the contact points between starting carbon particles. The stress concentration occurs at contact points that are located at preferential positions to the direction of compressive force [27], as will be described in Chapter 6. This process of graphitization is heterogeneous, in contrast with the more homogeneous process that takes place under atmospheric pressure. The 00l diffraction profiles of the samples in this process are observed to be composite, consisting of a broad line with a spacing of about 0.343 nm (T-component) and a sharp line with 0.336 nm (G-component). This effect contrasts with the gradual change of the 00l diffraction lines, i.e. gradual decreases in diffraction angle, 2θ, and width whilst keeping almost symmetrical profiles.

The collapsing of hollow particles mentioned above has to occur particle by particle, and this, in turn, depends strongly on the size of the particles and their orientation relative to neighboring particles, in order to generate the necessary concentrated stress. The hollow particles formed in carbons with a random nanotexture, for example glass-like carbon, are very small and contain large numbers of defects, so that they appear to be homogeneous, and so their 002 diffraction profile is usually very broad and weak, even after receiving heat treatment at 1500 °C under pressure. Because homogeneous and small-sized hollow particles can collapse abruptly, these types of carbons show an abrupt change to graphite in a narrow range of HTT, around 1600–1700 °C, and the composite diffraction profile of T- and G-components is scarcely observed. When carbons with an oriented texture, such as the coke PV-7, are subjected to high pressure, the orientation of the grains is strongly affected by the degree and direction of the stress concentration

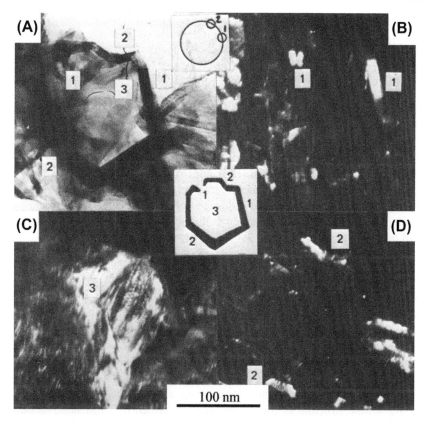

FIGURE 5.6 Transmission Electron Micrographs of a Particle Formed From a Glass-like Carbon Sphere After Heat Treatment at 1300 °C Under 0.5 GPa

(A) Bright-field image with selected-area diffraction pattern (inserted); (B and D) 002 dark-field images corresponding to positions **1** and **2**, respectively; and (C) 10 dark-field image of position **3**.

From [25]

that arises at the points of contact. Consequently, the hollow particles formed during the pre-graphitization step are inhomogeneous, depending on the size and processing conditions. As these hollow particles are collapsed and graphitized at slightly different HTTs, the diffraction profiles appear to be a composite of T- and G-components. For these carbons, retardation of the graphitization process was observed when a pre–heat treatment was performed under atmospheric pressure. This effect is explained as follows: the growth and partial improvement in stacking order of the crystallites occurred during pre–heat treatment in these types of carbons, and so, in order to have sufficient stress concentration at the contact points of grains, a higher temperature and/or higher pressure of heat treatment are required.

5.2 Graphitization in coexistence with minerals under pressure

In the previous section, carbons were shown to be graphitized very rapidly above 1600 °C and 0.3 GPa. No indication was found of the presence of a graphitic structure in the carbons below 1300 °C, even under 1 GPa [2]. However, the temperature-pressure conditions for the formation of natural graphite crystals have been geologically estimated to be several hundred degrees Celsius and several tenths of 1 GPa [4,5]. Therefore, the metamorphic rocks, in which the graphitic crystals are embedded, may have an accelerating effect on the graphitization process, in addition to the acceleration by pressure itself. Thus, experimental investigation was expanded to the heat treatment of carbon under pressure in coexistence with certain minerals [8,9,28–32]. Carbon samples were sandwiched between two disks of the mineral under a pressure of 0.3 GPa, and heated to different temperatures for different periods of time.

5.2.1 Coexistence with calcium compounds

In Figure 5.7, changes in the diffraction profile of the 002 line with HTT are compared in three cases: coexistence with $CaCO_3$ under a pressure of 0.3 GPa, without $CaCO_3$ under the same pressure, and coexistence with $CaCO_3$ under atmospheric pressure [28].

The appearance of a sharp peak corresponding to the G-component (d_{002} of about 0.336 nm) is observed only when the carbon is heat-treated in the presence of $CaCO_3$ under pressure. The G-component starts to appear at 1100 °C. The presence of flakes with a graphitic structure in these samples was verified by TEM observations. It

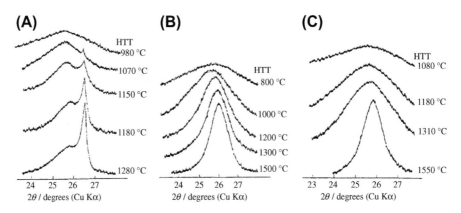

FIGURE 5.7 Changes in the 002 Diffraction Profile of PVC-derived Coke PV-7 with HTT

(A) In coexistence with $CaCO_3$ under 0.3 GPa, (B) without $CaCO_3$ under 0.3 GPa, and (C) in coexistence with $CaCO_3$ under atmospheric pressure of N_2 flow.

From [28]

should be emphasized that the formation of the G-component needs heat treatment above 1600 °C without $CaCO_3$ at the same pressure, of 0.3 GPa (see Figure 5.7B).

For the T-component, the value of d_{002} decreases gradually from the initial value of about 0.346 nm to about 0.344 nm, and the value of $Lc(002)$ increases slightly, up to 8 nm, with an increase in HTT. On the other hand, the d_{002} value for the G-component is practically constant at about 0.336 nm, but the $Lc(002)$ value increases significantly with increasing HTT. The formation of the G-component depends strongly on the residence time, suggesting that the transition from turbostratic to graphitic components is a rate process. The relative ratio of the G-component increases with increasing HTT and residence time.

The same experiment was conducted using $CaCO_3$ disks saturated with water (about 8 mass% at room temperature) [33]. The appearance of the G-component occurred at a lower temperature than when dried $CaCO_3$ was used. The same acceleration of the graphitization process for PV-7 was observed using natural limestone [9]. The acceleration of graphitization was more effective by natural limestone than by reagent-grade $CaCO_3$.

The graphitization of PV-7 under 0.3 GPa was conducted in coexistence with three calcium oxide disks with different degrees of crystallinity, which were prepared at three different calcination temperatures [30]. Changes in 002 diffraction profile with HTT are shown in Figure 5.8. In coexistence with CaO-9, which is prepared by calcination of $CaCO_3$ at 920 °C and has a relatively small crystallite size of 72 nm, the G-component first appears at 1100 °C, and it grows significantly with increasing HTT (Figure 5.8A). Close to 100% G-component is obtained after heat treatment at

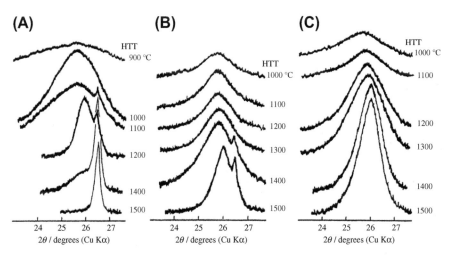

FIGURE 5.8 Changes in the 002 Diffraction Profile of PV-7 with HTT Under 0.3 GPa in Coexistence with CaO Calcined at Different Temperatures and with Different Crystallinities

CaO calcined at (A) 920 °C, (B) 1000 °C, and (C) 1470 °C.

From [30]

1500 °C. In coexistence with CaO-15, which is calcined at 1470 °C and has a large crystallite size of more than 100 nm, however, no G-component is detected even up to 1500 °C (Figure 5.8C). In the case of CaO-10, with an intermediate size of crystallite, the graphitization behavior is intermediate between that of CaO-9 and CaO-15 (Figure 5.8B). Changes in d_{002} and $Lc(002)$ with HTT were similar to those seen in the coexistence of $CaCO_3$; the variations in d_{002} and $Lc(002)$ of the T-component were small, about 0.343 and less than 10 nm, respectively, even after heat treatment at 1500 °C.

The effect of $Ca(OH)_2$ disks on the graphitization of PV-7 was studied under 0.3 GPa [29]. The results are summarized as changes in the 002 profile and content of the G-component with HTT (Figure 5.9). After heat treatment at 800 °C for 60 min, a sharp peak for the G-component is overlapped with the broad band of the T-component. Even though the G-component is a small amount (c. 2%), it is worth noting that the temperature required for the formation of the graphitic component in the coke is lowered to 800 °C by the coexistence of $Ca(OH)_2$, while heat treatment above 1000 °C was needed in the cases of calcium carbonate and oxide. These temperatures for the formation of the G-component are much lower than the temperature required without any minerals (above 1600 °C), as described in section 5.1. The content of the G-component increases with increasing residence time at 800 °C, suggesting that the formation reaction of the G-component is a rate process.

In the samples showing a sharp peak due to the G-component, flaky particles with a diffraction pattern of six-fold symmetry were detected under TEM. Even when the sample was heat-treated at a temperature as low as 600 °C for 60 min, a few graphitic flakes could be found under TEM, although no G-component was detected in an XRD. Heat treatment above 900 °C was difficult to carry out because the calcium hydroxide melted and reacted with the graphite heater during the experiment.

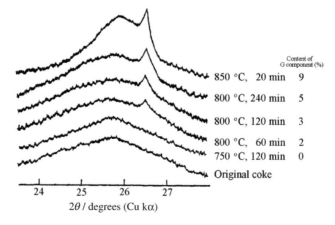

FIGURE 5.9 Change in the 002 Diffraction Profile with Heat Treatment Under 0.3 GPa in Coexistence with Ca(OH)$_2$

From [29]

5.2.2 **Coexistence with other minerals**

Heat treatment of PV-7 in coexistence with Al_2O_3, SiO_2, MgO, Na_2CO_3, and MgF_2 was carried out under 0.3 GPa [33].

With Al_2O_3 and SiO_2, a certain acceleration of graphitization was detected, although it was not very pronounced, the G-component being about 30 % with Al_2O_3 above 1200–1300 °C and only 10 % with SiO_2 after heat treatment at 1500 °C. In an X-ray powder pattern, aluminum carbide (Al_4C_3) was detected at the interface between the carbon and the Al_2O_3 disk, but no silicon carbide (SiC) was detected even after the high-temperature treatment.

With MgO, no G-component was observed on the 002 diffraction profile, but a small number of graphitic flakes were found under TEM. On MgF_2, graphitic flakes were only detected under TEM after heat treatment at 1500 °C under pressure.

In coexistence with Na_2CO_3, graphitic particles were detected under TEM after heat treatment as low as 500 °C. The number of particles seemed to increase with increasing HTT and residence time. However, heat treatment above 800 °C was not possible because of the melting and decomposition of Na_2CO_3.

5.2.3 **Mechanism for acceleration of graphitization**

In the course of graphitization under pressure, the 00l diffraction profile of carbon is a composite of turbostratic and graphitic components (T- and G-components), revealing a heterogeneous transformation of the carbon from the turbostratic to graphitic structure. In the case of the graphitization of carbons without any minerals under pressure, the appearance of the G-component, i.e. the beginning of graphitization, was found preferentially at and near the contact points between adjacent carbon particles where the stress is concentrated. In other words, change in d_{002} to 0.343 nm is accelerated by stress concentration. In the coexistence of $CaCO_3$, however, the G-component appears even before the d_{002} value of the T-component has reached 0.343 nm, as a consequence, at lower temperatures. It is reasonable to suppose that the occurrence of graphitization is accelerated by some catalytic effect of the coexisting $CaCO_3$.

After heat treatment with coexistence of calcium compounds, the sintered carbon blocks always contain a certain amount of calcium. Figure 5.10 shows the distribution of calcium in a resultant sintered disk along two directions [31]. Although the content of the G-component is a value averaged over a range of distance, as shown in Figures 5.10A and B, there appears to be some direct correspondence between the content of the G component and the amount of calcium present in the sintered disk. This suggests that calcium atoms accelerate the graphitization of carbon under pressure. This conclusion is valid for all the other calcium compounds examined. After heat treatment, the disks of the calcium compounds were found to be partially recrystallized [28]. The content of the G-component was closely related to the thickness of the recrystallized part in the $CaCO_3$ disks, irrespective of HTT and residence time.

The melting point of $CaCO_3$ is reported to be lowered by the addition of water [4]: the melting point of 1310 °C at 0.1 GPa is lowered to 1130 °C by 8% addition

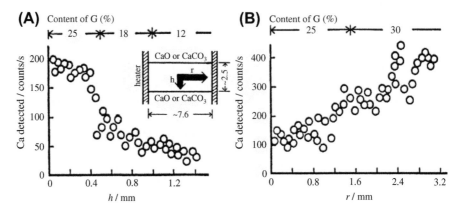

FIGURE 5.10 Distribution of Ca measured by electron microprobe analysis and the G-component in a Carbon Disk

(A) Distribution along the height, and (B) along the radius of the disk.

From [31]

of water. It was observed that the coke PV-7 started to be graphitized at about 900 °C when $CaCO_3$ containing about 8 mass% water was used. This is a reduction of temperature by more than 100 K compared with a dried calcium carbonate disk [28]. Since $Ca(OH)_2$ has a melting point at 840 °C under 0.3 GPa, heat treatment of the coke with $Ca(OH)_2$ was possible only below this temperature, but the graphitization of carbon was observed at a temperature as low as 600 °C [29].

All of the experimental results presented above suggest two possible routes for graphitization of carbon [33]. These can be expressed by using an oxide M_xO_y as follows.

At low temperatures:

$$M_xO_y + C \,(\text{turbostratic}) \rightarrow M_xO_{y-1} + CO \rightarrow M_xO_y + C \,(\text{graphitic}), \quad (5.1)$$

and at high temperatures:

$$mM_xO_y + (n+my)\,C\,(\text{turbostratic}) \rightarrow M_{mx}C_n \qquad (5.2)$$
$$+ myCO \rightarrow mM_xO_y + (n+my)\,C\,(\text{graphitic}).$$

On calcium compounds, these two routes can be rewritten as follows:

$$CaO + C\,(\text{turbostratic}) \rightarrow Ca + CO \rightarrow CaO + C\,(\text{graphitic}), \quad (5.3)$$

$$CaO + 3C\,(\text{turbostratic}) \rightarrow CaC_2 + CO \rightarrow CaO + 3C\,(\text{graphitic}). \quad (5.4)$$

The experimental facts indicate that calcium carbide was formed at the interface between carbon and calcium oxide after heat treatment above 1200 °C. On the other hand, calcium carbide was never detected below 1200 °C, and also it was difficult to decompose of CaC_2 below 1200 °C under pressure. Therefore, the first mechanism may be dominant below 1200 °C and the second above 1200 °C. CaO-9 with lower crystallinity reacts with carbon at a lower temperature than CaO disks with higher degrees of crystallinity [30]. When $CaCO_3$ and $Ca(OH)_2$ were used as the coexisting minerals, a nascent oxide was formed at temperatures around their melting points and this may react with carbon to form a reactive calcium metal. The formation of reactive CaO by the reaction between the starting CaF_2 and water vapor under high temperature and pressure was favorable for accelerating the graphitization process [33]. In the case of Al_2O_3, the two routes expressed by Equations 5.1 and 5.2 are thought to occur at temperatures up to 1500 °C under 0.3 GPa. In the cases of SiO_2 and MgF_2, only the first route, Equation 5.1, seems to be possible under 0.3 GPa. To form SiC, temperature higher than 1500 °C was needed under pressure. With MgF_2, MgO was not formed by the reaction with water vapor under high temperature and pressure. As a consequence, only minute amounts of graphite particles were found after the 1500 °C treatment. From these experimental results, it is concluded that graphite can be formed under mild conditions (below 1000 °C under 0.3 GPa) with the catalytic action of some metal species, such as Ca.

When $CaCO_3$ was used, the following reaction between CO_2 produced from the carbonate and the sample carbon is expected:

$$CO_2 + C \rightarrow CO. \tag{5.5}$$

The formed CO can join the reactions in Equations 5.3 and 5.4 to form graphite. The C^{13} isotope content in the G-component formed in the coexistence of limestone under 0.3 GPa could be used to establish the above scheme, because the carbon in limestone contains a much larger amount of C^{13} than in the original sample of PV-7. The results are summarized in Table 5.1 [33], where the negative value for δC^{13} means that the C^{13} content is below that of the standard gas. Table 5.1 shows that the carbons

Table 5.1 Content of Carbon Isotope and Graphitic Component

Sample	$[\delta C^{13}]_{obs}$	Graphitic Component	
		Content (%)	$[\delta C^{13}]_{calc}$
Original PV-7	−0.66	0	−
Carbon heat-treated at 1000 °C in coexistence with limestone	+0.00	15	3.7
Carbon heat-treated at 1250 °C in coexistence with limestone	+1.20	35	48
Carbon contained as carbonate group in the limestone	+20.14	−	−

obs, observed; calc, calculated.
$[\delta C^{13}] = \{[(C^{13}/C^{12})_{sample} - (C^{13}/C^{12})_{standard}]/(C^{13}/C^{12})_{standard}\} \times 1000.$
From [33]

heat-treated above 1000 °C have higher values of δC^{13} than the starting PV-7, and that the G-component has a higher δC^{13}-value, i.e. a higher concentration of C^{13}, than the original PV-7 after heat treatment in coexistence with limestone. This result suggests that the CO_2 formed by the decomposition of limestone reacts with the carbon sample according to Equation 5.5 to produce CO. Consequently, CO can take part in the reactions between CaO and the carbon sample (Equations 5.3 and 5.4).

Partial graphitization of anthracite was reported to occur as low as 600 °C under pressures of 0.8–1.0 GPa with a simple shear stress [34]. In this case, there may also have been some accelerating graphitization effect of ZrO_2, which was used to transfer the shear stress to the anthracite sample, although the authors did not mention it.

5.3 Stress graphitization in carbon/carbon composites
5.3.1 Acceleration of graphitization

Marked graphitization was found in the composites of carbon fibers/glass-like carbon, although the individual components, polyacrylonitrile (PAN)-based carbon fiber and glass-like carbon prepared from furfuryl alcohol condensates, were not graphitized [3].

In Figure 5.11A, 004 diffraction profiles of composites with different fiber fractions are compared to those of the component carbons, carbon fiber and glass-like carbon, after heat treatment at 2800 °C under atmospheric pressure. These X-ray profiles for the composites are very similar to those observed when carbons are heat-treated under high pressure, as described in the previous sections. Also, it is found that the d_{002} spacing for the component on the high-angle side for the composites is about 0.336 nm, which is the same as that of the G-component appearing under high pressures. On the other hand, PAN-based carbon fiber and glass-like carbon that have been individually processed under exactly the same conditions show only a broad band on the low-angle side, which is significantly different from the composites. It should be pointed out also that the asymmetry of the diffraction profile seems to be more pronounced in the composites with a lower fraction of fiber, revealing that the composite of 30 vol% fiber fraction contains a larger amount of graphitized component than the composite with 60 vol% of fibers. In Figure 5.11B, change in 002 diffraction profile with HTT is shown for the composite with fiber content 45 vol%. The 002 line gradually shifts to the high-angle side above 2100 °C and asymmetry of the profile becomes more evident with increase in HTT.

Optical microscopic observations were carried out on the polished cross-section of composite under crossed Nicols. Both glass-like carbon and PAN-based carbon fiber appear isotropic under optical microscopy, even after being heat-treated at high temperatures. However, the composite of carbon fibers with glass-like carbon matrix shows anisotropic texture, as shown in Figure 5.12. After heat treatment at 1000 °C, anisotropic regions develop at the fiber-matrix interfaces (Figure 5.12B), although the composite is isotropic after curing furfuryl alcohol condensates at 110 °C (Figure 5.12A). In these anisotropic regions, the basic structural units (BSUs) of small carbon layers were found to align parallel to the fiber surface and to the

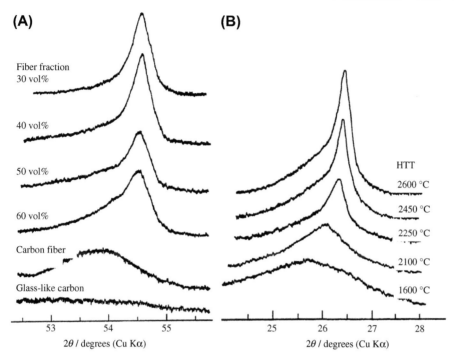

FIGURE 5.11 Changes in the Diffraction Profiles of Composites

(A) With fiber fraction after heat treatment at 2800 °C, and (B) with HTT on 45 vol% fiber fraction.

From [3]

fiber axis, from the analysis of pleochroism. The anisotropic regions were greatly enlarged by the heat treatment of these composites to high temperatures, such as 2450 °C (Figure 5.12C). Furthermore, a few fibers, which are isotropic, are distinctly seen in the enlarged anisotropic regions. The development of these anisotropic regions corresponds to the graphitic component's peak in the X-ray diffraction profile. After 2800 °C treatment, almost all the matrix becomes anisotropic, as shown in Figure 5.12D, and isotropic areas due to the cross-sections of carbon fiber are difficult to recognize, although there are a few.

A stress field is formed at the interface between the carbon fiber and its surrounding matrix in the carbon fiber/glass-like carbon composites during heat treatment, which tends to create a texture of concentric orientation of BSUs whose c-axis radiates from the center of the fibers, and consequently graphitization is accelerated in this field. The formation of oriented texture at the interface between carbon fiber and matrix was clearly shown by scanning electron microscopy (SEM) after etching [35]. It is worth noting that the formation of lamellae at the fiber-matrix interfaces in the composites is very similar to the anisotropy development at the contact points between carbon grains under high pressure, as described in the previous

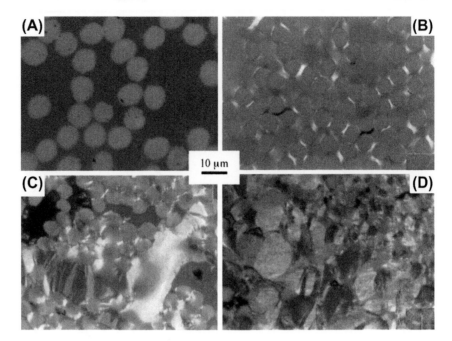

FIGURE 5.12 Polarized-light Micrographs Under Crossed Nicols of Carbon Fiber/Glass-like Carbon Composites

(A) After curing, and after heat treatment at (B) 1000 °C, (C) 2450 °C, and (D) 2800 °C.

From [3]

section (Figure 5.6). The lamellae appear to surround the fiber concentrically and this arrangement suggests the orientation of layer planes parallel to the fiber axis, which was proved by the measurement of magnetoresistance [3].

Composites prepared from flaky natural graphite by bonding with a glass-like carbon matrix were found to be graphitized after heat treatment at high temperature, by the measurement of bulk density, electrical resistivity, and averaged d_{002} [36]. The development of anisotropic areas was also observed on the composites prepared from glass-like carbon heat-treated at 1000 °C with furfuryl alcohol condensates as a binder, which was a precursor of glass-like carbon. Polarized-light micrographs are shown for the composites heat-treated at 1000 and 2800 °C in Figures 5.13A and 5.13B, respectively [37]. The formation of anisotropic areas in the matrix is easily recognized between the irregularly shaped particles.

Similar stress accumulation was also thought to occur in the carbon/carbon composites of glass-like carbon derived from polycarbodiimide (PCDI) with different carbon materials, such as graphite, carbon fiber, and glass-like carbon [38]. In the matrix around the filler particle, anisotropic texture was developed during heat treatment at 1700 °C, where carbon layers were in parallel to the surface of the filler. Wrinkling, cracking, and even fragmentation of the composites were observed.

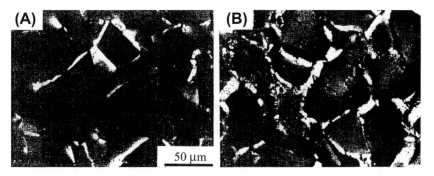

FIGURE 5.13 Polarized-light Micrographs for Natural Graphite Flake/Glass-like Carbon Composite After Heat Treatment

(A) At 1000 °C, and (B) at 2800 °C.

From [37]

A complicated texture, consisting of three layers with different orientation of BSUs, was formed in the glass-like carbon derived from PCDI [39].

5.3.2 Mechanism

With the aid of polarized-light microscopic observations, the anisotropic regions were clearly seen to be formed at the interfaces of the carbon fiber and the surrounding glass-like carbon matrix, and to expand gradually into the matrix with increase in HTT and residence time [40,41]. Most glass-like carbons are known to show a large amount of shrinkage in volume; about 47% for the furfuryl alcohol condensates during carbonization up to 1000 °C. On the other hand, the carbon fibers employed here are dimensionally stable because they have been previously carbonized around 1200 °C, so there was no shrinkage during the carbonization process of the composites. Therefore, it is reasonable to assume that the anisotropic regions are formed by the accumulation of stress, which is caused by the large difference in the volume shrinkage between the carbon fiber and the matrix glass-like carbon during carbonization. This assumes, of course, that there is a strong bonding between the fiber and the matrix during pyrolysis.

Stress accumulation was shown to occur in the composites of natural graphite (NG) flakes and glass-like carbon matrix derived from a phenol resin by measuring thermal expansion of the composites and by comparing with NG flake [42]. The results are reproduced in Figure 5.14, which shows the difference in the slopes, or coefficients of thermal expansion, between NG powder and the composites carbonized at 700–1000 °C. Large thermal expansion of NG powder is found to be constrained when it is embedded in a glass-like carbon matrix and heated up to 1000 °C. However, when the composite was heat-treated at 1800 °C, its slope is seen to be equivalent to the powder, which means that the stress accumulated is released by heat treatment up to 1800 °C. The stress accumulated at the NG boundary was estimated to reach a few tenths of 1 GPa.

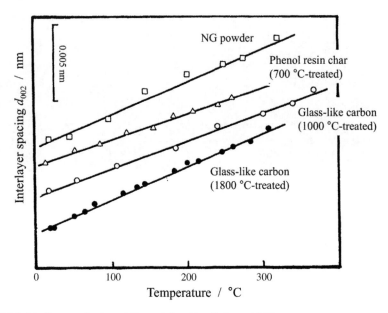

FIGURE 5.14 Changes in d_{002} of Natural Graphite Flakes with Temperature

From [42]

The stress accumulation at the boundary between carbon fiber and matrix was evaluated as a function of the distance from the surface of the carbon fiber by using a thick-walled-cylinder model for the matrix [43]. Experimental verification of this model was achieved by measuring the optical reflectance of a glass-like carbon matrix under crossed Nicols as a function of the distance from the fiber surface in a composite that had been heat-treated to 1000 °C. The results are shown in Figure 5.15. The calculated stress distribution appears to agree with that observed. Consequently, the maximum stress accumulated at the fiber-matrix interface was calculated to be 0.4 GPa during carbonization. From the shrinkage of the composites, it was also calculated that the residual stress at the boundary between fiber and matrix would be 0.7 GPa [43].

By taking into account the experimental fact that non-graphitizing carbons, such as glass-like carbons, can be graphitized under pressures above 0.3 GPa, the effect of stress accumulation is considered to be responsible for the graphitization of the glass-like carbon matrix in composites at high HTTs. This is one of the major possible explanations for the occurrence of the graphitization process, provided the accumulated stress is still sufficient at high temperatures for the process to proceed. An alternative explanation for the graphitization of the matrix in composites is a change from a random to an oriented texture of BSUs in the glass-like carbon matrix during carbonization up to 1000 °C, which is assisted by the stress accumulated in the region of the boundary between fiber and matrix [44]. In reality, the graphitization process may be some combination of these two processes, i.e. the orientation of the BSU is

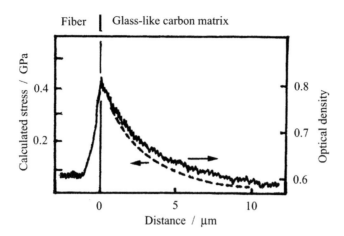

FIGURE 5.15 Stress Calculated From Optical Density as a Function of the Distance From the Interface Between the Carbon Fiber and Glass-like Carbon Matrix

From [43]

promoted by stress accumulation near the interface between carbon fiber and glass-like carbon matrix at moderate temperatures, and the oriented regions are graphitized at high temperatures with the assistance of the accumulated stress. Clearly, further experimental evidence is needed to define more precisely the details of the stress graphitization process in carbon/carbon composites and other carbonaceous materials.

5.4 Concluding remarks

5.4.1 Graphitization under pressure

Structural change from disordered turbostratic carbon to three-dimensionally ordered graphite, i.e. graphitization, is associated with a volume decrease. Therefore, it is reasonable to suppose that this process would be accelerated by external pressurization. This hypothesis has been experimentally verified because graphitization of turbostratic carbon occurs at temperatures above 1600 °C under applied pressures above 0.3 GPa, which is more than 1000 K lower than under atmospheric pressure. It was also clearly shown that this structural change in graphite is associated with and accelerated by internal stresses that are generated by interaction between the material's constituents at the contact points between grains or at the interface between fiber and matrix in carbon-carbon composites. In other words, it is graphitization assisted by stress (stress graphitization).

Remarkably, the acceleration of graphitization occurs for non-graphitizing carbons, which normally do not show any significant development of a graphitic structure even at HTTs above 3000 °C under atmospheric pressure. Under a pressure of 0.5 GPa, disordered glass-like carbons were completely converted to ordered graphite by heat treatment at 1600 °C.

An ordered nanotexture of BSUs is produced in regions where there is a sufficient accumulation of stress, e.g. at the contact points between grains. These regions are quickly converted to graphitic structure at high temperatures. Therefore, the process of structural change from carbon to graphite is heterogeneous, depending on the stress distribution and thermal conditions. This heterogeneous nature of structural change was observed by X-ray powder diffraction as composite profiles of 00l diffraction lines and by micrographs using SEM, TEM, and polarized-light microscopy.

The graphitization of carbons under high pressure was found to be accelerated by the coexistence of some minerals, typically calcium compounds. The formation of nascent calcium oxide was considered to be responsible for enhancing graphitization and the addition of water to calcium compounds, such as $CaCO_3$ and CaO, lowered the temperature for graphitization. With the coexistence of $Ca(OH)_2$, the flaky graphite particles were observed even after heating up to 800 °C, though the complete conversion to graphite was not attained.

5.4.2 Occurrence of graphite in nature

Many graphite ores occur naturally in the world. Some naturally occurring graphite crystals can have a very high degree of crystalline perfection. The thermal history of these natural graphite crystals has been geologically estimated from the minerals coexisting with them and is thought to have been very mild, at temperatures lower than 1000 °C and a pressure of about 0.5 GPa [4,5]. However, these estimated conditions for the formation of graphite crystals in nature are much milder than temperature-pressure conditions used to convert carbon with amorphous structure to crystalline graphite. Graphitization is the process to improve the stacking regularity, accompanied by growth of the size of the hexagonal carbon layers and increase in the number of stacked parallel layers, and has been studied in detail in relation to various conditions. Under atmospheric pressure without catalysts, however, it is known to require a temperature higher than 2500 °C. When some metals, such as iron and nickel, coexisted and acted as the catalysts, graphitization of carbon materials was found to be accelerated, but it still needed a temperature above 1000 °C [45,46].

As explained above, carbons have been graphitized rapidly at temperatures above 1600 °C and pressures above 0.3 GPa in the laboratory. However, these conditions for graphitization are much more severe than those geologically estimated for the occurrence of natural graphite [4,5]. Therefore, the metamorphic rocks, in which the graphite crystals are embedded, may have an accelerating effect on the graphite formation (graphitization), in addition to the acceleration by pressure. Graphitization of carbons was proved to be accelerated in coexistence of different minerals under a pressure of 0.3 GPa, as explained above. Different inorganic compounds, such as $CaCO_3$, CaO, $Ca(OH)_2$, CaF_2, Al_2O_3, and $NaCO_3$, including the natural mineral of limestone, have been used [9,28–33]. From these experimental results, it is concluded that graphite can be formed under mild conditions with the catalytic action of some metallic species, such as calcium. If we could realize extremely long residence time under these mild conditions, it seems that it would be possible to reproduce graphite crystals in the laboratory that are similar to those formed geologically in nature.

5.4.3 Stress graphitization in carbon/carbon composites

In carbon/carbon composites, stress can accumulate owing to the difference in thermal expansion and lead to the development of graphitic structure, as observed on various composites, such as carbon fiber with a glass-like carbon matrix, or natural graphite flakes with phenol resin. It has to be pointed out that this stress graphitization occurs by locally accumulated stress under atmospheric pressure. Most carbon/carbon composites are composed of filler carbon particles or fibers carbonized in advance and a matrix of resin. During preparation, the matrix resin shrinks by the carbonization, though the filler carbon does not show marked change, so that the stress induced by the disturbance of matrix shrinkage can change the nanotexture of the matrix, in the vicinity of contact, from random to oriented. The oriented regions in the matrix can easily be graphitized at high temperatures. Similar stress accumulation often occurs at the interfaces between different phases, even between filler coke particles and binder pitch. In most cases, however, no marked acceleration of graphitization is observed, probably because accumulated stress is not so high and also the stressed field is not a large area, as in the case of filler coke/binder pitch where the pitch (corresponding to the matrix in carbon fiber/carbon composites) is a very thin layer.

In the mixture of fine-sized MgO particles and pitch, in contrast, the development of micropores was observed after carbonization, in addition to mesopores formed by the replication of MgO particles [47], as described in Chapter 7. In this case, a strong disturbance in the flow of pitch during its pyrolysis is thought to result in the formation of microporous carbon on the MgO surface.

These results pose us the challenge of studying the interfaces between different phases, including carbon/carbon.

References

[1] Mrozowski S. Proceedings of the First and Second Conferences on Carbon. University of Buffalo: Waverley Press; 1951:31–45.
[2] Noda T, Kato H. Carbon 1965;3:289–97.
[3] Hishiyama Y, Inagaki M, Kimura S, et al. Carbon 1974;12:249–58.
[4] Wyllie PJ, Tuttle OF. Nature 1959;183:770.
[5] Buseck PR, Huang BJ. Geochim Cosmochim Acta 1985;49:2003–16.
[6] Inagaki M, Meyer RA. Chem Phys Carbon 1999;26:149–244.
[7] Inagaki M. Earth Sci Frontier 2005;12:171–81.
[8] Noda T. Carbon 1968;6:125–33.
[9] Noda T, Hirano S, Amanuma K, et al. Bull Chem Soc Jpn 1968;41:1245–8.
[10] Kamiya K, Mizutani M, Noda T, et al. Bull Chem Soc Jpn 1968;41:2169–72.
[11] Inagaki M. TANSO 1987; No.129: 68–80 [in Japanese].
[12] Noda T, Kamiya K, Inagaki M. Bull Chem Soc Jpn 1968;41:485–92.
[13] Kamiya K, Inagaki M, Noda T. Carbon 1971;9:287–9.
[14] Noda T, Inagaki M. Nature 1962;196:772–3.
[15] Noda T, Inagaki M, Sekiya T. Carbon 1965;3:175–80.
[16] Fischbach DB. Chem Phys Carbon 1971;7:1–05.

[17] Inagaki M, Oberlin A, Noda T. TANSO 1975; No.81: 68–72.
[18] Hishiyama Y, Kaburagi Y, Ono A, et al. Carbon 1980;18:427–32.
[19] Kamiya K, Inagaki M, Saito H. Bull Chem Soc Jpn 1969;42:1425–8.
[20] Inagaki M, Tamai Y, Naka S, et al. Carbon 1976;14:203–6.
[21] Inagaki M, Tamai Y, Naka S. TANSO 1973; No.75: 118–125 [in Japanese].
[22] Inagaki M, Tamai Y, Naka S, et al. Carbon 1974;12:639–43.
[23] Kamiya K, Mizutani M, Noda T, et al. Bull Chem Soc Jpn 1968;41:2169–72.
[24] Inagaki M, Oberlin S, de Fonton S. High Temp High Press 1977;9:453–60.
[25] de Fonton S, Oberlin A, Inagaki M. J Mat Sci 1980;5:909–17.
[26] Kamiya K, Yugo S, Inagaki M, et al. Bull Chem Soc Jpn 1968;41:2782–5.
[27] Inagaki M, Horii K, Naka S. Carbon 1975;13:97–101.
[28] Noda T, Inagaki M, Hirano S, et al. Kogyo Kagaku Zasshi 1969;72:643–8; [in Japanese].
[29] Noda T, Inagaki M, Hirano S, et al. Bull Chem Soc Jpn 1969;42:1738–40.
[30] Hirano S, Saito H, Inagaki M. Bull Chem Soc Jpn 1970;43:2599–603.
[31] Hirano S, Inagaki M, Saito H. Bull Chem Soc Jpn 1970;43:2624–5.
[32] Hirano S, Inagaki M, Saito H. Carbon 1979;17:395–8.
[33] Hirano S. Ph.D. Thesis 1970; Nagoya Univ.
[34] Bustin RM, Rouzaud J-N, Ross JV. Carbon 1995;33:679–91.
[35] Kimura S, Yasuda E. Zairyo Kagaku 1983;20:36–44; [in Japanese].
[36] Nakamura S, Ishii T, Yamada S. Symposium on Carbon 1964; IX-6.
[37] Inagaki M, Tamada K, Yamada S, TANSO. No. 1977;90:85–9; [in Japanese].
[38] Yamashita J, Shioya M, Hashimoto T, et al. Carbon 2001;39:129–33.
[39] Yamashita J, Shioya M, Hashimoto T, et al. Carbon 2001;39:119–27.
[40] Tanaka H, Kimura S, Tamada K. TANSO 1977; No. 91: 136–137 [in Japanese].
[41] Tanaka H, Kaburagi Y, Kimura S. J Mater Sci 1978;13:2555–9.
[42] Kamiya K, Inagaki M. Carbon 1973;11:429–30.
[43] Kimura S, Yasuda E, Tanaka H, et al. Yogyo Kyokai Shi 1975;83:122–7; [in Japanese].
[44] Yasuda A, Tanabe Y, Machino H, et al. TANSO 1987; No. 128: 7–11 [in Japanese].
[45] Inagaki M, Fujita K, Takeuchi Y, et al. Carbon 2001;39(6):921–9.
[46] Konno H, Fujita K, Habazaki H, et al. TANSO 2002; No.203: 113–116.
[47] Inagaki M, Kato M, Morishita T, et al. Carbon 2007;45:1121–4.

Glass-like Carbon: Its Activation and Graphitization

6

A carbon completely different from conventional carbon materials has been produced from thermosetting precursors, such as poly(furfuryl alcohol), phenol-formaldehyde, and cellulose [1]. It has a glass-like nature, with high hardness, isotropy in various properties, brittle conchoidal fracture, and gas impermeability, although lacks optical transparency. Despite the fact that these carbon materials have been given different trade names, such as *glassy carbon*, *cellulose carbon*, and *carbone vitreu*, they have almost the same characteristics. Therefore, these kinds of carbon materials are often called glass-like carbon.

Glass-like carbons are typical non-graphitizing carbons, which are poorly graphitized by heat treatment at high temperatures, as their X-ray powder patterns show, in Figure 6.1A. Even after 3000 °C treatment, the 002 line becomes only slightly sharp and a broad 004 line is detected, but the structure is still turbostratic because the 10 and 11 bands are unsymmetrical, interlayer spacing, d_{002}, being 0.341 nm. The characteristic conchoidal fractured surface of glass-like carbon is shown in Figure 6.1B.

In this chapter, oxidation of glass-like carbon spheres using air is reviewed, with discussion of the activation process, which has been often employed to develop and control the pore structure in carbon. On gas-impermeable glass-like carbon spheres, oxidation (i.e. activation) surely starts from the outermost surface of the spheres. Graphitization of glass-like carbon is also reviewed because it is non-graphitizing under normal conditions and needs melting for the conversion to graphite.

6.1 Activation of glass-like carbon

6.1.1 Glass-like carbon spheres

Starting material of phenol resin spheres for the production of glass-like carbon spheres were prepared through the polymerization of phenol resin suspended in water using calcium fluoride [2]. A typical process is as follows: an aqueous solution of phenol and formaldehyde with hexamethylenetetramine and calcium chloride is mixed with sodium fluoride under stirring, and then heated slowly to 85 °C to get spherical particles of resol-type phenol resin, which are covered with

Prerequisite for readers: Chapter 3.3 (Non-graphitizing and glass-like carbons) in *Carbon Materials Science and Engineering: From Fundamentals to Applications*, Tsinghua University Press.

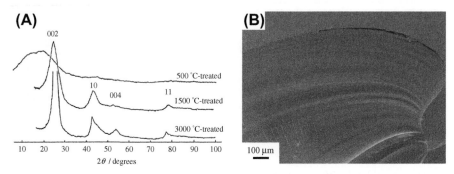

FIGURE 6.1 Characteristics of Glass-like Carbon

(A) Change in X-ray diffraction pattern with heat-treatment temperature, and (B) characteristic conchoidal fractured surface

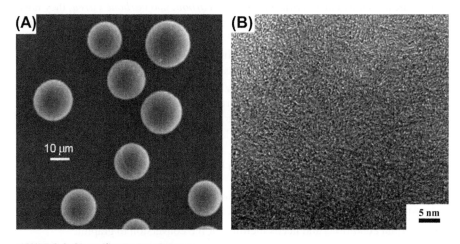

FIGURE 6.2 Glass-like Carbon Spheres

(A) Scanning electron microscopy (SEM) image, and (B) 002 lattice fringe image

From [3]

calcium fluoride and suspended in water. After cooling the suspension, the precipitates are separated from aqueous solution and then washed with water. Spherical particles with average sizes ranging from 5 to 800 μm are easily obtained. Glass-like carbon spheres are prepared from the phenol resin spheres by heat treatment at high temperature. The recommended procedure for getting glass-like carbon spheres [2] is as follows: heating to 900 °C with a rate of 50 K/h, maintaining at 900 °C for 3 h, and cooling to room temperature at a rate of 50 K/h in an inert atmosphere. After carbonization under these conditions, spherical morphology is conserved, as shown in Figure 6.2A. No cracks or fissures were observed on the surface under high magnification. Carbon yield through this carbonization process was about 63 mass% after 1000 °C carbonization. These heating conditions are not as severe as is necessary for the production of glass-like carbon blocks with either a large size or a complicated form, and can be reproduced even in a laboratory. The glass-like carbon nanotexture can be confirmed under high-resolution transmission electron microscopy (TEM), as in the 002 lattice fringe image shown in Figure 6.2B.

Glass-like carbon spheres with a particle size of c. 10 μm have been activated at different temperatures between 355 and 430 °C for different periods from 1 to 100 h in a flow of dry air with a rate of 50 m^3/min [3,4].

6.1.2 Activation in a flow of dry air

For activation of the glass-like carbon spheres, air oxidation was selected, for the following two reasons: the oxidation reaction was expected to start homogeneously from the surface of spheres, and air is the cheapest oxidizing agent and has a constant composition, if humidity is controlled. Activation was performed through the oxidation by air, either dry or wet. Oxidation of carbon spheres was followed by measurements of oxidation yield and pore-structure parameters from adsorption/desorption isotherm data at 77 K, through different analysis procedures.

Oxidation yields at different temperatures are plotted as a function of oxidation time, t, in logarithmic scale in Figure 6.3A [3,4]. The curves were then able to be put together by shifting along the abscissa at a reference temperature to get a so-called "master curve". Since some degree of scattering in the experimental points cannot be avoided, the master curve has to be obtained by shifting each experimental point at different oxidation temperatures along the abscissa (logarithm of oxidation time) to be consistent with the points measured at the reference temperature. The master curve at a reference temperature of 400 °C for the oxidation in dry air is shown in Figure 6.3B, with experimental points shifted. All experimental points can be superimposed on a smooth curve, i.e. the master curve. A plot of the shift factor against the inverse of oxidation temperature, T (an Arrhenius plot), gives a linear relationship, as shown in Figure 6.3C. These experimental results suggest that the oxidation of the carbon spheres is a single rate process and the conversion between oxidation temperature and residence time is possible in the present temperature range (355–430 °C). The slope of the Arrhenius plot gives an apparent activation energy, ΔE, of about 150 kJ/mol.

The Brunauer-Emmett-Teller method (BET surface area), S_{BET}, is plotted as a function of oxidation temperature and time in Figure 6.4A. The S_{BET} of the carbon spheres after oxidation at 355 °C for 10 h readily reaches about 400 m^2/g and increases with increasing oxidation time. At the highest temperature, 430 °C, it reaches about 800 m^2/g even after only 10 h of oxidation, and peaks after 30 h oxidation at 1000 m^2/g, but then decreases with increasing oxidation time. Intermediate oxidation temperatures resulted in intermediate change in S_{BET}, as shown by the dotted line for 400 °C in the figure. The same procedure and the same values of shift factors as for the oxidation yield were applied to obtain the master curve at 400 °C

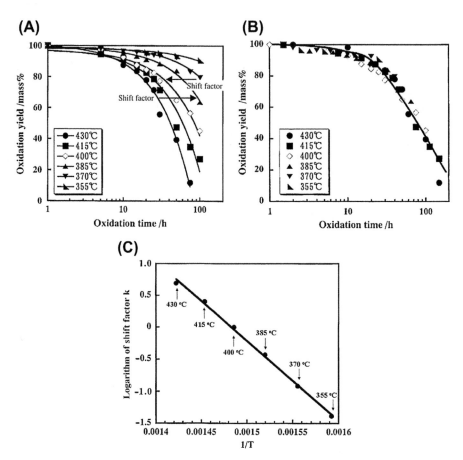

FIGURE 6.3 Oxidation of Glass-like Carbon Spheres

(A) Oxidation yield at different temperatures, (B) master curve at 400 °C, and (C)
Arrhenius plot of shift factors

From [3]

for S_{BET} (Figure 6.4B). The scattering in the data is a little larger than that for oxidation yield (Figure 6.3B), but a master curve can be clearly defined.

Three surface areas determined by α_s plot analysis—total surface area, S_{total},
external surface area, S_{ext}, and microporous surface area, S_{micro}—were analyzed by
the same procedure using the same shift factors as for oxidation yield. The master
curves obtained are shown in Figures 6.5A to 6.5C, together with the experimental
points shifted from each oxidation temperature. Micropore volume, V_{micro}, can be
determined by α_s plot analysis and its master curve is shown in Figure 6.6A. In
Figure 6.6B, V_{micro}, evaluated by using the density functional theory (DFT), and volumes separated into two subclasses, ultramicropore having the size less than 0.7 nm
and supermicropore having 0.7–2 nm size, are shown as master curves.

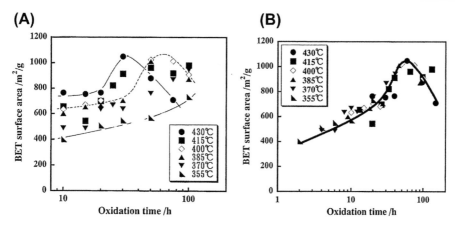

FIGURE 6.4 Change in S_{BET} of the Glass-like Carbon Spheres with Oxidation Time at Different Temperatures

(A) Change with time at different temperatures, and (B) master curve at 400 °C using the same shift factors as for oxidation yield

From [4]

Master curves obtained from α_s analysis and the DFT method show the same tendency, although coincidence in the absolute values cannot be expected. At the beginning of oxidation, V_{micro} increases gradually with the increase in oxidation time, which is thought to be because of the increase in ultramicropore volume. Beyond 10 h oxidation, V_{micro} starts to increase rapidly, where ultramicropores decrease gradually and supermicropores increase. After oxidation for more than 70 h, V_{micro} decreases rapidly, which is mainly because of the rapid decrease in ultramicropore volume. From the cumulative curve obtained by the BJH method, V_{meso} is calculated and its master curve is shown in Figure 6.6C. The changing tendency in V_{meso} in Figure 6.6C is very similar to that in S_{ext} in Figure 6.5B, which is reasonable because the principal part of S_{ext} is due to mesopores.

In order to understand the development of pores at the very beginning of air oxidation, the pore-size distribution in the micropore range was calculated by the DFT method. Pore-size distributions measured on carbon spheres oxidized for 2.5, 20, and 75 h at 400 °C are shown in Figure 6.7, which correspond to the beginning of micropore development, the starting point of increasing V_{micro}, and the maximum point in V_{micro}, as shown in Figure 6.6A, respectively.

On carbon spheres oxidized at 400 °C for 2.5 h, pores less than 0.4 nm seem to be predominant, even though these pores cannot be measured; in other words, most of the pores existing originally in the carbon spheres are not yet opened and only entrances with size less than 0.4 nm have been formed. After 20 h of oxidation at 400 °C, the population of pores with sizes around 0.6 and 1.2 nm increases, and the V_{micro} increases from about 0.17 m³/g to about 0.28 m³/g (Figure 6.6A), which may suggest the opening of closed pores. In this sample, there is also a small population of pores of 0.9-nm size. The pore-size distribution curve observed on the

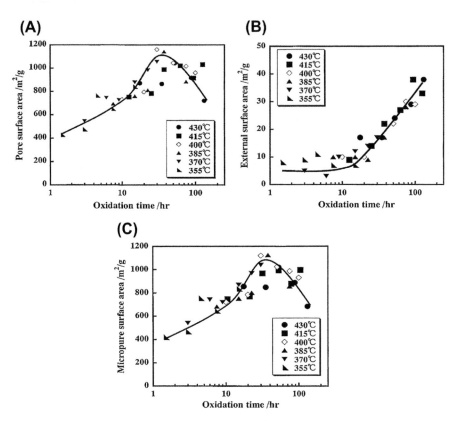

FIGURE 6.5 Surface Area Master Curves for the Oxidation of Glass-like Carbon Spheres

(A) S_{total}, (B) S_{ext}, and (C) S_{micro}, determined by α_s plot

From [4]

sample oxidized at 400 °C for 20 h suggests a large population of pores smaller than 0.4 nm, larger than for the samples oxidized for 2.5 h. This result may suggest that all pores originally exist in carbon spheres, i.e. in glass-like carbon nanotexture, are not yet opened after 20 h oxidation at 400 °C. After oxidation for 75 h, however, the population of pores smaller than 0.4 nm becomes almost zero; in other words, all closed pores are opened, and the population of pores with a size of around 0.6 nm decreases, but that of the pores with 1.2-nm size increases markedly. In this sample, the presence of pores with a size of about 0.9 nm is clearly observed. At 400 °C, a trend to increase the volume of pores with sizes of around 0.6 and 1.2 nm is clearly observed, and pores less than 0.4 nm seem to decrease. At 430 °C, the same trend could be recognized on pore-size distributions obtained through DFT analysis. After 75 h oxidation at 430 °C (corresponding to oxidation for 150 h at 400 °C), however, pore volume corresponding to pores with 0.6- and 1.2-nm sizes decreased, probably because pores were widened to more than 2 nm by oxidation.

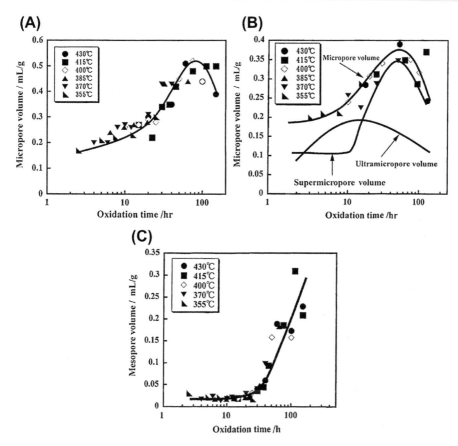

FIGURE 6.6 Micropore Volume Master Curves for Oxidation of Glass-like Carbon Spheres

(A) V_{micro} determined by α_s plot, (B) V_{micro} with ultramicropore and supermicropore volumes determined by the DFT method, and (C) mesopore volume, V_{meso}, determined by the Barrett-Joyner-Halenda method (BJH)

From [4]

6.1.3 Activation in a flow of wet air

The process of oxidation of glass-like carbon spheres using wet air can also be expressed as a master curve of oxidation yield at a reference temperature of 400 °C [5], as shown in Figure 6.8A. The Arrhenius plot of shift factors for each oxidation temperature was linear, and gave an apparent activation energy, ΔE, of about 200 kJ/mol. By using the conversion factors used for oxidation yield, it was also possible to obtain master curves of various pore-structure parameters.

In Figures 6.8B and 6.8C, pore-structure parameters are plotted against oxidation time, but not on a logarithmic scale as in the figures shown above. The V_{total} (Figure 6.8B) increases rapidly with increasing residence time at 400 °C and peaks

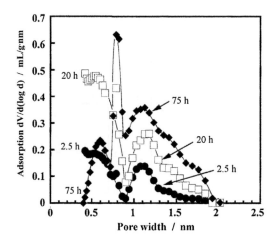

FIGURE 6.7 Pore-size distributions of spheres oxidized for different periods at 400 °C

From [4]

at about 0.6 m^3/g after 40 h of oxidation. Passing through the maximum, V_{total} tends to decrease gradually with further increase in residence time. S_{BET} changes in very similar way to V_{total}, giving a maximum of about 1000 m^2/g. Both V_{micro} and V_{meso} increase rapidly up to about 40 h oxidation, but show slightly different behavior beyond 40 h, V_{micro} keeping almost constant but V_{meso} decreasing (Figure 6.8C). The decrease in V_{total} by oxidation longer than 40 h seems to be due to the decrease in V_{meso} by widening of mesopores to macropores.

6.1.4 Activation process

The shift factors used for oxidation yield and also for the various pore parameters gave the apparent activation energy of about 150 kJ/mol for oxidation in dry air [3,4]. Taking into account the fact that the formation energies of CO_2 and CO gases are about 394 and 111 kJ/mol, respectively, the obtained value of apparent activation energy is in between these two formation energies. In wet air, an apparent activation energy of about 200 kJ/mol was obtained. From the analyses of gasification processes, similar values of activation energy have been reported for chars [6]. The activation energy determined here is only an apparent value, but it is considered to be the coefficient for the conversion between temperature and time for oxidation. The values of activation energy obtained by the present shifting procedure are thought to depend on the particle size and morphology, as well as oxidation conditions, such as the oxidizing agent and its concentration and flow rate, etc.

It has to be pointed out here that it is difficult to compare the absolute values of pore parameters that are determined by different methods of analysis, because these methods are based on different assumptions [7,8]. However, relative changes in various pore parameters with residence time at a reference temperature are possible

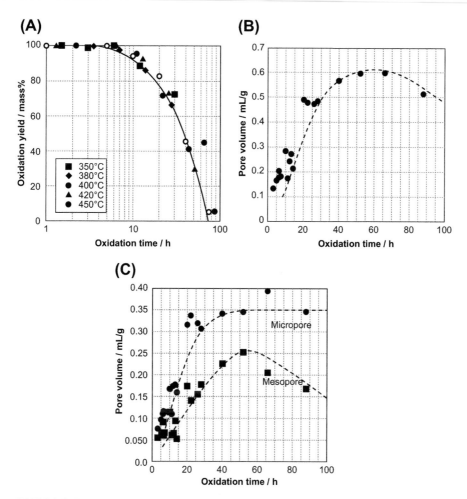

FIGURE 6.8 Pore-structure Parameter Changes with Oxidation in Wet Air for Glass-like Carbon Spheres

Master curves of (A) oxidation yield, (B) V_{total}, and (C) V_{micro} and V_{meso}. In (A) experimental points determined at different oxidation temperatures and then shifted to compose master curve at 400 °C are shown

From [5]

to compare using the master curves. The master curve of S_{BET} is, in general, very similar to that of S_{total} determined by α_s plot, and also very similar to that of S_{micro} determined by α_s plot, which is reasonable because S_{BET} is governed mostly by the presence of micropores. The S_{BET} reached its maximum after 65 h of oxidation at 400 °C, but the S_{micro} after 30 h of oxidation, shifting to a shorter oxidation time, which is reasonable considering that mesopores also contribute to S_{BET}. The development of mesopores started at about 30 h of oxidation, as observed on the change in V_{meso} by the BJH method. The S_{ext} determined by α_s plot has a very similar changing tendency with oxidation time to that of V_{meso}. The master curves of V_{micro} determined by α_s plot and the DFT method are also similar to each other in showing a maximum, as do the curves for S_{micro} and S_{BET}. Maxima observed for S_{BET}, S_{total}, and S_{micro} are reasonably understood as a result of the competition between widening and collapsing of pores and subsequent pore-surface-area loss.

In Figure 6.9, master curves for dry-air oxidation at 400 °C are reproduced for V_{micro} determined by α_s plot, ultramicropore and supermicropore volumes by the DFT method, and V_{meso} by the BJH method, in order to make comparison easier, together with some scanning electron microscopy (SEM) images showing the surface morphology of the spheres. At the beginning of oxidation, i.e. up to 10 h oxidation, the main process is the formation of ultramicropores, which is thought to be due to the opening of closed pores originally present in the glass-like carbon nanotexture. From 10 h up to about 60 h, ultramicropores decrease but supermicropores and also mesopores increase with increasing oxidation time. After 65 h, micropores decrease rapidly but mesopore volume increases slightly, which results in the decrease in surface areas measured. SEM observation of the sphere surface seems to agree with this pore development sequence, as shown in Figure 6.9, and it gives some information on macropores, which

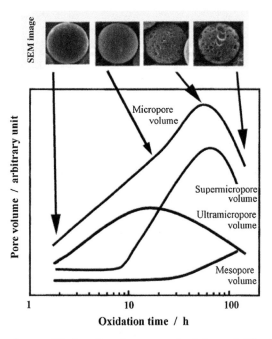

FIGURE 6.9 Master Curves of Various Pore Parameters for Spheres Oxidized at 400 °C in Dry Air, with Typical SEM Images of the Surface

From [4]

could not be measured by the gas adsorption/desorption analyses. After a few hours of oxidation at 400 °C, no change on the surface of sphere was detected in SEM images. After 20 h oxidation, only a few pit-like holes were observed, where supermicropores and mesopores were developing. After passing through the maximum of V_{micro}, the surface of spheres became rough due to the formation of macropores on the surface.

The changes in pore volumes suggest that the development of pores in the glass-like carbon spheres occurs stepwise: formation of ultramicropores by opening of closed pores pre-existing in the glass-like carbon nanotexture followed by gradual enlargement and collapse of ultramicropores to supermicropores and then to meso-pores, and finally to macropores.

6.1.5 Direct observation of micropores

Micropores formed by air activation on the surface of glass-like carbon spheres were directly observed by scanning tunneling microscopy (STM) [9–11]. Commercially available phenol-resin-derived carbon spheres with a glass-like carbon nanotexture were carbonized at 1000 °C in either a CO_2 atmosphere or in N_2, and named APT and APS, respectively. From an STM image observed using a Pt/Ir tip with a current of about 2 nA and a bias voltage between 20 and 800 mV (Figure 6.10A), a counter map as shown in Figure 6.10B is obtained, where the height difference in nm, with respect to the deepest zone of zero-level, is indicated for each counter line. After differentiat-ing pores from depressions and trenches in the STM images, pore-size distribution can be obtained. STM observation shows only the entrance of the pores formed on the surface of sphere, but it gives very useful information for understanding the acti-vation process because oxidation starts at the surface of glass-like carbon spheres. In Table 6.1, a summary of pore-structure parameters is given together with S_{BET} for

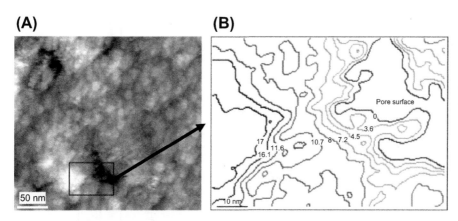

(A)

(B)

FIGURE 6.10 Imaging of Micropores

(A) Typical STM image of oxidized glass-like carbon sphere, and (B) counter map of a squared part

From [9]

Table 6.1 Pore-structure Parameters Determined by STM on Four Types of Glass-like Carbon Spheres

Sample (Area Studied)	S_{BET}	Pore Size	<0.5 nm	0.5–0.9 nm (Ultramicropores)	0.9–1.4 nm	1.4–1.8 nm (Supermicropores)	1.8–2.2 nm	2.2–4 nm (Mesopores)
APS (26,700 nm²)	2	Mean size	0.25	0.60	–	–	–	–
		Number (%)	54 (96)	2 (4)	–	–	–	–
		Density	2023	75	–	–	–	–
APS-44 (26,970 nm²)	980	Mean size	0.43	0.70	1.10	1.54	1.97	2.7
		Number (%)	10 (6.6)	44 (29.3)	42 (28)	24 (16)	14 (9.3)	16 (10.6)
		Density	371	1631	1557	890	519	593
APT (23,492 nm²)	310	Mean size	0.36	0.66	1.08	1.60	2.00	2.70
		Number (%)	28 (28.6)	19 (19.4)	16 (16.3)	7 (7.1)	11 (11.2)	17 (17.3)
		Density	1192	809	681	298	468	724
APT-44 (24,200 nm²)	1200	Mean size	0.49	0.75	1.00	1.55	2.00	2.90
		Number (%)	1 (0.5)	51 (29)	66 (37.9)	36 (20.3)	15 (8.5)	8 (4.5)
		Density	41	2107	2727	1488	620	331

S_{BET}, BET surface area in m²/g; mean size, in nm (standard deviation less than 0.2 nm); number, number of pores in studied area (%); and density, number of pores per μm². From [9].

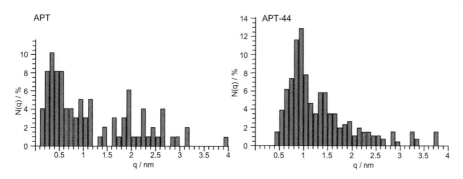

FIGURE 6.11 Pore-size Distributions Determined by STM Observation for Two Types of Glass-like Carbon Spheres

From [10]

the pristine spheres, APT, and APS, and for those oxidized in boiling 13 vol% nitric acid for 3 h and then in air at 400 °C for 4 h, named APT-44 and APS-44, respectively. In the table, pore sizes determined by STM were divided into six classes; the sizes 0.5–0.9 nm and 0.9–1.4 nm corresponding to ultramicropores, 1.4–1.8 nm and 1.8–2.2 nm to supermicropores, and 2.2–4 nm to mesopores.

APT carbonized in CO_2 has many micropores, even mesopores, on the surface, although the pores are mainly of sizes less than 0.5 nm on the surface of APS carbonized in N_2. By oxidation, pore size is markedly broadened, as can be understood by comparing the pristine to the oxidized spheres, APS to APS-44 and APT to APT-44. A marked shift in pore size distribution to larger sizes can be seen in Figure 6.11 on APT and APT-44.

The results of pore analysis based on STM observation reveal that pore size is broadened by oxidation (activation); in other words, small pores are sacrificed for the formation of larger pores by oxidation.

6.1.6 Two-step activation

Based on the oxidation process on carbon spheres, a two-step activation process was proposed [12]: creation of mesopores or macropores on the surface of glass-like carbon spheres (first-step activation) by air oxidation at a high temperature for a short time, followed by opening of the closed pores in the glass-like carbon matrix and increase of micropores (second-step activation) by air oxidation at a low temperature for a long time.

In Figure 6.12, adsorption/desorption isotherms are shown for the carbon spheres oxidized in one-step and two-step processes, respectively. Two isotherms for the spheres oxidized at 500 °C for 3 h and for those oxidized at 415 °C for 40 h are compared with the isotherm for spheres oxidized in a two-step process, the first step at 500 °C for 3 h followed by the second step at 415 °C for 30 h. By applying two-step activation, micropores are thought to be developed efficiently, because the increased adsorption in a low relative pressure range indicates the formation of a larger amount

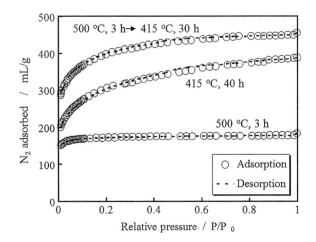

FIGURE 6.12 Nitrogen Adsorption/Desorption Isotherms for Carbon Spheres Oxidized Under Different Conditions

From [12]

of micropores by the two steps of activation at 500 °C and then 415 °C for a total residence time of 33 h than by the one-step activation at 415 °C for 40 h.

In Figures 6.13A–6.13C, pore-structure parameters S_{BET}, S_{micro}, and V_{micro} are plotted against oxidation yield, Y, for the one-step and two-step activation processes. Oxidation yield, Y, becomes the lower with the more severe oxidation conditions (higher temperature for longer residence time). The relationships of pore-structure parameters with Y can be approximated by straight lines, which are broken at around 50 mass% of Y, where each pore-structure parameter gives a maximum, as shown in Figure 6.13. Above Y of about 50 mass%, the experimental points obtained by the two-step activation are always higher for the three pore parameters than those by the one-step activation.

The results show that at a given S_{micro} and V_{micro}, the two-step activation results in higher Y by c. 10%; in other words, at a given oxidation yield, larger S_{micro} by c. 10% and larger V_{micro} by c. 20% can be obtained.

6.2 Graphitization of glass-like carbons
6.2.1 Graphitization through melting

Carbon melts at 4000 K under a pressure of about 10 MPa, giving graphite with exactly the same lattice constants as natural graphite, i.e. $c_0 = 0.67078$ and $a_0 = 0.2461$ nm [13]. Glass-like carbons, as typical non-graphitizing carbon, were found also to give graphite, though they were not graphitized even by heat treatment at the very high temperature of 3000 °C [14].

A rod of glass-like carbon with a diameter of 4 mm, which was heat-treated at 2000 °C, was melted by directly passing electric current under the pressure of

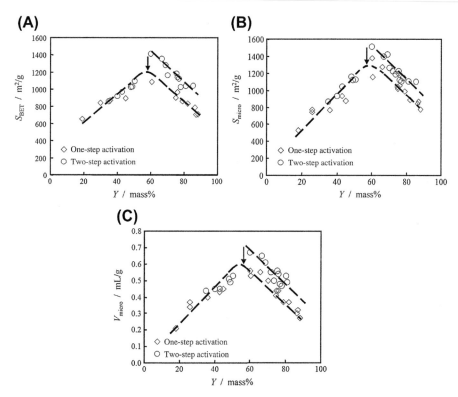

FIGURE 6.13 Relationships of Pore Parameters to Oxidation Yield, *Y*, by One-step and Two-step Activation

(A) S_{BET}, (B) S_{micro}, and (C) V_{micro}

From [12]

11 MPa of argon. The melting of the glass-like carbon started in the inner part of the middle of the rod and made a molten mass in a spindle form along the rod axis, although coke-based carbon gave the molten carbon as a ball attached to a crater-type cavity. Frequently, the molten carbon in glass-like carbon spouted out through a break in the wall, which was not melted yet, and a narrow but long cavity with a spindle form remained in the rod. The spouted molten carbon was scattered as many balls smaller than 0.5 mm in diameter. When a rod 5 mm in diameter was used, the spouted molten carbon had a cochleate appearance. Minute fragments of the molten carbon were also found on the inside wall of the cavity.

These small balls and cochleate particles were soft and of metallic luster. Their lattice constants were the same as those of natural graphite. Their X-ray diffraction patterns provided only the pattern due to hexagonal graphite, with no pattern of rhombohedral modification. By grinding in a mortar, however, a trace of the rhombohedral modification was detected.

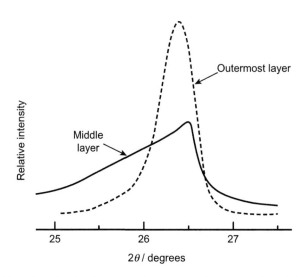

FIGURE 6.14 002 Diffraction Profiles of the Crater Wall After the Melting of Glass-like Carbon

From [14]

When the rod did not melt down and showed no change in appearance after the passage of electric current, a cavity of about 10 mm long and about 1.5 mm in diameter was found in the central part of the rod. The wall of the cavity was apparently composed of three layers. The innermost layer was the molten carbon, being soft and of metallic luster and having the same lattice constants as natural graphite. The middle layer was very hard and showed a conchoidal fracture, characteristic of glass-like carbon. The 002 diffraction line of this layer was asymmetrical, as shown in Figure 6.14 (solid line), showing that only partial graphitization had occurred even after heating at temperatures very near the melting point of carbon. These experimental results illustrate the fact that glass-like carbon does not lose its characteristics even at temperatures very near its melting point. In coke-based carbons, the inner wall of the crater formed by melting had the lattice constant $c_0 = 0.6714$ nm, a little larger than that of natural graphite [14]. The outermost layer of the rod showed a symmetrical 002 diffraction profile, as shown in Figure 6.14 (dotted line). This layer was very thin, about 0.05 mm thickness, and so soft that it could be whittled by a razor blade, but no metallic luster. Its lattice constant, c_0, was measured to be 0.6748 nm, much larger than that of natural graphite. This layer seemed to be graphitized at a temperature near the melting point under an acceleration by a minute amount of oxygen in the atmosphere, because it was in direct contact with the ambient argon gas.

6.2.2 Graphitization under high pressure

Graphitization of glass-like carbons was possible under a pressure of 0.5 GPa at a temperature around 1600–1700 °C [15–17]. In Figure 6.15, change in the 002 diffraction profile with temperature under 0.5 GPa is shown on carbon beads with glass-like nano-texture [15]. After heat treatment above 1600 °C, the 002 profile becomes sharp and

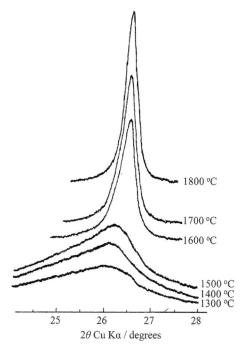

25   26   27   28

2θ Cu Kα / degrees

1800 °C
1700 °C
1600 °C
1500 °C
1400 °C
1300 °C

FIGURE 6.15 Change in the 002 Diffraction Profile of Glass-like Carbon Spheres at Different Temperatures Under a Pressure of 0.5 GPa

From [16]

shifts to a higher diffraction angle, indicating that the graphitization of glass-like carbon occurs abruptly. Almost the same results were obtained for a glass-like carbon powder consisting of irregular particles [16]. Above 1600 °C, an interlayer spacing, d_{002}, of 0.336 nm was obtained.

The graphitization starts preferentially at the contact points between the glass-like carbon spheres, where stress is concentrated upon pressurization. In Figure 6.16, polarized-light micrographs are shown on the cross-section of a sintered block obtained at different temperatures under 0.5 GPa [15]. Bright areas showing optical anisotropy are observed even after 1300 °C treatment, indicating the occurrence of stress concentration. Bright areas grow and also increase in number with increasing temperature, where anisotropic carbon layers are preferentially oriented to make their layers perpendicular to the direction of stress [17]. Above 1600 °C, the whole area of the cross-section changes to be anisotropic.

Glass-like carbons can be graphitized at temperatures above 1600 °C under high pressure of 0.5 GPa, but it is characteristic for glass-like carbon, including other non-graphitizing carbons with random nanotexture, to occur very abruptly, although graphitizing carbons with oriented nanotexture are graphitized gradually under pressure.

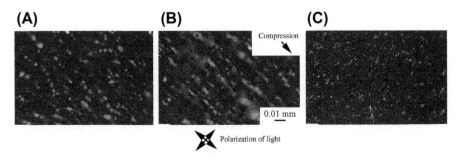

FIGURE 6.16 Polarized-light Micrographs of Glass-like Carbon Spheres under 0.5 GPa at Different Temperatures

(A) 1300 °C, (B) 1500 °C, and (C) 1600 °C

From [15]

6.2.3 Graphitization in C/C composites

Composites of carbon fibers with a glass-like carbon matrix were prepared at 1000 °C and heat-treated up to 2800 °C, in which polyacrylonitrile (PAN)-based carbon fibers were aligned unidirectionally in the glass-like carbon matrix formed from furfuryl alcohol condensates [18].

In Figure 6.17, the X-ray diffraction profiles of the 004 diffraction line of composites with different fiber fractions are compared to those for the component carbons, carbon fiber and glass-like carbon. These X-ray profiles for the C/Cs are very similar to those observed when carbons are heat-treated under pressure, as described in the previous section. Also, the d_{002} spacing for the component on the high-angle side for the composites is about 0.336 nm, which is the same as that observed for carbon that has been graphitized under high pressure.

On the other hand, component samples of the PAN-based carbon fiber and the glass-like carbon that have been individually processed under exactly the same conditions show only a broad band on the low-angle side, which is significantly different compared to the composites. This is because they both have very low graphitizability. The 004 diffraction profile of the composites indicates that there is development of graphite structure, which appears to be accelerated by the interaction between these two components, carbon fiber and glass-like carbon, in the composites during the high-temperature treatment, as explained in Chapter 5 using polarized-light optical micrographs (Figure 5.12). Asymmetry of the X-ray profile seems to be more pronounced in the composites with lower fiber content. This means that composites whose fiber fraction is 30 vol% contain a larger amount of graphitized matrix than those with 60 vol% of fibers. This is because of stress graphitization, which is attributed to the graphitization of the matrix by the interaction between the fibers and the matrix at elevated temperatures.

The reason why this graphitization occurs is that a stress field is created in C/Cs during heat treatment between each fiber and its surrounding matrix. The formation

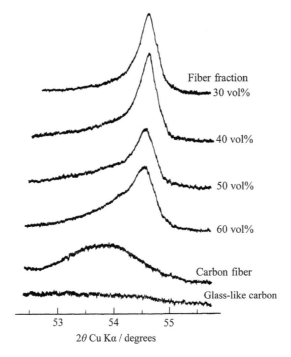

FIGURE 6.17 004 Diffraction Profiles of Carbon Fiber/Glass-like Carbon Composites with Different Fiber Fractions, in Comparison with the Component Carbon Fiber and Glass-like Carbon, After Heat Treatment at 2800 °C

From [18]

and orientation of graphite layers around the fiber can be clearly demonstrated by SEM on a similar carbon fiber/glass-like carbon composite after a slight etching by oxidation, as shown in Figure 6.18 [19]. This stress field tends to create a texture of concentric orientation of basic structural units (BSUs) whose c-axis radiates from the center of the fibers. On X-ray diffraction, graphitization was observed by a gradual shift of the 002 diffraction line to the high-angle side above 2100 °C and the asymmetry of its profile becomes more evident with increase of heat-treatment temperature.

The magnetoresistance ($\Delta\rho/\rho$) has been measured on these types of composites in order to evaluate the degree of graphitization and orientation of carbon layers along the fiber axis. The dependence of the maximum transverse magnetoresistance, $(\Delta\rho/\rho)_{max}$ [18], on the magnetic field is shown in Figure 6.19. $(\Delta\rho/\rho)_{max}$ of the composite changes its sign from negative to positive at a temperature between 2250 and 2450 °C, suggesting that this is the temperature where the growth of the graphitic structure starts to occur. The values of $(\Delta\rho/\rho)_{max}$ for the composite after 2800 °C treatment are comparable to the values of nuclear-grade high-density isotropic graphite blocks, revealing a very high degree of graphitization. In contrast, there

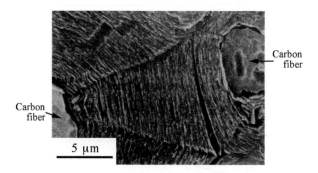

FIGURE 6.18 SEM Image of the Cross-section of the Carbon Fiber/Glass-like Carbon Composite After Slight Etching

Courtesy of Prof. E. Yasuda of Tokyo Institute of Technology, Japan

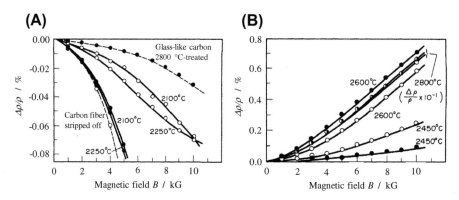

FIGURE 6.19 Magnetic Field Dependences of Magnetoresistance, $\Delta\rho/\rho$, on Carbon Fiber/ Glass-like Carbon Composites Heat-treated at Different Temperatures, in Comparison with the Component Glass-like Carbon and a Carbon Fiber Stripped Off the Composite After Heat Treatment at 2800 °C

From [18]

is no indication of graphitization for the glass-like carbon, as it retains a negative $(\Delta\rho/\rho)_{max}$ (Figure 6.19A) even after being heat-treated to above 2800 °C. In order to verify that the graphitization effect occurred in the matrix, the magnetoresistance was measured on a single fiber that was stripped off the composite after the heat treatment at 2800 °C. It is clearly shown in Figure 6.19A that the fibers have negative values of magnetoresistance, i.e. they are not graphitized even after 2800 °C treatment.

6.3 Concluding remarks

For the structure of glass-like carbon prepared from various precursors, the model of the aggregated, minute closed shells consisting of carbon hexagonal layers [20–23] explains their characteristic properties, such as low bulk density, gas impermeability, and brittleness [1]. Two other models have been proposed: where two kinds of domains of sp^2 and sp^3 carbons are linked by oxygen bridges [24], and the three-dimensional entanglement of two-dimensional sp^2 carbon ribbons [25]. However, the model of aggregation of minute closed shells [22] is believed to be the most reasonable. The key process for the production of these glass-like carbons is the slow heating during carbonization of precursor resins, slow enough to secure complete shrinkage at the beginning of carbonization of the carbonaceous materials formed through pyrolysis of precursors, as described above.

Kinetic studies of oxidation of the glass-like carbon spheres in air have given a fundamental understanding of the activation process.

1. The activation process can be expressed by the master curve, based on the possible conversion between oxidation temperature and time.
2. The master curve can be obtained for various parameters, such as oxidation yield and various pore structure parameters, by using the same apparent activation energy.
3. Activation has been experimentally shown start by the opening of closed pores, which had been formed in the glass-like carbon matrix in advance, forming micropores.
4. Pores developed by enlarging and collapsing of micropores to macropores, via mesopores.

Based on these results, a new activation procedure, two-step activation, was proposed, which was effective for achieving comparative pore structure with a higher oxidation yield in a shorter oxidation time than in one-step activation. This fundamental understanding relates to carbon spheres with glass-like carbon nanotexture, where oxidation starts principally from the surface, not the inside of the sphere; its basic reaction is simple formation of CO and CO_2 gases, and it occurs homogeneously, at least in the beginning.

Since glass-like carbon is a typical non-graphitizing carbon, its graphitization is possible only under very severe conditions, by melting, under pressures as high as 0.5 GPa, and under marked stress concentration in carbon/carbon composites. By melting, it can be converted to graphite spheres with high crystallinity under an argon pressure of 10 MPa. However, the wall of the crater behind the molten part is not graphitized even though it experiences a high temperature near the carbon melting point (about 3800 °C) under 10 MPa; in other words, the random nanotexture of glass-like carbon does not change even at a high temperatures near the melting point. It can be graphitized under very high pressure, as high as 0.5 GPa. At the interface between the filler of carbon fiber and the matrix of glass-like carbon, marked stress concentration occurs, which can induce the structural change of glass-like carbon to graphite at the high temperature of 2800 °C. These experimental results reveal that the random nanotexture formed in glass-like carbon has to be destroyed in order to develop the graphite structure; in other words, the nanotexture established upon the carbonization of organic precursors governs predominantly the structural change at high temperatures.

References

[1] Noda T, Inagaki M, Yamada S. J Non-Cryst Solids 1969;1:285–302.

[2] Okada T, Echigo Y. Plastics 1987;38:81–6 [in Japanese].

[3] Nishikawa T, Inagaki M. Adv Sci Technol 2005;23:827–37.

[4] Inagaki M, Nishikawa T, Oshida K, et al. Adv Sci Technol 2006;24:55–64.

[5] Inagaki M, Suwa T. Mol Cryst Liq Cryst 2002;386:197–203.

[6] Raghunathan K, Yang RYK. Ind Eng Chem Res 1989;28:518–23.

[7] Sing KSW. Porosity in Carbons. In: Patrick JW, editor. Edward Arnold; 1995. p. 49–66.

[8] Hanzawa Y, Kaneko K, Carbon Alloys, Yasuda E, et al. Elsevier 2003:319–34.

[9] Vignal V, Morawski AW, Konno, et al. J Mater Res 1999;14:1102–12.

[10] Inagaki M, Vignal V, Konno H, et al. J Mater Res 1999;14:3152–7.

[11] Inagaki M, Sunahara M, Shindo A, et al. J Mater Res 1999;14:3208–10.

[12] Wang L, Fujita M, Toyoda M, et al. New Carbon Mater 2007;22:102–8.

[13] Noda T, Matsuoka H. Kougyou Kagaku Zasshi 1960;63:465–7 [in Japanese].

[14] Noda T, Inagaki M. Bull Chem Soc Jpn 1964;37:1709–10.

[15] Inagaki M, Horii K, Naka S. Carbon 1975;13:97–101.

[16] Inagaki M, Oberlin A, de Fonton S. High Temp High Press 1977;9:453–60.

[17] Kamiya K, Inagaki M. Carbon 1981;19:45–9.

[18] Hishiyama Y, Inagaki M, Kimura S, et al. Carbon 1974;12:249–58.

[19] Yasuda E, Kimura S, Shibusa V. Jap Soc Compo Mater 1983;9:2–7 [in Japanese].

[20] Noda T, Inagaki M. Bull Chem Soc Jpn 1964;37:1534–8.

[21] Noda T, Yamada S, Inagaki M. Bull Chem Soc Jpn 1968;41:3023–4.

[22] Shiraishi M. Introduction to Carbon Materials. Carbon Society Japan 1984:29–40 [in Japanese].

[23] Oberlin A. Chem Phys Carbon vol. 22. Marcel Dekker 1989. p. 1–143

[24] Kakinoki J. Acta Cryst 1965;18:578.

[25] Jenkins GM, Kawamura K. Nature 1971;231:175–6.

Template Carbonization: Morphology and Pore Control

Novel carbon materials have been attracting the attention of materials scientists and engineers since the 1960s: very hard and amorphous glass-like carbons, pyrolytic carbons consisting of well-oriented carbon layers, fibrous carbons derived from different polymers, amorphous diamond-like carbons, closed cages consisting of carbon atoms (fullerenes), tubular carbons with nano-sized diameters (carbon nanotubes), and graphene consisting of single carbon layers have been developed. Development of new types of carbon materials has been supported by innovations in various techniques for the preparation of these materials; for example, chemical vapor deposition, exfoliation, and intercalation. Carbonization of various precursors using different materials as templates, or template carbonization, is one of these new, innovative techniques.

Template carbonization is the process to replicate the structure and texture of the template into the carbon in nanoscale, and has been successfully applied to prepare carbon flakes, carbon nanotubes and nanofibers, and nanoporous carbons. It was first proposed to control the morphology of carbon materials, and then used to prepare nanoporous carbons with controlled size and alignment of pores by employing various templates. In Table 7.1, the templates often discussed in the literature are listed together with the characteristics of the resultant carbons, dividing the application of template carbonization into two purposes, morphology and pore-structure control.

Review articles about template carbonization have been published from different points of view [1–8], although most of them are focused on pore control in carbon materials. The carbons prepared through template carbonization have been used in various applications and so they have been referred to in a number of reviews on specific applications. In this section, template carbonization is divided into the processes aiming to control morphology and those to control pore structure of the resultant carbon materials. The prospects of the template-carbonization processes are discussed, including the possibility of applying these techniques to large-scale production.

Prerequisite for readers: Chapters 2.3 (Nanotexture development in carbon materials (carbonization)) and 2.4 (Novel techniques for carbonization) in *Carbon Materials Science and Engineering: From Fundamentals to Applications*, Tsinghua University Press.

Table 7.1 Template Carbonization

Template	Resultant Carbon
Morphological control	
Inorganic layered compounds (clays)	Lamellar morphology
Anodic aluminum oxide (AAO) films	Tubular morphology
Organic foams (polyurethane foams)	Foamy morphology
Pore-structure control	
Zeolites	Microporous, ordered
Mesoporous silicas	Mesoporous, ordered
MgO	Mesoporous
Block copolymer surfactants	Mesoporous, ordered
Metal-organic frameworks (MOFs)	Mesoporous

7.1 Template carbonization for morphological control

7.1.1 Inorganic layered compounds

The template-carbonization technique was first developed for the preparation of thin oriented graphite films using two-dimensional spaces in layered compounds, such as montmorillonite (MONT) and taeniolite (TAEN) [9–15]. The fundamental scheme of this template method is as in the following example. Firstly, Ca^{2+}-exchanged MONT was exposed to the vapor of a carbon precursor, acrylonitrile (AN), for 24 h at room temperature. The MONT-AN complex was irradiated by γ-ray at room temperature in order to polymerize the intercalated AN to polyacrylonitrile (PAN), and then heat-treated at 700 °C in N_2 flow to carbonize the PAN. The heat-treated complex was put into 46% HF solution at 0 °C under rigorous stirring and then refluxed in 37% HCl solution to recover the resultant carbon, of which the yield was 0.055 g/g-clay. The mass gain of the MONT template after exposure to AN vapor was measured to be 0.14 g/g-clay, which was confirmed to be due to the intercalation of a monolayer of AN into the gallery between the lamellae of MONT. The resultant carbon contained a relatively large amount of foreign atoms, 2.5 mass% H, 16 mass% N, and 15.6 mass% O, but was composed of flaky particles [9,10]. Its flaky morphology was maintained during the treatment at high temperatures, even though a marked change in structure was associated with this treatment, which indicated the development of graphitic structure (graphitization), while the carbon precursor itself, AN, did not show appreciable growth of graphite structure by high-temperature treatment [11]. The flakes heat-treated to 2800 °C showed a lattice fringe image under transmission electron microscopy (TEM) similar to graphite, and a negligibly small D-band in the Raman spectrum. In Figure 7.1, lattice fringe images are shown for the flakes heat-treated at 1400, 2100, and 2800 °C. Many layers are already stacked in parallel after the treatment at 1400 °C, but regular graphitic stacking is realized only above 2500 °C.

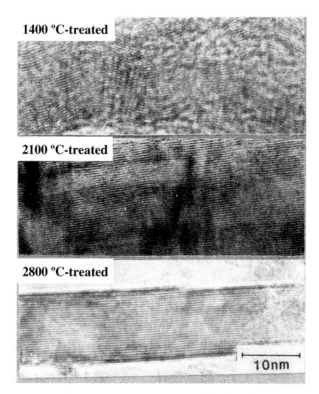

FIGURE 7.1 002 Lattice Fringe Images for Carbon Flakes Prepared Through Templating and Heat-treating at Different Temperatures

Courtesy of Prof. T. Kyotani of Tohoku University, Japan

Furfuryl alcohol (FA) and vinyl acetate (VAC) have been successfully used as carbon precursors [12,13]. Other layered compounds, saponite (SAPO) and TAEN, have also been used as templates [14,15]. All these complexes of carbon precursor/template clay gave carbon flakes with high graphitizability. Since the templates used were easily obtained as films, graphite films were prepared from VAC [16]. Carbon flakes thus prepared had thickness of about 40 nm and it was possible to form them into pellets by simple compression without any binder [17]. They reacted with potassium vapor to form a first-stage intercalation compound [18].

7.1.2 Anodic aluminum oxide films

Tubular carbons were prepared by carbon deposition at 800 °C in a flow of a mixture of N_2 and propylene (2.5%) on the inner wall of nano-sized channels in an anodic aluminum oxide (AAO) film [19,20]; the process being a kind of chemical vapor infiltration (CVI). The template AAO film was dissolved either with 46% HF at room temperature or with a 10 mol/L NaOH aqueous solution at 150 °C in an autoclave.

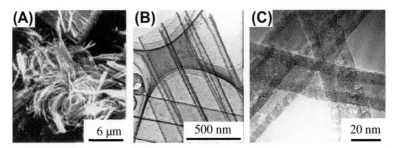

FIGURE 7.2 Carbon Tubules Prepared by Template Carbonization Using Anodic Aluminum Oxide Film

(A) SEM image, (B and C) TEM images

Courtesy of Prof. T. Kyotani of Tohoku University, Japan

Scanning electron microscopy (SEM) and TEM images of carbon tubules, which were prepared by using a laboratory-made AAO film with 30 or 200 nm diameter channels, are shown in Figure 7.2. In the as-prepared tubules, carbon layers are small and do not yet have a perfect orientation along the tube axis, as shown in Figure 7.2C, but they grow markedly and align perfectly along the tube axis after heat treatment at the high temperature of 2800 °C (Figure 7.2B), the resultant carbons being comparable to multi-walled carbon nanotubes synthesized by arc-discharging. The control of channel size (diameter and length) of template oxide films has been well established by changing the electrolyte, current density, temperature, and time for oxidation of aluminum foil. Therefore, the diameter and length of tubular carbons depend on the diameter of channels and the thickness of the template film, respectively, and the wall thickness of tubular carbons can be controlled by the deposition conditions. It has to be pointed out that the tubes prepared via this process have very uniform diameter and length, and align along the channels of the AAO film, which are perpendicular to the film surface.

Ni, Co, or Fe metal, which were deposited at the bottom of the channels of the AAO template by AC electrolysis and the following reduction in a flow of CO, served as a catalyst for CVI into the channels to form tubular carbons [20,21]. The resultant tubular carbons were aligned with their tubular axes parallel, and so could form a membrane with a porosity of 60% and thickness of 60 μm [22,23]. The membranes thus prepared were tested for Li intercalation/deintercalation, electro-catalyzed O_2 reduction, and methanol oxidation, after deposition of Pt, Pt/Ru, and Fe [22]. Highly ordered arrays of tubular carbons with a diameter of 32 nm and length of 6 μm were prepared by the same process using a Co catalyst at the bottom of the channels of an AAO film, and by CVI of acetylene at 650 °C [23,24]. By using AAO films with Y-branched channels of 60 nm in diameter, tubular carbons with a Y-junction were prepared through CVI of acetylene at 650 °C [25]. By repeating the anodization and carbon deposition from acetylene in the channels of the template, a tube-in-a-tube and linearly joined tubular carbons were prepared [26].

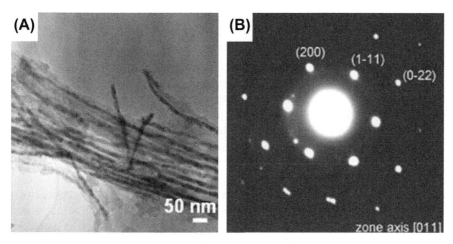

FIGURE 7.3 Permalloy-filled Carbon Tubules

(A) TEM image, and (B) selected-area electron diffraction pattern

From [36]

This template-carbonization technique has many advantages for control of the structure of tubular carbons, in addition to high uniformity in diameter and length of tubes. Another advantage of this technique is the fact that the tubular carbons are fixed inside the channels of the template aluminum oxide, and one end is open. This gives a possibility to relatively easily fill the inside of the tubes with either metals or metal oxides to prepare nanowires; the carbon tubules filled with foreign material can then be separated from the template AAO film by dissolution. Filling with highly crystallized Pt was carried out; a carbon-deposited AAO film was impregnated into an ethanol solution of H_2PtCl_6 at room temperature and then heat-treated at 500 °C in a H_2 flow [27]. Pt and Pt/Ru alloy were deposited on the walls of tubules as nanoparticles, about 7 and 1.6 nm, by immersing into aqueous solution of H_2PtCl_6 and H_2PtCl_6 + $RuCl_3$, respectively, followed by heating at 580 °C in a flow of H_2 [28]. Tubules were filled with crystalline Fe_3O_4 through metal-organic chemical vapor deposition (MOCVD) of ferrocene ($Fe(C_5H_5)_2$), the filling efficiency being more than 20% [29]. Ni nanowire formed through MOCVD of nickelocene ($Ni(C_5H_5)_2$) in the tubule was thought to be a single crystal with a uniform diameter of 4 nm, of which the (111) planes are preferentially parallel to the tube axis [30]. NiO nanoribbons (4 × 20 × 800 nm) [31] and also $Ni(OH)_2$ single crystals [32,33] were successfully used to fill the tubes. Filling of tubules with Permalloy (Fe-Ni alloys), about 40 nm in diameter and 900 nm in length, was also carried out by electroplating in an electrolyte [34–36]. These Permalloy-filled nanotubes were shown to be effective for the suppression of electromagnetic noise emission [36]. In Figure 7.3, a TEM image of Permalloy-filled nanotubes and its selected-area electron diffraction pattern are shown, revealing the formation of a single crystal [36].

The inner surface of these tubular carbons was also able to be modified by fluorination at a temperature of 50–200 °C [37–39] and oxidized by immersing into 20% HNO_3 under refluxing conditions [40], because the inner surfaces of the tubes were exposed to the atmosphere but the outer surface was completely covered by the template Al_2O_3.

The one-end-closed tubular carbons separated from the AAO template (nano-test-tubes) were found to be water-dispersible without any post-treatment, when length was not longer than 5 μm [41]. These nano-test-tubes were easily realigned to a monolayer film at the liquid-liquid interface by dispersing in an aqueous ethanol/toluene solution, which was transferable to solid substrate [42,43]. The tubes filled with Permalloy, however, could be dispersed in water, even when their length was around 1 μm, probably owing to the magnetic interaction between the tubes. Upon oxidation of the outer surface of the Permalloy-filled test tubes in H_2O_2, water dispersibility was improved, which might make it possible to apply these tubes as a carrier for a magnetic drug-delivery system [44]. It was experimentally confirmed that a dye, as a model drug, sealed into these tubes was able to dissolve out [45] and also that adsorption capacity of the tubes for proteins depends upon both surface nature and channel size [46]. Aligned nano-test-tubes with co-axial double walls, consisting of N-doped and undoped carbons [47,48] and also of N-doped and B-doped carbons [49,50] were successfully synthesized by using propylene, acetonitrile, and/or benzene with BCl_3 as precursors. Carbon-coated AAO films, of which the channel walls were coated by carbon and then fluorinated, were able to permeate water molecules preferentially from a water/ethanol mixture [51].

As well as CVI of organic gases, carbon deposition inside the channels of AAO film has also been carried out through the impregnation of mesophase pitch [52,53], hexa(4-dodedylphenyl)-*peri*-hexabenzocoronene [54], poly(vinyl chloride) (PVC), and poly(vinyl alcohol) (PVA) [55], and carbon nanofibers with platelet nanotexture were obtained. Nanofibers with three different nanotextures were prepared in the channels of AAO film [56]: nanofibers with platelet nanotexture by using PVC (sample A in Figure 7.4A), nanotubes formed by CVI of acetylene followed by filling with PVC decomposition (sample B), and nanotubes filled with Ni (sample C). Electrocatalytic activity for the redox reaction of ferri/ferro hexacyanide of these three nanofibers is shown in Figure 7.4B, plotting the difference in the peak position between the oxidation and reduction peaks, ΔE_p, against the sweeping rate of cyclic voltammetry. Nanofibers from sample A, with platelet nanotexture, after heat treatment at 1200 °C show activity comparable with Pt, but those from samples B and C have low reversibility, lower than Pt, even after heat treatment. The estimated electron-transfer rate showed a large difference between samples A and B, which was thought to be owing to the difference in nanotexture on the surface of nanofibers, as schematically shown in Figure 7.4A.

7.1.3 Organic foams

Carbon foams have been prepared using either polyurethane (PU) or melamine foam as a template with impregnation of poly(amide acid), followed by imidization and

(A)

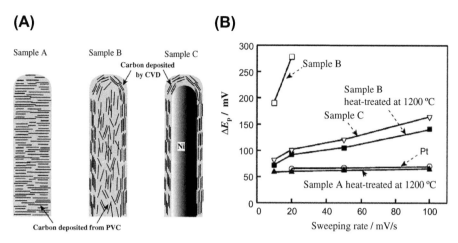

(B)

FIGURE 7.4 Carbon Nanofibers Prepared using an AAO Film as Template

(A) Three different nanotextures, and (B) electrocatalytic activity in comparison with Pt wire, plotting ΔE_p against sweeping rate

From [56]

carbonization [57,58]. The templates, PU and melamine, gave only a small amount of carbon residues after carbonization, but functioned as a template to form a polyimide foam, which was easily converted to carbon foam at high temperatures. Because of the presence of macropores, activation using air was effectively performed to increase the micropores, which resulted in the acceleration of water-vapor adsorption [57]. By using fluorinated polyimide as the carbon precursor, microporous carbon foams were prepared by a one-step process of carbonization, having S_{BET} of 1540 m^2/g and V_{micro} of 0.63 cm^3/g [58]. FA [59] and petroleum pitch [60] have been used as carbon precursors to impregnate into PU-based foam for the fabrication of carbon foam by carbonization. The pore structure in the template PU foam was controlled by adding small particles of clay, although some decomposition residues from the clay mineral remained in the carbon foams [59]. Pitch impregnation into PU foam was performed using water-slurry-dispersed small pitch particles with a surfactant [60]. Carbon foams consisting of mesoporous walls have been prepared from resol-type phenol-formaldehyde by using PU foam and block copolymer surfactant F127 [61].

7.2 Template carbonization for pore-structure control

7.2.1 Zeolites

Zeolites, which are microporous and so widely used as adsorbents, are aluminosilicate minerals with a framework consisting of tetrahedra of AlO_4 and SiO_4. In their micropores, various cations, such as H^+, Na^+, K^+, Ca^{2+}, and Mg^{2+}, can be accommodated and easily exchanged for other cations in solution. Three types of zeolites,

Y, β, and L, which have different frameworks and consequently different pore interconnections, have been used for template carbonization [62].

CVI of propylene in the H-form zeolite Y at 700 °C was effective in resulting in a high S_{BET} up to 2200 m²/g, consisting of both micropores and mesopores, after washing in HF and HCl solutions. However, the deposition of acrylonitrile (AN) vapor under its saturated pressure, followed by polymerization under γ-ray irradiation, and also the liquid impregnation of FA under reduced pressure did not confer high S_{BET} on the resultant carbons [63]. Impregnation, polymerization, pyrolysis, and carbonization in the pores of different zeolites have been studied with PAN [64] and phenol-formaldehyde (PF) [65]. Carbon filling into zeolite channels has also been done by CVI of propylene [66,67]; the resultant carbon being porous, with S_{BET} of 1380 m²/g and total pore volume, V_{total}, of 0.60 cm³/g [67]. Coupling of impregnation of FA with CVI of propylene into Na-form zeolite Y was demonstrated to be effective to synthesize highly microporous carbons, having S_{BET} of 3600 m²/g, micropore volume, V_{micro}, of 1.5 cm³/g, and negligibly small mesopore volume, V_{meso} [68,69]. Micropores in these carbons were thought to be aligned regularly because they showed a sharp peak around 6° in 2θ in the X-ray diffraction (XRD) pattern (Cu Kα) and highly ordered lattice fringes under TEM, as shown in Figure 7.5, which revealed a structural regularity with a periodicity of about 1.4 nm, similar to the template zeolite [69]. Detailed studies showed that: (1) the coupling of impregnation of polymer with CVI of organic gas (two-step filling) and (2) the heat treatment at the high temperature of 900 °C in the zeolite channels were essential to form highly microporous carbons [70]. Through this two-step filling, almost complete filling of zeolite channels was possible, which seemed to result in the formation of a sufficiently rigid carbon framework even after removal from the zeolite channels.

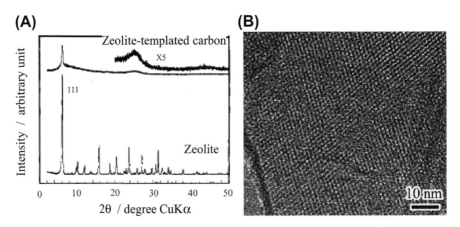

FIGURE 7.5 Zeolite-templated Carbon

(A) XRD pattern in comparison with the template zeolite, and (B) high-resolution TEM image

Courtesy of Prof. T. Kyotani of Tohoku University, Japan

Since the zeolite-templated carbon thus obtained had a very high microporosity, its pore structure was studied by various analytical methods of N_2 adsorption/desorption isotherms [71]. Its high microporous surface area, S_{micro}, of 3700 m^2/g, high micropore volume, V_{micro}, of 1.8 cm^3/g, and sharp pore-size distribution in the range of 1.0–1.5 nm were confirmed.

The zeolite-templated carbon was concluded to be formed by the assembly of single non-stacked nanometer-sized graphene fragments, these graphene sheets being curved like buckybowls owing to the steric hindrance of the template nanochannels, and their edges being bound by oxygen-containing functional groups [72]. This curved graphene and periodical nanoporous structure could be preserved up to 380 °C [73]. The presence of curved graphene sheets in these carbons was thought to be responsible for their high-temperature ferromagnetism [74].

The two-step filling procedure, coupled with heat treatment at 700–800 °C, was applied to other zeolite templates, zeolite β, ZSM-5, mordenite, and zeolite L [75]. A relatively high S_{BET} of about 2000 m^2/g was obtained from zeolite β, but this could not surpass the S_{BET} value obtained from zeolite Y. By using NH_4-form zeolite Y, microporous carbons with nitrogen-containing functional groups were obtained by the impregnation of PF, followed by carbonization above 900 °C [76]. S_{BET} increased with increasing heat-treatment temperature (HTT) from 1700 m^2/g after 900 °C-treatment to 3700 m^2/g after treatment at 1100 °C. Particle size of the template zeolite was shown to have an influence on the pore structure of the resultant carbons: small particles of zeolite gave a carbon with a little higher S_{BET}, higher V_{total}, and higher carbonization yield than large particles [77]. A laboratory-prepared zeolite, EMC-2, which had interconnected cages along the a-axis and straight channels along the c-axis, could be replicated by two-step filling with FA and propylene to result in a carbon with a high S_{BET} of 4000 m^2/g and V_{micro} of 1.8 cm^3/g [78]. CVI of acetylene at 600 °C, followed by CVI of propylene at 700–800 °C and heat treatment at 900 °C, gave carbons with a long-range regularity and uniformity of micropores, better than the carbons prepared by the above-mentioned two-step filling process [79].

The zeolite-templated carbons were able to be pelletized without any binder at 300 °C and 147 MPa, their densities being 0.7–0.9 g/cm^3, although no pellets could be obtained from commercially available activated carbons [80]. By this densification, the volumetric surface area increased from 1100 to 1300 m^2/cm^3 and average micropore size decreased with increasing pelletizing pressure.

Adsorption of various gases was studied on these microporous zeolite-templated carbons [81–87]. Nitrogen-doped carbon, which was prepared using impregnation of FA followed by CVD of acetonitrile, showed higher affinity for H_2O molecules than the nitrogen-free carbon [81]. Adsorption of hydrogen was studied experimentally [82–85] and a hydrogen uptake of 2.2 mass% at 30 °C and 34 MPa was reported [84]. The ordered micropore structure was preserved by potassium adsorption at 380 °C, although it was destroyed by bromine adsorption. The magnetic properties of zeolite-templated carbons were found to be sensitive to the adsorption of these gases, including He [73,86]. ^{4}He molecules formed a layer of about 1.4 atomic thickness in a carbon consisting of a three-dimensional arrangement of micropores with 1.4 nm periodicity, micropore diameter being 1.2 nm [87]. Zeolite-templated carbon loaded with Pt was reported to have higher specific activity for room-temperature methanol oxidation than the commercial catalyst [88]. Application of zeolite-templated carbons to the electrodes of electric double-layer capacitors has been examined in many papers [89–93].

7.2.2 Mesoporous silicas

Silicas with ordered mesoporous structure, which were formed by templating self-assembly of surfactants, including MCM-48, MCM-41, and SBA-15, were successfully used to form ordered mesoporous carbons. In Table 7.2, mesoporous silicas used as templates and the resultant carbons are listed with representative pore parameters. Reviews focusing on silica-templated carbonization have also been published [94,95].

The pores (channels) in the template silicas are replicated in the carbons by either impregnating or CVI of a carbon precursor, followed by carbonization and removal of the templates, mostly with HF. In Figure 7.6, the powder XRD patterns at very low

Table 7.2 Silica Templates and Resultant Carbons

Silica Template		Mesoporous Carbon Synthesized		
Notation	Symmetry	Notation	Symmetry (XRD)	Pore Parameters
MCM-48	Cubic, *Ia3d*, inter-penetrating and non-interpenetrating channel system	CMK-1	($2\theta = 1.6$ and $2.7°$) $I4_1/a$ or $Ia3d$	$S_{BET} = 1700$, $w_{meso} = 3.4$, $V_{meso} = 1.0$
SBA-15	Hexagonal, *P6mm*, channels interconnected by micropores	CMK-3 –	2D-hexagonal	$W_{channel} = 5.9$, $S_{BET} = 1823$, $V_{total} = 2.23$, $w_{meso} = 5.8$
MCM-41	Hexagonal, *P6mm*, uniform channels without interconnection	–	Random nanochannels	Microporous, $S_{BET} = 1100$
MSU-H	Hexagonal, ordered worm-like holes	–	2D-hexagonal, wormholes	$S_{BET} = 1228$, $w_{meso} = 3.9$, $V_{total} = 1.26$
Ordered mesopores		–	Ordered, $2\theta = 2.2°$	$S_{BET} = 1056$, bimodal $w_{meso} = 2.0$, $w_{micro} = 0.6$

CVI, chemical vapor infiltration; FA, furfuryl alcohol; I, impregnation in solution; PF, phenol/formaldehyde resin; RF, resorcinol/formaldehyde resin; S_{BET}, BET surface area (m²/g); V_{meso}, mesopore volume (mL/g); V_{micro}, micropore volume (mL/g); V_{total}, total pore volume (mL/g); w_{meso}, average width of mesopores (nm); w_{micro}, average width of micropores (nm), $w_{channel}$, average width of channels (nm)

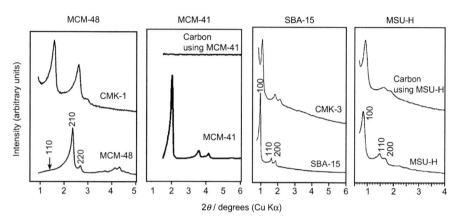

FIGURE 7.6 XRD Patterns at Low Diffraction Angle for the Template Silica and the Resultant Carbon

diffraction angle are compared for different silica templates and the carbons synthesized with them, to show correspondence in pore-structure symmetry.

When MCM-48 was used as a template, the resultant carbons did not replicate the symmetry of the template exactly, having a lower symmetry. After the carbonization of MCM-48 impregnated by sucrose with sulfuric acid from aqueous solution, the template silica was dissolved out in an aqueous solution of NaOH and ethanol to isolate porous carbon [96]. Other sugars, glucose and xylose, have been successfully used as carbon precursors. The resultant carbon, denoted CMK-1, had S_{BET} of 1380 m^2/g and ordered mesopores 3.0 nm in width, together with micropores of 0.5–0.8 nm width, V_{meso} of 1.1 cm^3/g, and V_{micro} of 0.3 cm^3/g. Ordered pore structure was confirmed by the diffraction peaks at 1.6 and 2.7° in 2θ(Cu Kα), which was different from the symmetry of the template MCM-48 (*Ia3d*) (see Figure 7.6). In most cases, the pore symmetry in the resultant carbons was either *I4₁32*, *I4/a* or lower, probably because of collapsing after the removal of template [96–99]. In order to use PF resin as the carbon precursor, aluminum was implanted onto the channel walls of mesopores in template silicas such as MCM-48 and MCM-41, to generate strong acid catalytic sites for polymerization of the resin [97,98]. Silylation of the pore surface of MCM-48 was effective to get highly ordered mesopores in the resultant carbon [99]. Pore-structure symmetry *Ia3d* in the template MCM-48 was able to be replicated in the carbon by the repetition of an impregnation/drying step [100]. By applying CVI of acetylene at 800 °C to fill the pores in the template MCM-48, the symmetry of the template could be preserved in the resultant carbon even after carbonization at 900 °C and removal of the template [101]. Mesoporous MCM-48-type silicas with different pore sizes were successfully used as templates for the preparation of CMK-1-type ordered mesoporous carbons [102]. In order to synthesize ordered mesoporous carbon, the repeated impregnation/carbonization of sucrose to fill the pores in the template completely, addition of the optimum amount of sulfuric acid, and carbonization at a temperature above 600 °C were essential [103].

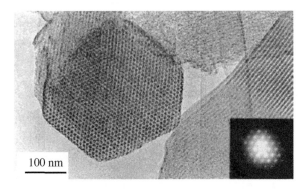

FIGURE 7.7 Lattice Fringe Image and Diffraction Pattern (Inserted) for the Carbon (CMK-3) Prepared Using Silica SBA-15 as a Template

From [105]

The carbon prepared using MCM-41 as a template contained disordered micropores, probably owing to the collapse of the template framework upon its removal, and consisted of carbon nanowires randomly separated [102], as shown by XRD in Figure 7.6. However, MCM-41 was able to give self-supporting carbon nanowire arrays, probably because adjacent nanowires are connected by irregular rods [104]. By using Al-implanted MCM-41 with leaf-like morphology and FA, carbon nanowires with a diameter of 4–5 nm were packed side-by-side without any regularity.

Carbon with ordered mesopores in the same symmetry as the template SBA-15 was obtained by impregnation of sucrose aqueous solution with sulfuric acid in aqueous solution, and was designated CMK-3 [105], as shown in Figure 7.6. The TEM images of CMK-3 are shown in Figure 7.7. It had an S_{BET} of 1520 m^2/g and V_{total} of 1.3 cm^3/g, due to the channels of 4.5 nm in diameter. The carbon obtained using FA consisted of channels with an inner diameter of 5.9 nm and mesopores formed between the adjacent channels [103]. To obtain ordered mesoporous carbons, the calcination temperature for template SBA-15 had to be below 880 °C, because SBA-15 calcined at 970 °C gave a disordered pore arrangement [106]. The size of mesopores in the carbon increased from 2.2 to 3.3 nm with increasing thickness of the wall of template SBA-15 from 1.4 to 2.2 nm, which was controlled by changing the ratio of hexadecyltrimethylammonium bromide/polyoxyethylene hexadecyl ether-type surfactants [107]. By using SBA-15 with rod-like morphology under hydrothermal conditions, CMK-3 rods were obtained, having S_{BET} of 1823 m^2/g, V_{total} of 2.23 cm^3/g, and mesopore width of 5.8 nm [108]. From a powder mixture of the SBA-15/FA composite and NaCl particles of about 200 nm, with compression under a pressure of 1000 kg/m^3, a CMK-3 monolith was obtained, which consisted of mesopores of 3.7 nm in width and macropores of 550 nm in width [109]. By CVI of acetonitrile into the mesopores of SBA-15 at 950–1100 °C, ordered mesoporous N-doped carbons were prepared [110]. Ordered mesoporous carbons were synthesized from FA by using Al-implanted SBA-15 with different Si/Al ratios of 5–80 [111,112]. Heat treatment of the mixture of SBA-15 with FA at 180 °C under hydrothermal conditions gave mesoporous carbon after dissolution of silica in a 4 mol/L aqueous solution of NH_4HF_2 [113].

The mesoporous silica MSU-H, which had a porous framework similar to that of SBA-15, was also used as a template for the preparation of mesoporous carbons, preserving its pore structure symmetry [114], as shown in Figure 7.6. The carbons obtained from FA with paratoluene sulfonic acid at 800 °C using MSU-H of different mesopore sizes from 2.4 to 27 nm had mesopore sizes from 2 to 10 nm [115]. By controlling the amount of carbon precursor impregnated, mesoporous carbons with unimodal (2.9 nm in width) and bimodal (2.9 and 14 nm in width) pore structures could be prepared from FA [115,116] and also from phenol resin [116]. The size of mesopores in the carbon was controlled from 3.8 to 10.5 nm by infiltration of a mixture of sucrose and different amounts of boric acid (0–25 mol%) into MSU-H [117]. A two-dimensional hexagonal array of mesopores in the template was retained in the resultant carbons, of which S_{BET} and V_{total} decreased from 1337 to 848 m^2/g and from 1.60 to 1.25 cm^3/g, respectively, with increasing content of boric acid from 0 to 25 mol%.

Mesophase pitch was able to be used as a carbon precursor for the ordered mesoporous silica templates, but its impregnation and carbonization above 750 °C had to be done in an autoclave [118]. A petroleum pitch with a low softening point between 114 and 122 °C was able to be impregnated into the templates MCM-48 and SBA-15 at 302 °C under atmospheric pressure [119]. The mesoporous carbon prepared from the pitch was thermally stable up to 1400 °C.

Silica-templated mesoporous carbons have been developed with potential applications as adsorbents for large molecules [120–123], hydrogen [124–128], and CO_2 [129]; catalyst supports [130,131]; and also as electrodes for electric double-layer capacitors [97,98,132–145]. For hydrogen storage, the activation of mesoporous carbons either by KOH [124] or CO_2 [125] was required, and nitrogen enrichment was also performed on CMK-3 [126–128].

7.2.3 **MgO**

Nanoporous carbons have been prepared using MgO particles as an inorganic template [146–154] and the results were reviewed in two papers: one focusing on the procedure of pore control [155] and the other on applications related to porous structure [156]. A mixture of the MgO precursor, which gives nano-sized MgO particles after pyrolysis, and the carbon precursor was heat-treated at 900 °C in an inert atmosphere. From the carbon-coated MgO particles thus obtained, template MgO was dissolved out using a diluted acid at room temperature to isolate the carbon formed. MgO was selected as a template mainly because of its chemical and thermal stabilities, structural and compositional stabilities, lack of reaction with carbon precursors up to the carbonization temperature of carbon precursors, and easy removal in diluted acidic solution.

Different MgO precursors have been used: MgO itself; magnesium acetate, $Mg(CH_3COO)_2$; citrate, $Mg_3(C_6H_5O_7)_2$; gluconate, $Mg(C_{11}H_{22}O_{14})$; hydroxycarbonate, $3MgCO_3 \cdot Mg(OH)_2$; and $Mg(OH)_2$. PVA was used as the carbon precursor in most of this work, and both thermoplastic and thermosetting precursors, such as pitches, poly(ethylene tetraphthalate) (PET), poly(amic acid), poly(vinyl pyrrolidone) (PVP), polyacrylamide (PAA), phenol, asphalt, and starch, have been used. Mixing of two precursors was performed in different ratios either in powder (powder mixing) or in solution (solution mixing). After carbonization at 900 °C, the products were obtained as powder with no marked aggregation of particles being observed, when the MgO/carbon precursor ratios were larger than 5/5.

When commercially available MgO powder with particle size of about 100 nm was used with PVA in the ratio of 7/3, microporous carbon with S_{micro} 800 m^2/g was obtained [147]. From the same MgO/PVA ratio of 7/3, however, mesoporous carbons were obtained in systems using Mg acetate, citrate, and gluconate, particularly with solution mixing in the Mg acetate/PVA system. In Figure 7.8A, S_{BET} is plotted for the carbons obtained from the mixture of Mg acetate and PVA prepared via powder and solution mixing [148]. The solution mixing resulted in very high S_{BET}, as high as 1800 m^2/g, much higher than for powder mixing. As shown in Figure 7.8B, pore-size distribution in the carbon prepared via solution mixing shows a high population of pore sizes of about 13 nm, although a broad distribution is obtained via powder mixing. For Mg citrate/PVA and Mg gluconate/PVA systems, however, no marked differences in surface area or pore-size distribution were observed between the carbons prepared via powder- and solution-mixing processes.

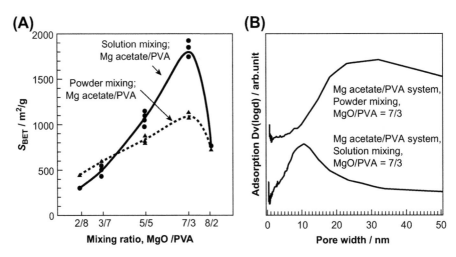

FIGURE 7.8 Effect of Mixing Process of Mg Acetate and PVA to Prepare MgO-templated Carbon

(A) S_{BET}, and (B) pore-size distribution

From [150]

With the Mg citrate/PVA system, the carbons obtained are mesopore-rich. The pore-size distribution of these carbons showed a sharp maximum at about 5 nm and the S_{meso} and V_{meso} determined by the Barrett-Joyner-Halenda method (BJH) reached 1600 m²/g and 1.7 cm³/g, respectively. With the Mg gluconate/PVA system, the changes in pore-structure parameters with mixing ratio were a little different from those with the Mg citrate/PVA system [150]. In Figure 7.9A, S_{total}, S_{meso}, and S_{micro} determined by the BJH method are plotted against the MgO/PVA ratio for the Mg gluconate/PVA system. From Mg gluconate itself (MgO/PVA = 10/0), S_{total} of the resultant carbon reaches 1300 m²/g and consists mainly of micropores, S_{micro} being about 900 m²/g and V_{micro} about 0.6 cm³/g, though S_{meso} and V_{meso} are small in comparison with other carbons prepared in this system. For the carbon prepared from an MgO/PVA of 7/3 with this system, S_{micro} is comparable to S_{meso}. With decreasing MgO/PVA ratio from 7/3, mesopores become predominant. Pore-size distribution in the mesopore region is shown for these carbons in Figure 7.9B. The mixtures of Mg gluconate and PVA gave a size distribution sharply peaking at 2–4 nm, the pore volume of this size range decreasing with decreasing MgO/PVA ratio.

A coal tar pitch has been used as a carbon precursor, and gave mesoporous carbons [152]. The Mg acetate/pitch system gave carbons with mesopores centered at around 13 nm and the Mg citrate/pitch system resulted in mesopores centered at around 5 nm, even though the two precursors are mixed as powder. The high wettability of pitches to the surface of oxides, MgO in the present case, seems to be the main reason for the narrow pore-size distribution. When polyimide was used as a carbon precursor, however, the dispersion of MgO precursor particles in the organic solution of poly(amic acid) was important [155]. With PET, repeated fusion

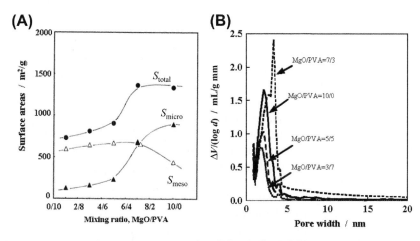

FIGURE 7.9 Changes in Pore Parameters with Mixing Ratio MgO/PVA to Prepared MgO-templated Carbon in the Mg Gluconate/PVA System

(A) Surface areas S_{total}, S_{micro}, and S_{meso}, and (B) pore-size distribution

From [151]

and crushing of the MgO precursor/PET mixture were needed to obtain high surface areas in the resultant carbons [153].

The crystallite size calculated from the full width at half maximum intensity of the 101 X-ray diffraction line for MgO and the average size of MgO particles observed under TEM were almost the same as the pore size at the population maximum of the size distribution determined from the N_2 gas adsorption isotherm. This result suggests that nano-sized MgO left pores of almost the same size in the carbon after its dissolution. MgO particles formed by the pyrolysis of MgO precursors, Mg acetate, citrate, and gluconate, are homogeneous in size, and as a consequence, a very narrow pore-size distribution in the resultant carbons is observed.

It has been experimentally demonstrated that template MgO can be recycled [155]. Acetic and citric acids were selected to dissolve out MgO from carbon-coated MgO. The recovered Mg acetate and citrate aqueous solutions were mixed with PVA again and subjected to the same procedure to produce MgO-templated porous carbon. Over five cycles, almost the same S_{BET} and carbon yield were obtained, revealing that 100% of MgO can be recycled, by supplying new carbon precursor in every cycle. It has to be pointed out that citric acid itself is the solvent for MgO as well as the carbon precursor.

Porous carbon nanofibers have been prepared by electrospinning of a *N,N'*-dimethylformamide (DMF) solution of PAN and $MgCl_2$ in different ratios, followed by stabilization at 250 °C, carbonization at 1050 °C, and dissolution of MgO in 1 mol/L H_2SO_4 [157]. $MgCl_2$ changed to MgOHCl during the stabilization process and to MgO during carbonization. The nanofibers prepared were microporous. From natural magnesite (mainly magnesium carbonate) and PET, nanoporous carbon, of which S_{micro} and S_{meso} were about 380 m²/g, was obtained at 650 °C [158]. By using $MgCO_3$, $Mg(OH)_2$, magnesite, and MgO, mesoporous carbons with S_{BET} of 600–2000 m²/g were prepared and tested as adsorbents for various pollutants in water [159–163]. Needle-like nanoparticles of $Mg(OH)_2$ prepared by a surfactant-mediated solution process entrained tunnel-like mesopores in the resultant carbons [161–163].

MgO-templated mesoporous carbons have been used as electrode materials for electric double-layer capacitors (EDLCs) [149,154,161–169]. Using nitrogen-containing organic compounds as carbon precursors in this procedure, nitrogen-doped nanoporous carbon was prepared, which gave better EDLC performance [154]. By using these mesoporous carbons as either one of the electrodes of asymmetric EDLCs, the roles of micropores and mesopores and the predominant contribution of the negative electrode to capacitive performance were demonstrated [165–167]. Carbon nanofibers prepared with a $MgCl_2$/PAN system have been applied to hydrogen storage [157]. MgO-templated carbon containing nanoparticles of Sn metal in its pores was prepared from the mixture of MgO, SnO_2, and PVA by one-step heat treatment at 900 °C, and successfully applied to the anode of lithium-ion rechargeable batteries [170]. The space neighboring an Sn nanoparticle, which was caused by the dissolution of an MgO particle, is thought to absorb a marked expansion due to alloying of Sn with Li, and the carbon shell disturbs the movement of the Sn particle due to alloying/de-alloying during charge/discharge cycles of the batteries. Porous carbon loaded with MgO before its complete dissolution was able to remove SO_2 gas in air [171]. MgO/CaO-loaded porous carbon prepared from the mixture of natural dolomite and PET has been tested as an adsorbent for CO_2 [172].

A similar process using a $Ni(OH)_2$ template with phenol in an ethanol solution was proposed to prepare mesoporous carbons, and the obtained carbon was reported to give high energy and power densities to EDLC in both aqueous and organic electrolytes [173]. Barium citrate gave porous carbons containing both micropores and mesopores, but much smaller S_{BET} than the carbon prepared from Mg(II) citrate [168,174]. Nitrogen-containing porous carbons have been prepared from melamine-formaldehyde resin using $CaCO_3$ as the template, S_{BET} reaching 800 m^2/g [175].

7.2.4 Block copolymer surfactants (soft templates)

Mesoporous carbons with ordered pore structure have been successfully prepared by using some surfactants as organic templates [176–189]. The organic templates used in this process may be called "soft templates," in contrast to inorganic templates ("hard templates"), such as zeolites, silicas, and MgO. The key to this process was the direct use of the self-assembly of the surfactant block copolymers as templates, although the same surfactants were used for preparation of the mesoporous silicas, which were used as templates to synthesize nanoporous carbons, as explained above. In most cases of soft-template carbonization, the solvent-evaporation-induced self-assembly (EISA) method was applied to the mixture of a carbon precursor and a surfactant.

Carbon films with hexagonal arrays of channels (cylindrical mesopores) perpendicular to the film surface were synthesized by using a self-assembly of diblock copolymer, polystylene-*block*-poly(4-vinylpridine) (PS-P4VP), as a template and resorcinol-formaldehyde (RF) as a carbon precursor [176]. A solution of PS-P4VP and resorcinol was cast onto a silica substrate to form a film, in which most of the resorcinol molecules were located in the P4VP domain owing to the hydrogen-bond association between the basic P4VP blocks and the acidic resorcinol monomers. By the controlled evaporation of the solvent DMF in DMF/benzene vapor at 80 °C, a highly ordered nanotexture of the film was obtained, where the PS domain became the cylinder directed perpendicularly to the film surface. This film was exposed to formaldehyde vapor to form a highly cross-linked phenol resin in the P4VP domain and then carbonized at 800 °C. Channels were formed perpendicularly in the carbon film, of which the diameter was 33.7 nm and the wall thickness about 9.0 nm. Crack-free carbon films with thickness up to 1 μm and size up to 6 cm^2 were prepared. By using the same diblock copolymer and a phenol resin with hexamethylenetetramine, the products after pyrolysis at 600 °C had disordered pores [177].

A mixture of commercially available triblock copolymer, poly(ethylene oxide)-*b*-poly(propylene oxide)-*b*-poly(ethylene oxide) (PEO_{106}-PPO_{70}-PEO_{106}, Pluronic F127), resorcinol, triethyl orthoacetate, and formaldehyde in water/ethanol/HCl mixed solution was spin-coated on a silicon substrate and then carbonized up to 800 °C to obtain mesoporous carbon films [178]. The periodic texture of the resorcinol-formaldehyde/triethyl orthoacetate (carbon precursor) and F127 was established in the original organic films by EISA with a spacing of 9.2 nm and this was retained even after carbonization at 800 °C. The surfactant F127 was decomposed at 400 °C to form mesopores. The carbon films prepared at 800 °C had S_{BET} of 1354 m^2/g, V_{total} of 0.743 cm^3/g, and mesopore width of 5.9 nm. By changing the concentration of the surfactant F127 in the starting solution, a carbon with hexagonal arrays of channels (phenol/F127 = 1/0.012) and one with cubic arrays of mesopores (phenol/F127 = 1/0.005–1/0.006) could be prepared [180]. The former consisted of channels with diameter of 2.9 nm, S_{BET} of 968 m^2/g, and V_{total} 0.56 cm^3/g, and the latter of mesopores with a size of about 3.7 nm, S_{BET} of 778 m^2/g, and V_{total} 0.44 cm^3/g. As shown in Figures 7.10A and 7.10B, N$_2$ adsorption/desorption isotherms for the two mesoporous materials, starting polymers, and resultant carbons, are type IV with a pronounced hysteresis and have sharply peaking pore-size distribution. By using phloroglucinol-formaldehyde with F127, mesoporous carbons were prepared in monolith, fiber, and film morphologies under much faster and milder conditions than by using PF and RF [181].

By using another triblock copolymer, Pluronic P123 (PEO$_{20}$-PPO$_{70}$-PEO$_{20}$), with resols in aqueous solution, carbons with bicontinuous mesopores were obtained [182]. A resin of resorcinol/phloroglucinol-formaldehyde was successfully used for a carbon precursor for the process using a P123 template. Another kind of triblock copolymer, of acrylonitrile (AN) with |-butyl acrylate (BA), (AN)$_{45}$-(BA)$_{530}$-(AN)$_{45}$,

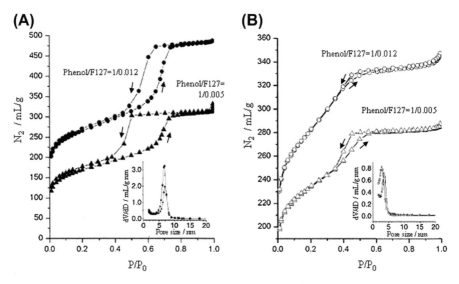

FIGURE 7.10 N$_2$ Adsorption/Desorption Isotherms for Mesoporous Polymers and Resultant Mesoporous Carbons with Pore-size Distribution (Inserted)

(A) Polymers, and (B) carbons prepared at 900 °C

From [180]

was carbonized on a cleaved mica or silicon wafer after stabilization at 200–230 °C, but no information on pore structure was presented [183].

Different pore structures in carbon materials have been obtained by simply adjusting the ratio of phenol to template surfactant: two-dimensional hexagonal, three-dimensional bicontinuous, body-centered cubic, and lamellar [184–187]. Even after carbonization at 1200 °C, carbons with channels (about 2.8 nm in diameter) in hexagonal arrangement were obtained from the compositions in the range of phenol/F127 molar ratio of 1/0.010–1/0.015 and phenol/P123 of 1/0.007–1/0.016 [184]. TEM images of two orthogonal faces of the resultant carbon are shown in Figures 7.11A and 7.11B. From compositions in the range of phenol/F127 of 1/0.003–1/0.008 and phenol/F108 (PEO_{132}-PPO_{50}-PEO_{132}) of 1/0.005–1/0.010, carbon with mesopores (diameter 3.8 nm) arranged in cubic symmetry was

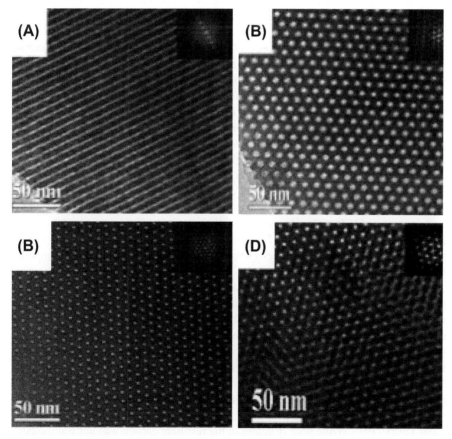

FIGURE 7.11 TEM Images of Ordered Mesoporous Carbons

(A and B) With hexagonal symmetry, (C and D) with cubic symmetry

From [184]

synthesized, as shown in Figures 7.11C and 7.11D. Carbon with bicontinuous mesopores arranged in cubic symmetry was obtained from compositions in the very narrow range of phenol/P123 of 1/0.018–1/0.019, of which mean mesopore size was 2.3 nm. By increasing the phenol/P123 ratio to 1/0.022–1/0.027, lamellar mesostructure was formed.

Stability of the hexagonal symmetry arrays of channels in the composite films of PF and triblock copolymer and in the carbon films after removal of template at 350 °C has been discussed, taking into account the constraint due to wetting of thin film with the substrate and the contraction stress during template removal [188]. The removal of the template at 350 °C from the composite films templated by F127 changed the pore structure from two-dimensional hexagonal symmetry to disordered alignment. However, the films templated by P123, which had a shorter PEO segment than F127, was able to give well-ordered channels after removal of the template. The composition phenol/P123 to give ordered channels was markedly narrowed for thin films in comparison to bulk powders, with a range of 1/0.010–1/0.013 for the film but 1/0.008–1/0.016 for the powder.

Carbons with ordered macropores, of which the walls contained ordered mesopores, have been synthesized via a dual-templating technique using polymethyl methacrylate (PMMA) and triblock copolymer as templates [189]. A solution of PF and F127 was added into sedimented colloidal PMMA spheres, followed by evaporation of the solvent, cross-linking, and heating at 450 °C to produce phenol resin monolith, which was converted to carbon monolith at 900 °C. The sizes of macropores formed by the PMMA template and of mesopores by the F127 template were 342–404 and around 3 nm, respectively, the latter being in either face-centered cubic or two-dimensional hexagonal symmetry.

The frameworks in the mesoporous carbons obtained by this template method were shown to be relatively rigid, even though S_{BET} and V_{total} decreased gradually with increasing HTT; 607 m^2/g and 0.58 cm^3/g after 850 °C-treatment and 230 m^2/g and 0.30 cm^3/g after 2600 °C [187]. The mesoporous carbon thus obtained could be activated using KOH to create micropores with only a slight sacrifice of mesopores [190,191].

Ordered mesoporous TiC/C composites were prepared by using resol-type phenol, titanium citrate, and F127 in an ethanol/water mixture, followed by carbothermal reduction up to 1000 °C [192]. Crystalline TiC particles of 4–7 nm in size were confined in the carbon pore walls and were able to enhance the oxidation resistance of the carbon open framework. Starting from phenol, tetraethyl orthosilicate (TEOS), and colloidal silica with F127, porous carbons, consisting of macropores of 20 or 50 nm in size, mesopores of about 12 nm, and micropores of about 2 nm were prepared; macropores were obtained by dissolution of colloidal silica, mesopores by a soft-templating process using F127, and micropores by dissolution of silica formed from TEOS [193]. While pore structure in the resultant carbon was not significantly affected by embedding silica nanoparticles using TEOS, the incorporation of alumina nanoparticles (~50 nm) caused a gradual reduction in mesoporosity with increasing loading of alumina [194].

7.2.5 **Metal-organic frameworks**

Porous metal-organic frameworks (MOFs) have attracted much attention as adsorbents for hydrogen and methane storage, as explained in Chapter 15. They also offer a potential as templates to synthesize nanoporous carbon materials, because they have nano-scaled cavities and open channels. One MOF, MOF-5 ($Zn_4O(OOCC_6H_4COO)_3$) which has a three-dimensional intersecting channel system (diameter 1.8 nm), was exposed to FA vapor at 150 °C, FA being polymerized in the channels of MOF-5, and then heat-treated at 1000 °C in Ar to carbonize polymerized FA [195]. Template MOF-5 decomposed at 425–525 °C and left ZnO, which was reduced to metallic Zn at around 800 °C and then vaporized above its boiling point, 908 °C, leaving carbon alone. The resultant carbon was mesoporous and had a high S_{BET} and V_{total} of 2872 m^2/g and 2.06 cm^3/g. A strong effect of the carbonization temperature was observed, showing that a temperature 1000 °C was needed to get a high surface area and consequently high performance in EDLC [196]. By using glycerol as the carbon precursor, a similar mesoporous carbon has been obtained [197]. MOF-5 itself can leave microporous carbon with S_{BET} of 1800 m^2/g, V_{total} of 2.87 cm^3/g, and V_{micro} of 0.92 cm^3/g, but the filling of its channels by either phenol or carbon tetrachloride/ethylenediamine resulted in less porous carbons [198].

By using a zeolite-type MOF (ZIP-8, $Zn(C_4H_5N_2)_2$) as a template, which acted also as the carbon precursor, and FA as an additional carbon precursor, nanoporous carbon was obtained, S_{BET} reaching 3405 m^2/g and V_{total} 2.58 cm^3/g after carbonization at 1000 °C [199]. The scheme of the preparation procedure is shown in Figure 7.12A and nitrogen adsorption/desorption isotherms are shown for the porous carbons prepared at 800 and 1000 °C in Figure 7.12B. By gelation of the mixture of RF and MOF at 80 °C, followed by carbonization at 950 °C, porous carbon with S_{BET}

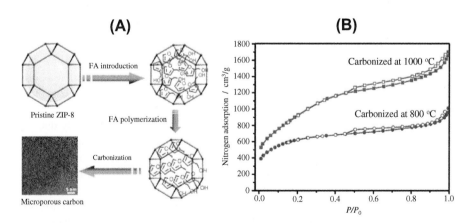

(A)   **(B)**

FA introduction

Pristine ZIP-8   FA polymerization

Carbonization

Microporous carbon

Carbonized at 1000 °C

Carbonized at 800 °C

Nitrogen adsorption / cm^3/g

P/P_0

FIGURE 7.12 Porous Carbon Prepared Using a Zeolite-type MOF as Template

(A) Scheme of preparation, and (B) nitrogen adsorption/desorption isotherms

From [199]

of 2368 m^2/g was obtained [200]. Zn metal used in the MOF was reported to act as an activation reagent during carbonization.

7.2.6 Other templates

In 1987, glass-like carbon foams were prepared by using sintered NaCl as a template [201]. This might be the first work on template carbonization. Template NaCl was removed by careful washing in water and 0.4 mol/L HNO$_3$, and then the carbon was freeze-dried. The carbon foam thus prepared was composed of macropores with a size of approximately 8 μm and had a bulk density in a range of 0.035 to 0.075 g/cm^3. A Li-form of taeniolite was intercalated with either hydroxylaluminum (Al$_2$(OH)$_5$Cl·2.5H$_2$O) or hydroxylaluminum-zirconium (Al$_{1.2}$Zr$_{0.3}$Cl·5H$_2$O) and then saturated with an 80% benzene solution of FA, followed by polymerization of FA and carbonization at 700 °C [202]. The resultant carbons were microporous, having molecular sieving properties. Porous clay heterostructures prepared from natural montmorillonite by intercalation of different surfactants, octyl-, decyl-, and dodecyl-amines, and TEOS have also been used as templates to prepare nanoporous carbons from FA [203]. The carbon obtained by using decyl-amine and carbonizing at 700 °C gave the highest S_{BET} of 1469 m^2/g and a volume of 2-nm pores of 0.68 cm^3/g. Similar porous carbons have been obtained by using porous clay heterostructures loaded by different metals and sucrose [204].

Colloidal silica has also been used as a template to synthesize porous carbons [205–216]. A porous carbon with three-dimensional periodical alignment was prepared from a phenolic resin and a silica template, which was prepared by sintering silica spheres with uniform diameter in the range of 150–300 nm [206]. A mixture of silica colloid stabilized by cetyltrimethylammonium bromide and RF gave a mesoporous carbon with S_{BET} of 1512 m^2/g and V_{meso} of 3.6 cm^3/g [207]. Through a sol-gel process using tetraethoxy silane in the presence of FA, followed by carbonization at 800 °C, a carbon with S_{BET} of 1170 m^2/g and V_{total} of 1.27 cm^3/g, composed of 67% mesopores centered at 4 nm width, was prepared [208]. From a mixture of TEOS and sucrose with or without colloidal silica, unimodal or bimodal mesoporous carbons were prepared [209]. The mixture of mesophase pitch and colloidal silica resulted in mesoporous carbons, after penetrating colloidal particles into pitch at 260 °C and then carbonizing at 900 °C (referred as colloid-imprinted carbon) [210,211]. V_{total} of the resultant carbons was 1.0–1.6 cm^3/g, mainly due to mesopores of about 13–24 nm in width. Colloidal spherical silicas of 8 nm in diameter and elongated silica particles (5–20 nm in diameter and 40–300 nm in length) were used as a template by mixing with RF, followed by polymerization and carbonization at 850 °C, leading to mesoporous carbons [213]. The resultant carbons had S_{BET} of 600–900 m^2/g and V_{total} of 1.0–2.0 cm^3/g. By using a silica template, carbons with fully interconnected mesopores of about 4.3 nm in width and macropores of 0.5–30 μm in width were prepared in a monolith (5 mm in diameter and 1 cm in length) [214]. Microporous carbon monolith was prepared by carbonization of the mixture of cyclodextrin and tetramethyl orthosilicate with a sulfuric acid catalyst, which had S_{BET} of 1970 m²/g, V_{total} of 1.0 cm³/g, and micropore width centered at 1.6 nm [215]. Silica-templated carbon films with a size larger than 15×25 mm² were prepared by spin-coating of an acidic aqueous solution of sucrose with TEOS onto a silicon wafer, followed by carbonization at 400 °C and dissolution of the silica template [212]. The resultant carbon films had S_{BET} of 2600 m²/g and V_{total} of 1.4 cm³/g, mainly consisting of mesopores with width centered at about 2.4 nm. Onion-like carbon-silica composite vesicles were prepared from aqueous emulsion of a low-molecular-weight resol-type phenol resin as a carbon precursor, TEOS as a silica source, the surfactant F127 as a template, and 1,3,5-trimethylbenzene as an organic co-solvent [217]. These vesicles could be converted to onion-like mesoporous carbon vesicles having bimodal pores of 4–23 and 66–82 nm widths. Mesocellular carbon foams with pore sizes larger than 20 nm have been prepared by selecting mesocellular aluminosilicates as templates and PF as the carbon precursor [218]. Porous carbons composed of ordered three-dimensionally interconnected macropores with uniform mesoporous walls were synthesized from divinylbenzene by using polystyrene spheres for the macropore template and silica particles for the mesopore template [219]. By acid-catalyzed polymerization of RF in the presence of TEOS, colloidal silica, and triblock copolymer, highly porous carbons were prepared after carbonization and dissolution of silicas [220]. Primary mesopores were derived from a block copolymer template, spherical mesopores from silica, and micropores from TEOS. Thin carbon nanosheets that were transparent under electron microscopy have been obtained from the mixture of FA and halloysite, a kind of natural kaolinite, after carbonization at 700 °C and leaching with HF solution [221,222]. The nanosheet contained mesopores of about 20 nm in size.

7.3 Concluding remarks

Template carbonization using clays might be worthwhile considering as a route to prepare graphene [223], even though much experimental study of both templates and carbon precursors is needed. In the present procedure, the carbonization temperature is limited to 700 °C because the layer structure of the template MONT is only stable below this temperature. As a consequence, the resultant carbon still contains a large amount of hydrogen, nitrogen, and/or oxygen, which might leave lattice defects after removal by further carbonization up to about 1500 °C [11]. In order to synthesize graphene by this route, therefore, a new template that can resist high temperatures, hopefully to around 1500 °C, has to be developed to keep the carbon precursor in the interlayer space of the template until completion of its carbonization. Also, carbon precursors that have a high carbon yield after carbonization are needed.

Preparation of tubular carbons using AAO films as the template has been considered as one of the routes to synthesize multi-walled carbon nanotubes with uniform diameter and length, also being vertically aligned. Filling of metals and metal oxides into tubular carbons formed in AAO templates is easily carried out without touching the outer surfaces, more easily than in the case of carbon nanotubes prepared by other techniques, such as arc-discharging and catalytic CVD. Enclosing various drugs together with magnetic metal particles inside the tubular carbons formed in AAO templates can be done easily, and the resultant tubules may be applied to a magnetic drug-delivery system after separation from the template [44]. The drawbacks of this technique for practical application seem to be a low yield of the product and the necessity to use either high-concentration HF or NaOH solution for removing the template. These drawbacks may be compensated by the excellent functionality of the resultant carbon tubules. Even so, much work is required on the development of more appropriate template materials and carbon precursors.

Templating processes for the control of pore structure in carbon have various advantages and disadvantages in comparison with the conventional activation process. In the activation process, pores, particularly micropores, are created by the gasification of matrix carbon to CO and/or CO_2 through oxidation—in other words, by losing carbon—and the formation of mesopores is done by sacrificing smaller micropores that are formed either during carbonization of the precursor or through activation, resulting in lowering the final yield of porous carbon. Formation of ordered micropores and mesopores is very difficult to realize in carbon through the activation process. The templating processes, on the other hand, can create pores with uniform size and morphology, even in an ordered arrangement, as explained above. However, it has to be pointed out again that most of the templating processes are complicated and costly. Phenol resin, PAN, PVA, and pitch have been employed as carbon precursors in templating processes, but biomasses have rarely been used, although they are important carbon precursors for the mass production of activated carbons.

In Table 7.3, different template-carbonization processes for pore-structure control are compared, listing the carbon precursors, resultant nanoporous carbons, and template performance.

The advantage of zeolite-template carbonization is the formation of ordered micropores in the resultant carbons. For exact replication of the zeolite template, complete filling of pores in the template with carbon and sufficient carbonization in the zeolite channels were required [68,70]. Two-step filling by liquid impregnation and CVI, and heat treatment to 900 °C in the channels of zeolite Y gave a very high S_{BET} of 3600 m²/g and ordered micropores of a volume of 2.0 cm³/g. The zeolite template has to be sacrificed in microporous carbon synthesis. It will be necessary to develop applications in which the functions of the zeolite-templated microporous carbons are high enough to compensate for the cost of this multi-stepped process.

Template carbonization using ordered mesoporous silicas can give ordered mesopores in carbons, which inherit the mesoporous structure of the templates. Complete filling of the mesopores in the template is also required in this method, which has been done either by repeated impregnation or by combining the impregnation and CVI processes. When the filling of the carbon precursor was not enough, an ordered framework in the carbon cannot be retained after removal of the silica template.

Table 7.3 Comparison of Template-carbonization Methods for Pore Control in Carbon

	Templates	Carbon Prepared			Template Performance	
		Carbon Precursors	Pore Structure	Pore Volume and Surface Area	Removal	Cyclability
Inorganic						
Zeolites	Zeolite Y, β, and L, ZSM-5	Propylene, CVI + FA, impregnation	Micropores ordered	3700 m²/g, 1.8 cm³/g 4000 m²/g, 1.8 cm³/g	HF or NaOH	No
Mesoporous silicas	MCM-48, MCM-41, SBA-1, and SBA-15	Sucrose, FA, PF and MP, impregnation Acetylene, CVI	Mesopores, ordered or disordered	1130 m²/g~1.0 cm³/g 1520 m²/g, 1.3 cm³/g	HF or NaOH	No
Colloidal silicas	Organic silicates and inorganic silica	Mixing with RF, AN,	Micropores and mesopores disordered	1512 m²/g, 3.6 cm³/g 2600 m²/g, 1.4 cm³/g	HF	No
MgO	MgO, Mg acetate and citrate, Mg(OH)₂	Mixing with PVA, coal tar pitches, PET, PIs	Mesopores disordered	1800 m²/g 1600 m²/g, 1.7 cm³/g	Acetic or citric acids	Easy to recycle
Organic						
Surfactants	Diblock and triblock copolymers	Mixing with RF, EOA, PGF, AN	Mesopores ordered or disordered	1354 m²/g, 0.74 cm³/g, 968 m²/g, 0.56 cm³/g	Not needed	No
Organic foams	Urethane and melamine foams	impregnation of PIs	Macropores and micropores	1540 m²/g, 0.63 cm³/g	Not needed	No

AN, acrylonitrile; CVI, chemical vapor infiltration; EOA, triethyl orthoacetate; FA, furfuryl alcohol; MP, mesophase pitch; PET, poly(ethylene tetraphthalate); PF, phenol formaldehyde; PGF, phloroglucinol-formaldehyde; PIs, polyimides; PVA, poly(vinyl alcohol); RF, resorcinol-formaldehyde

In some cases, micropores are created in the walls of mesopores. The advantage of this process using mesoporous silicas is the formation of ordered mesopores in the resultant carbons. However, it has to be pointed out that ordered mesoporous carbons can only be obtained with the necessity of sacrificing both surfactants and silicas. Also, for the removal of the template silica, either corrosive HF or NaOH in an autoclave has to be used.

Soft-template methods, most of them using either diblock or triblock copolymers, have an advantage of giving mesopores ordered in different symmetries and also in random arrays. The process is a simplification of the silica-templating process for mesoporous carbons. However, the carbon precursors are limited to thermosetting resins, the procedure is still complicated, and the surfactant template has to be sacrificed for the synthesis of mesoporous carbons. It may be key for further development into large-scale production to expand the possible carbon precursors and also to create a certain amount of micropores in addition to the mesopores.

In comparison with the template carbonization processes discussed above, MgO template carbonization has the following advantages:

1. Template MgO is easily removed by diluted non-corrosive acid.
2. MgO can be recycled.
3. Size and volume of mesopores are tunable by selecting the MgO precursor [155].

Pore size formed in MgO-templated carbons is very uniform, compared to other processes. Therefore, the MgO template carbonization process has a good possibility of application to the production of mesoporous carbons on a large scale. In 2010, it was announced that a few hundred grams of nanoporous carbon in a batch could be prepared by the MgO-template method [156] and this has since been commercialized (see http://www.toyotanso.co.jp/Products/Newly_developed_Porous_carbon_en.html). Also, one of the advantages of this process is the use of thermoplastic precursors, such as PVA and pitches, without any stabilization process. A disadvantage of this method is that mesopores can not be obtained in ordered arrangement. However, it has to be pointed out that pores in the carbon do not need to be ordered for many of the applications of nanoporous carbons. MgO-templated carbon was shown to be able to keep its pore structure and to have relatively high electrical conductivity after heat treatment at a temperature as high as 2200 °C [224].

Pore structures in the carbons prepared from FA using various templates—a clay (bentonite), β-zeolite, and Al-implanted MCM-48—were discussed, focusing on their application for gas separation and storage [225]. Electrochemical performances for energy storage were compared in the carbons from AN, FA, pyrene, and vinyl acetate, using various zeolites and montmorillonite as templates [226]. Since 2007, a process has been developed where two templates are applied on one carbon precursor (bimodal-template carbonization) to fulfill the requirements of certain applications [227–230].

References

[1] Kyotani T. Carbon 2000;38:269–86.
[2] Sakintunas B, Yueruem Y. Ind Eng Chem Res 2005;44:2893–902.
[3] Kyotani T. Bull Chem Soc Jpn 2006;79:1322–37.
[4] Pandolfo AG, Hollenkamp AH. J Power Sources 2006;157:11–27.
[5] Liang C, Li Z, Dai S. Angew Chem Int Ed 2008;47:3696–717.
[6] Stein A, Wang Z, Fierke MA. Adv Mater 2009;21:265–93.
[7] Inagaki M, Orikasa H, Morishita T. RSC Adv 2011;1:1620–40.
[8] Nishihara H, Kyotani T. Adv Mater 2012;24:4473–98.
[9] Kyotani T, Sonobe N, Tomita A. Nature 1988;331:331–3.
[10] Sonobe N, Kyotani T, Tomita A. Carbon 1988;26:573–8.
[11] Sonobe N, Kyotani T, Hishiyama Y, et al. J Phys Chem 1988;92:7029–34.
[12] Sonobe N, Kyotani T, Tomita A. Carbon 1990;28:483–8.
[13] Sonobe N, Kyotani T, Tomita A. Carbon 1991;29:61–7.
[14] Kyotani T, Sonobe N, Tomita A. TANSO 1992;1992(No.155):301–6.
[15] Kyotani T, Yamada H, Sonobe N, et al. Carbon 1994;32:627–35.
[16] Kyotani T, Mori T, Tomita A. Chem Mater 1994;6:2138–42.
[17] Sonobe N, Kyotani T, Tomita A, et al. TANSO 1990; No.141: 38–44 [in Japanese].
[18] Kyotani T, Suzuki K, Sonobe N, et al. Carbon 1993;31:149–53.
[19] Kyotani T, Tsai L, Tomita A. Chem Mater 1995;7:1427–8.
[20] Li J, Moskovits M, Haslett TL. Chem Mater 1998;10:1963–7.
[21] Che G, Lakshmi BB, Marth CR, et al. Chem Mater 1998;10:260–7.
[22] Che G, Lakshmi BB, Fisher ER, et al. Nature 1998;393:346–9.
[23] Suh JS, Lee JS. Appl Phys Lett 1999;75:2047–9.
[24] Li J, Papadopoulos C, Xu JM, et al. Appl Phys Lett 1999;75:367–9.
[25] Li J, Papadopoulos C, Xu J. Nature 1999;402:253–4.
[26] Lee JS, Gu GH, Kim H, et al. Chem Mater 2001;13:2387–91.
[27] Kyotani T, Tsai L, Tomita A. Chem Commun 1997;1997:701–2.
[28] Che G, Lakshml BB, Martin CR, et al. Langmuir 1999;15:750–8.
[29] Pradhan BK, Kyotani T, Tomita A. Chem Commun 1999;1999:1317–8.
[30] Pradhan BK, Toba T, Kyotani T, et al. Chem Mater 1998;10:2510–5.
[31] Matsui K, Pradhan BK, Kyotani T, et al. J Phys Chem B 2001;105:5682–8.
[32] Matsui K, Kyotani T, Tomita A. Adv Mater 2002;14:1216–9.
[33] Orikasa H, Karoji J, Matsui K, et al. Dalton Trans 2007;34:3757–62.
[34] Kim KH, Orikasa H, Kyotani T, et al. IEEE Trans Magn 2005;41:4075–7.
[35] Wang XH, Orikasa H, Inokuma N, et al. J Mater Chem 2007;17:986–91.
[36] Kim KH, Yamaguchi M, Orikasa H, et al. Solid State Commun 2006;140:491–4.
[37] Hattori Y, Watanabe Y, Kawasaki S, et al. Carbon 1999;37:1033–8.
[38] Touhara H, Okino F. Carbon 2000;38:241–67.
[39] Touhara H, Inahara J, Mizuno T, et al. J Fluor Chem 2002;114:181–8.
[40] Kyotani T, Nakazaki S, Xu WH, et al. Carbon 2001;39:781–4.
[41] Orikasa H, Inokuma N, Okubo S, et al. Chem Mater 2006;18:1036–40.
[42] Matsui J, Iko M, Inokuma N, et al. Chem Lett 2006;35:42–3.
[43] Matsui J, Yamamoto K, Inokuma N, et al. J Mater Chem 2007;17:3806–11.
[44] Orikasa H, Inokuma N, Ittisanronnachai S, et al. Chem Commun 2008;2008:2215–7.
[45] Ittisanronnachai S, Orikasa H, Inokuma N, et al. Carbon 2008;46:1361–3.
[46] Vijayaraj M, Gadiou R, Anselme K, et al. Adv Funct Mater 2010;20:2489–99.

[47] Xu WH, Kyotani T, Pradhan BK, et al. Adv Mater 2003;15:1087–90.
[48] Yang QH, Xu WH, Tomita A, et al. Chem Mater 2005;17:2940–5.
[49] Yang Q, Xu W, Tomita A, et al. J Am Chem Soc 2005;127:8956–7.
[50] Yang QH, Hou PX, Unno M, et al. Nano Lett 2005;5:2465–9.
[51] Kyotani T, Xu WH, Yokoyama Y, et al. J Membrane Sci 2002;196:231–9.
[52] Jian K, Shim HS, Schwartzman A, et al. Adv Mater 2003;15:164–7.
[53] Chan C, Crawford G, Gao Y, et al. Carbon 2005;43:2431–40.
[54] Zhi L, Wu J, Li J, et al. Angew Chem Int Ed 2005;44:2120–3.
[55] Konno H, Sato S, Habazaki H, et al. Carbon 2004;42:2756–9.
[56] Orikasa H, Akahane T, Okada M, et al. J Mater Chem 2009;19:4615–21.
[57] Inagaki M, Morishita T, Kuno A, et al. Carbon 2004;42:497–502.
[58] Ohta N, Nishi Y, Morishita T, et al. New Carbon Mater 2008;23:216–20.
[59] Harikrishnan G, Patro TU, Khakhar DV. Carbon 2007;45:531–5.
[60] Chen Y, Chen B, Shi X, et al. Carbon 2007;45:2132–4.
[61] Xue C, Tu B, Zhao D. Nano Res 2009;2:242–53.
[62] Kyotani T, Tomita A. J Jpn Petrol Inst 2002;45:261–70.
[63] Kyotani T, Nagai T, Inoue S, et al. Chem Mater 1997;9:609–15.
[64] Enzel P, Bein T. Chem Mater 1992;4:819–24.
[65] Johnson SA, Brigham ES, Ollivier PJ, et al. Chem Mater 1997;9:2448–58.
[66] Cordero T, Thrower PA, Radivic LR. Carbon 1992;3:365–74.
[67] Rodriguez-Mirasol J, Cordero T, Radovic LR, et al. Chem Mater 1998;10:550–8.
[68] Ma ZX, Kyotani T, Tomita A. Chem Commun 2000;2000:2365–6.
[69] Ma ZX, Kyotani T, Liu Z, et al. Chem Mater 2001;13:4413–5.
[70] Ma Z, Kyotani T, Tomita A. Carbon 2002;40:2367–74.
[71] Matsuoka K, Yamagishi Y, Yamazaki T, et al. Carbon 2005;43:876–9.
[72] Nishiyama H, Yang QH, Hou PX, et al. Carbon 2009;47:1220–30.
[73] Takai K, Suzuki T, Enoki T, et al. J Phys Chem Solids 2010;71:565–8.
[74] Kopelevich Y, da Silva RR, Torres JHS, et al. Phys Rev B 2003;68; 092408.
[75] Kyotani T, Ma Z, Tomita A. Carbon 2003;41:1451–9.
[76] Su F, Zhao XS, Lv L, et al. Carbon 2004;42:2821–31.
[77] Garsuch A, Klepel O, Sattler RR, et al. Carbon 2006;44:593–6.
[78] Gaslain FOM, Parmentier J, Valtchev VP, et al. Chem Commun 2006;2006:991–3.
[79] Hou PX, Yamazaki T, Orikasa H, et al. Carbon 2005;43:2624–7.
[80] Hou PX, Orikasa H, Itoi H, et al. Carbon 2007;45:2011–8.
[81] Hou PX, Orikasa H, Yamazaki T, et al. Chem Mater 2005;17:5187–93.
[82] Yang Z, Xia Y, Sun X, Mokaya R. J Phys Chem B 2006;110:18424–31.
[83] Yang Z, Xia Y, Mokaya R. J Am Chem Soc 2007;129:1673–9.
[84] Nishihara H, Hou PX, Li LX, et al. J Phys Chem C 2009;113(8):3189–96.
[85] Wang H, Gao Q, Hu J, et al. Carbon 2009;47:2259–68.
[86] Takai K, Suzuki T, Enoki T, et al. Phys Rev B 2010;81:205420.
[87] Nakashima Y, Matsushita T, Hieda M, et al. J Low Temp Phys 2011;162:565–72.
[88] Su F, Zeng J, Yu Y, et al. Carbon 2005;43:2366–73.
[89] Ania CO, Khomenko V, Raymundo-Pinero E, et al. Adv Funct Mater 2007; 17:1828–36.
[90] Nishihara H, Itoi H, Kogure T, et al. Chem Eur J 2009;15:5355–63.
[91] Portet C, Yang Z, Gogotsi KY, et al. J Electrochem Soc 2009;156:A1–6.
[92] Kwon T, Nishihara H, Itoi H, et al. Langmuir 2009;25:11961–6.
[93] Itoi H, Nishihara H, Kogure T, et al. J Am Chem Soc 2011;133:1165–7.

[94] Ryoo R, Joo SH, Kruk M, et al. Adv Mater 2001;13:677–81.
[95] Lee J, Han S, Hyeon T. J Mater Chem 2004;14:478–86.
[96] Ryoo R, Joo SH, Jun S. J Phys Chem B 1999;103:7743–6.
[97] Lee J, Yoon S, Hyeon T, et al. Chem Comm 1999;1999:2177–8.
[98] Lee J, Yoon S, Oh SM, et al. Adv Mater 2000;12:359–62.
[99] Yoon SB, Kim JY, Yu J S. Chem Commun 2001;2001:559–60.
[100] Yang H, Shi Q, Liu X, et al. Chem Commun 2002;2002:2842–3.
[101] Kaneda M, Tsubakiyama T, Carisson A, et al. J Phys Chem B 2002;106:1256–66.
[102] Kruk M, Jaroniec M, Ryoo R, et al. J Phys Chem B 2000;104:7960–8.
[103] Joo SH, Jun S, Ryoo R. Microp Mesop Mater 2001;44–45:153–8.
[104] Tian B, Che S, Liu Z, et al. Chem Commun 2003;2003:2726–7.
[105] Jun S, Joo SH, Ryoo R, et al. J Am Chem Soc 2000;122:10712–3.
[106] Shin HJ, Ryoo R, Kruk M, et al. Chem Commun 2001;2001:349–50.
[107] Lee JS, Joo SH, Ryoo R. J Am Chem Soc 2002;124:1156.
[108] Yu C, Fan J, Tian B, et al. Adv Mater 2002;14:1742–5.
[109] Lu AH, Li WC, Schmidt W, et al. Microp Mesop Mater 2006;95:187–92.
[110] Xia Y, Mokaya R. Adv Mater 2004;76:1553–8.
[111] Darmstadt H, Roy C, Kaliaguine S, et al. Chem Mater 2003;15:3300–7.
[112] Lu AH, Li WC, Schmidt W, et al. Carbon 2004;42:2939–48.
[113] Titirici MM, Thomas A, Antonietti M. J Mater Chem 2007;17:3412–8.
[114] Kim SS, Pinnavaia TJ. Chem Commun 2001;2001:2418–9.
[115] Alvarez S, Fuertes AB. Carbon 2004;42:433–6.
[116] Fuertes AB, Nevskaia DM. Microp Mesop Mater 2003;62:177–90.
[117] Lee HI, Kim JH, You DJ, et al. Adv Mater 2008;20:757–62.
[118] Kim TW, Park IS, Ryoo R. Angew Chem Int Ed 2003;42:4375–9.
[119] Vix-Guterl C, Saadallah S, Vidal L, et al. J Mater Chem 2003;13:22535–9.
[120] Han S, Sohn K, Hyeon T. Chem Mater 2000;12:3337–41.
[121] Vinu A, Streb C, Murugesan V, et al. J Phys Chem B 2003;107:8297–9.
[122] Choi M, Ryoo R. Nat Mater 2003;2:473–6.
[123] Wang DW, Li F, Lu GQ, et al. Carbon 2008;46:1593–9.
[124] Vix-Guterl C, Frackowiak E, Jurewicz K, et al. Carbon 2005;43:1293–302.
[125] Xia K, Gao Q, Wu C, et al. Carbon 2007;45:1989–96.
[126] Zheng Z, Gao Q, Jiang J. Carbon 2010;48:2968–73.
[127] Giraudet S, Zhu Z, Yao X, et al. J Phys Chem C 2010;114:8639–45.
[128] Giraudet S, Zhu Z. Carbon 2011;49:398–405.
[129] Pevida C, Drage TC, Snape CE. Carbon 2008;46:1464–74.
[130] Joo SH, Choi SJ, Oh I, et al. Nature 2001;412:169–72.
[131] Lu AH, Li WC, Hou Z, et al. Chem Commun 2007;2007:1038–40.
[132] Han S, Lee KT, Oh M, et al. Carbon 2003;41:1049–56.
[133] Yoon S, Lee J, Hyeon T, et al. J Electrochem Soc 2000;147:2507–12.
[134] Zhou H, Zhu S, Hibino M, et al. J Power Sources 2003;122:219–23.
[135] Jurewicz K, Vix-Guterl C, Frackowiak E, et al. J Phys Chem Solids 2004;65:287–93.
[136] Vix-Guterl C, Saadallah S, Jurewicz K, et al. Mater Sci Eng B 2004;108:148–55.
[137] Fuertes AB, Pico F, Rojo JM. J Power Sources 2004;133:329–36.
[138] Fuertes AB, Lota G, Centeno TA, et al. Electrochim Acta 2005;50:2799–805.
[139] Centeno TA, Sevilla M, Fuertes AB, et al. Carbon 2005;43:3012–5.
[140] Li L, Song H, Chen X. Electrochim Acta 2006;51:5715–20.
[141] Alvarez S, Blanco-Lopez MC, Miranda-Ordieres AJ, et al. Carbon 2005;43:866–70.

[142] Xing S, Qiao Z, Ding RG, et al. Carbon 2006;44:216–24.
[143] Sevilla M, Alvarez S, Centeno TA, et al. Electrochim Acta 2007;52:3207–15.
[144] Xia K, Gao Q, Jiang J, et al. Carbon 2008;46:1718–26.
[145] Banham D, Feng F, Burt J, et al. Carbon 2010;48:1056–63.
[146] Inagaki M, Kobayashi S, Kojin F, et al. Carbon 2004;42:3153–8.
[147] Morishita T, Suzuki T, Nishikara T, et al. TANSO 2005; No. 219: 226–231 [in Japanese].
[148] Morishita T, Suzuki T, Nishikawa T, et al. TANSO 2006; No. 223: 220–226 [in Japanese].
[149] Morishita T, Soneda Y, Tsumura T, et al. Carbon 2006;44:2360–7.
[150] Morishita T, Ishihara K, Kato M, et al. Carbon 2007;45:209–11.
[151] Morishita T, Ishihara K, Kato M, et al. TANSO 2007; No. 226: 19–24 [in Japanese].
[152] Inagaki M, Kato M, Morishita T, et al. Carbon 2007;45:1121–4.
[153] Przepiórski J, Karolczyk J, Takeda K, et al. Ind Eng Chem Res 2009;48:7110–6.
[154] Konno H, Onishi H, Yoshizawa N, et al. J Power Sources 2010;195:667–73.
[155] Morishita T, Tsumura T, Toyoda M, et al. Carbon 2010;49:2690–707.
[156] Morishita T, Wang L, Tsumura T, et al. TANSO 2010; No. 242: 60–68 [in Japanese].
[157] Jung M-J, Im JS, Jeong E, et al. Carbon Lett 2009;10:217–20.
[158] Przepiórski J, Czyżewski A, Kapica J, et al. Pol J Chem Technol 2011;13:42–6.
[159] Karolczyk J, Janus M, Przepiorski J. Pol J Chem Technol 2012;14:95–9.
[160] Zhang WF, Huang ZH, Guo Z, et al. Mater Lett 2010;64:1868–70.
[161] Zhang WF, Huang ZH, Cao GP, et al. J Power Sources 2012;204:230–5.
[162] Zhang WF, Huang ZH, Zhou CJ, et al. J Mater Chem 2012;22:7158–63.
[163] Zhang WF, Huang ZH, Cao GP, et al. J Phys Chem Solids 2012;73:1428–31.
[164] Fernandez JA, Morishita T, Toyoda M, et al. J Power Sources 2008;175:675–9.
[165] Wang L, Morishita T, Toyoda M, et al. Electrochim Acta 2007;53:882–6.
[166] Wang L, Toyoda M, Inagaki M. Adsorpt Sci Technol 2008;46:491–5.
[167] Wang L, Inagaki M, Toyoda M. TANSO 2009; [No. 240]: 230–238 [in Japanese].
[168] Zhou J, Yuan X, Xing W, et al. New Carbon Mater 2010;25:370–5.
[169] Wang Y, Wang C. New Carbon Mater 2010;25:376–81.
[170] Morishita T, Hirabayashi M, Nishioka Y, et al. J Power Sources 2006;160:638–44.
[171] Przepiórski J, Czyżewski A, Kapica J, et al. Chem Eng J 2012. http://dx.doi.org/10.1016/j.cej.2012.02.087.
[172] Przepiorski J, Czyzewski A, Pietrzak R, et al. J Therm Anal Calorim 2013;111: 357–64.
[173] Wang DW, Li F, Liu M, et al. Angew Chem Int Ed 2008;47:373–6.
[174] Zhou J, Yuan X, Xing W, et al. Carbon 2010;48:2765–72.
[175] Yang G, Han H, Li T, et al. Carbon 2012;50:3753–65.
[176] Liang C, Hong K, Guiochon GA, et al. Angew Chem Int Ed 2004;43:5785–9.
[177] Kosonen H, Valkama S, Nykaenen A, et al. Adv Mater 2006;18:201–5.
[178] Tanaka S, Nishiyama N, Egashira Y, et al. Chem Commun 2005;2005:2125–7.
[179] Meng Y, Gu D, Zhang F, et al. Angew Chem Int Ed 2005;44:7053–9.
[180] Zhang FQ, Meng Y, Gu D, et al. J Am Chem Soc 2005;127:13508–9.
[181] Tanaka S, Katayama Y, Tate MP, et al. J Mater Chem 2007;17:3639–45.
[182] Liang CD, Dai S. J Am Chem Soc 2006;128:5316–7.
[183] Kowalewski T, Tsarevsky NV, Matyjaszewski K. J Am Chem Soc 2002;124:10632–3.
[184] Meng Y, Gu D, Zhang F, et al. Chem Mater 2006;18:4447–64.
[185] Zhang F, Meng Y, Gu D, et al. Chem Mater 2006;18:5279–88.

[186] Jin J, Nishiyama N, Egashira Y, et al. Microp Mesop Mater 2009;118:218–23.
[187] Wang X, Liang C, Dai S. Langmuir 2008;24:7500–5.
[188] Song L, Feng D, Fredin NJ, et al. ACS Nano 2010;4:189–98.
[189] Wang Z, Kiesel ER, Stein A. J Mater Chem 2008;18:2194–200.
[190] Gorka J, Zawislak A, Choma J, et al. Carbon 2008;46:1159–61.
[191] Jin J, Tanaka S, Egashira Y, et al. Carbon 2010;48:1985–9.
[192] Yu T, Deng Y, Wang L, et al. Adv Mater 2007;19:2301–6.
[193] Gorka J, Jaroniec M. Carbon 2011;49:154–60.
[194] Gorka J, Jaroniec M. J Phys Chem C 2008;112:11657–60.
[195] Liu B, Shioyama H, Akita T, et al. J Am Chem Soc 2008;130:5390–1.
[196] Liu B, Shioyama H, Jiang H, et al. Carbon 2010;48:456–63.
[197] Yuan D, Chen J, Tan S, et al. Electrochem Commun 2009;11:1191–4.
[198] Hu J, Wang H, Gao Q, et al. Carbon 2010;48:3599–606.
[199] Jiang HL, Liu B, Lan YQ, et al. J Am Chem Soc 2011;133:11854–7.
[200] Deng HG, Jin SL, Zhan L, et al. New Carbon Mater 2012;27:194–9.
[201] Pekala R, Hopper RW. J Mater Sci 1987;22:1840–4.
[202] Bandoz TJ, Jagiello J, Putyera K, et al. Chem Mater 1996;8:2023–9.
[203] Santos C, Andrade M, Vieira AL, et al. Carbon 2010;48:4049–56.
[204] Nguyen-Thanh D, Bandosz TJ. Microp Mesop Mater 2006;92:47–55.
[205] Gilbert MT, Knox JH, Kaur B. Chromatographia 1982;16:138–46.
[206] Zakhidov AA, Baughman RH, Iqbal Z, et al. Science 1998;282:897–901.
[207] Han S, Hyeon T. Chem Commun 1999;1999:1955–6.
[208] Kawashima D, Aihara T, Kobayashi Y, et al. Chem Mater 2000;12:3397–401.
[209] Pang J, Hu Q, Wu Z, et al. Microp Mesop Mater 2004;74:7–8.
[210] Li Z, Jaroniec M. J Am Chem Soc 2001;123:9208–9.
[211] Li Z, Jaroniec M. Chem Mater 2003;15:1327–33.
[212] Pang J, Li X, Wang D, et al. Adv Mater 2004;16:884–6.
[213] Han S, Lee KT, Oh M, et al. Carbon 2003;41:1049–56.
[214] Taguchi A, Small JH, Linden M. Adv Mater 2003;15:1209–11.
[215] Han B-H, Zhou W, Sayant A. J Am Chem Soc 2003;125:3444–5.
[216] Gierszal KP, Jaroniec M. J Am Chem Soc 2006;128:10026–7.
[217] Gu D, Bongard H, Deng Y, et al. Adv Mater 2010;22:833–7.
[218] Lee J, Sohn K, Hyeon T. J Am Chem Soc 2001;123:5146–7.
[219] Chai GS, Shin IS, Yu J S. Adv Mater 2004;16:2057–61.
[220] Jaronieca M, Gorka J, Choma J, et al. Carbon 2009;47:3034–40.
[221] Wang AP, Kang F, Huang ZH, et al. Clays Clay Miner 2006;54:485–90.
[222] Wang A, Kang F, Huang Z, et al. Microp Mesop Mater 2008;108:318–24.
[223] Inagaki M, Kim YA, Endo M. J Mater Chem 2011;21:3280–94.
[224] Orikasa H, Morishita T. TANSO 2012; No. 254: 153–159.
[225] Barara-Rodrigues PM, Mays TJ, Moggridge GD. Carbon 2003;41:2231–46.
[226] Meyers C, Shah SD, Patel SC, et al. J Phys Chem 2001;105:2143–52.
[227] Zheng M, Cao J, Ke X, et al. Carbon 2007;45:1111–3.
[228] Zhao G, He J, Zhang C, et al. J Phys Chem C 2008;112:1028–33.
[229] Wang K, Wang Y, Wang Y, et al. J Phys Chem C 2009;113:1093–7.
[230] Wang Y, Zheng M, Lu H, et al. Nanoscale Res Lett 2010;5:913–6.

Carbon Nanofibers Via Electrospinning

Fibrous carbon materials have for some time been attracting the attention of scientists and engineers. In the 1960s, carbon fibers were developed as an important industrial material for modern science and technology, and have been produced from various carbon precursors via a melt-spinning process. Polyacrylonitrile (PAN) has been used as the principal precursor, associated with various modifications in processing, such as the use of some additives, oxidative stabilization of as-spun PAN fibers at a low temperature, and stretching during stabilization and carbonization. Isotropic pitches, anisotropic mesophase pitches, and phenolic resins have also been the precursors for carbon fibers. The catalytic chemical vapor deposition (CVD) process has also produced carbon fibers (vapor-grown carbon fiber, VGCF), which are different in structure and properties from those produced via melt-spinning. In the center of VGCFs, thin tubes consisting of straight carbon layers were found [1], which were later reported to be formed by arc-discharging [2] and named carbon nanotubes (CNTs) [3,4]. The process used for the production of VGCFs was successfully applied to synthesize CNTs [5]. The diameters of fibrous carbon materials are in the nanometer range, such as 7 nm for single-walled CNTs, in contrast to the diameters of carbon fibers, which are often a few micrometers. Fibrous carbon materials with diameters in the range of 10^1–10^3 nm, i.e. intermediate between carbon nanotubes and carbon fibers, will here be called "carbon nanofibers." Each group of fibrous carbons has variety in nanotexture, as summarized in Figure 8.1: CNTs consisting of single-, double-, and multi-walled tubes, in addition to having different chiralities in single-walled ones; carbon nanofibers having a variety of nanotextures, herringbone, platelet, and tubular types; carbon fibers having different structures mainly governed by the precursors used, PAN, pitches, phenols, benzene vapor, etc.

Electrospinning has been used to produce nanofibers of various polymers, with diameters from a few tens of nanometers to a few micrometers, as different forms of nonwoven mats (webs), yarns, etc. Electrospinning is a relatively simple and low-cost strategy to produce continuous nanofibers from polymer solutions or melts. Some electrospun polymer nanofibers have been successfully converted to carbon nanofibers by heat treatment in an inert atmosphere [6]. The carbon nanofibers produced via electrospinning are continuous fibrous materials with a uniform diameter

---

Prerequisite for readers: Chapter 3.4 (Fibrous carbons) in *Carbon Materials Science and Engineering: From Fundamentals to Applications*, Tsinghua University Press.

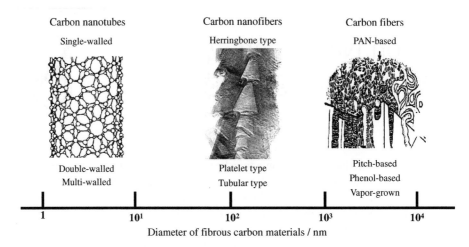

FIGURE 8.1 Fibrous Carbon Materials in Different Sizes

and without impurity metals, although the nanofibers synthesized by CVD have variable diameters and contain metallic impurities.

In this chapter, the carbon nanofibers prepared via electrospinning and carbonization are covered, focusing on their structure and properties in relation to their applications.

8.1 Carbon nanofibers synthesized via electrospinning

Various organic polymers have been used to synthesize nanofibers by electrospinning, but the organic polymers available as carbon precursors are relatively limited. PAN and pitches have frequently been used, probably because both of them are used also in the production of commercial carbon fibers. In addition, polyimides (PIs), poly(vinyl alcohol) (PVA), polybenzimidazol (PBI), phenolic resin, and lignin have been used.

In order to convert electrospun polymer nanofibers to carbon nanofibers, a carbonization process at around 1000 °C has to be applied. In principle, any polymer with a carbon backbone can potentially be used as a precursor. For carbon precursors, such as PAN and pitches, a so-called stabilization process before carbonization is essential to keep the fibrous morphology, of which the fundamental reaction is oxidation. The poor spinnability of pitches was solved by mixing with a polymer having good spinnability, such as PAN. During stabilization and carbonization of polymer nanofibers, they show significant weight loss and shrinkage, resulting in the decrease of fiber diameter.

8.1.1 Polyacrylonitrile

Polyacrylonitrile (PAN) has been frequently selected as a carbon precursor for the preparation of carbon nanofibers via electrospinning. Carbon nanofibers, which were

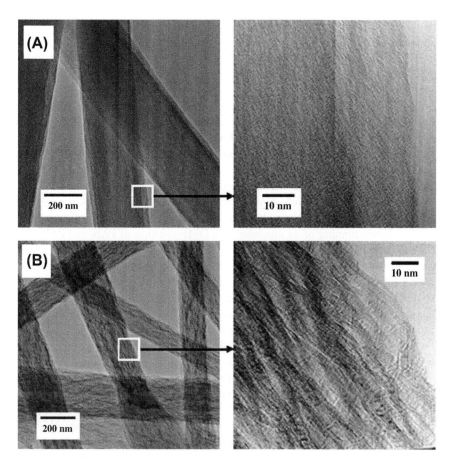

FIGURE 8.2 TEM Images of Electrospun PAN-based Carbon Nanofibers After Heat Treatment

(A) 1000 °C-treated, and (B) 2200 °C-treated

From [8]

prepared from an 8 mass% PAN/*N,N*-dimethylacetamide (DMF) solution by carbonization at 1100 °C, had an average diameter of 110 nm and interlayer spacing, d_{002}, of 0.368 nm [7]. From scanning and transmission electron microscopy (SEM and TEM) observations, the fiber was concluded to have skin-core heterogeneity; carbon layers were oriented predominantly parallel to the fiber surface in the skin. PAN-based carbon nanofiber bundles have been prepared from a 10 mass% PAN/DMF solution by the addition of 5 mass% acetone and 0.01 mass% dodecylethyldimethylammonium bromide. They were collected on the rim of a rotating disc covered with Al foil and then subjected to heat treatment [8]. The diameter of nanofibers composing the bundles was approximately 330 nm for as-spun fibers, 250 nm for 1000 °C-treated ones, and 220 nm for 1800 °C-treated ones. TEM images are shown in Figure 8.2 for 1000 °C-treated and 2200 °C-treated nanofibers, the latter having d_{002} of 0.344 nm. Aiming at better alignment of the basic structural units (BSUs) of hexagonal carbon layers along the fiber axis, multi-walled carbon nanotubes (MWCNTs) were embedded into electrospun PAN-based carbon nanofibers, although improvement was observed only around the MWCNTs [9]. On a PAN-based nanofiber web after activation by steam at 800 °C, adsorption of benzene vapor was studied at 343–423 K under a pressure of 4.0 KPa and high adsorption was confirmed, in comparison with an activated carbon fiber [10]. The electrospinnability, environmentally benign nature, and commercial viability have recently been reviewed, with a focus on PAN [11].

Core-shell polymeric fibers electrospun through a double capillary, PAN/DMF solution in the outer capillary and poly(methyl methacrylate) (PMMA) in the inner capillary, resulted in hollow carbon nanofibers after carbonization at 1100 °C, because PAN gave a relatively high carbonization yield but PMMA a negligibly small amount of carbon residues [12]. Similar hollow nanofibers have been synthesized by electrospinning of an emulsion-like DMF solution of PAN and PMMA in different ratios through a single capillary, followed by carbonization at 1000 °C and heat treatment up to 2800 °C [13]. In Figure 8.3, SEM and TEM images are shown for the nanofibers prepared from the mixture of PAN/PMMA = 5/5. By changing the PAN/PMMA ratio, the content of mesopores could be controlled; mesopore volume, V_{meso}, changed from 0.18 cm^3/g for a 9/1 ratio to 0.47 cm^3/g for a 5/5 ratio, although micropore volume, V_{micro}, was almost constant at 0.34 cm^3/g [13].

Mixing of poly(vinylpyrrolidone) (PVP) into PAN has also been employed to control the pore structure in nanofibers [14]. The Brunauer-Emmett-Teller (BET) surface area (S_{BET}) and total pore volume (V_{total}) of the carbon nanofibers obtained at 1000 °C changed from 237 to 571 m^2/g, and from 0.10 to 0.19 cm^3/g, respectively, with increasing PVP content. PVP and PAN solutions were separately fed into the spinneret to form side-by-side bicomponent nanofibers, which were comparable with PAN/PVP blended nanofibers after carbonization at 970 °C and activation at 850 °C in CO_2 [15].

Electrical conductivity was measured to be 4.9 S/cm for 800 °C-carbonized PAN-based carbon nanofibers [16], and the 1000 °C-treated nanofibers showed a large negative magnetoresistance, −0.75 at a temperature of 1.9 K under a magnetic field of 9 T [17]. Although fibrous morphology survived under the heat treatment, it has to be taken into account that these nanofibers are not stabilized before carbonization when these properties are compared with those of commercial PAN-based carbon fibers. The PAN-based carbon nanofiber bundles prepared by stabilization and carbonization at 2200 °C showed a conductivity of 840 S/cm in parallel to fiber axis, but 61 S/cm perpendicular to the fiber axis [8]. Electrospun PAN-based carbon nanofibers changed their electrical conductivity on exposure to NO gas after H_3PO_4 activation [18].

Mechanical properties have been measured on single fibers of PAN-based carbon nanofibers [7]. Averaged bending modulus was 63 GPa and Weibull fracture stress was 640 MPa with failure probability 63%. Tensile strength and Young's modulus measured on the bundles of electrospun PAN-based carbon nanofibers were 542 MPa and 58 GPa, respectively [8]. The mechanical properties reported

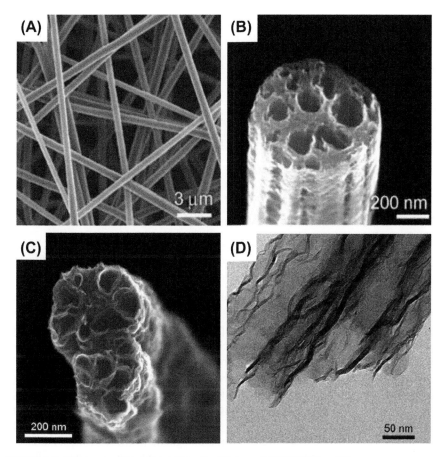

FIGURE 8.3 Electrospun Nanofibers From the Mixture of PAN/PMMA = 5/5

SEM images of (A) as-spun, (B) 1000 °C-carbonized, and (C) 2800 °C-treated fibers, and (D) TEM image of 2800 °C-treated fiber

From [13]

on the electrospun carbon nanofibers are much inferior to commercially available PAN-based carbon fibers. Since oxidative stabilization of PAN fibers is known to be the most important unit-process for PAN-based carbon fibers, optimization of stabilization conditions for electrospun PAN nanofibers has to be studied in detail. Relatively high tensile strength and Young's modulus were reported on electrospun single nanofibers prepared from 9 mass% PAN/DMF solution, followed by stabilization in air and carbonization at 1700 °C [19]. Tensile strength depended strongly on heat-treatment temperature (HTT), showing a maximum of 3.5 GPa at 1400 °C, and Young's modulus increased with increasing HTT, giving 181 GPa at 1700 °C, as shown in Figure 8.4. These changes in mechanical properties with heat treatment were explained by the growth of crystallite in nanofibers. However, it has to

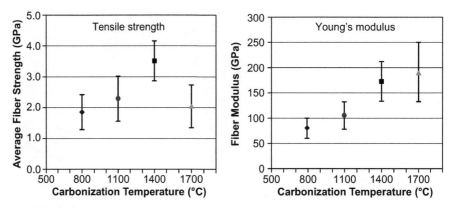

FIGURE 8.4 Dependence of Tensile Strength and Young's Modulus of Carbon Nanofibers on Carbonization Temperature

From [19]

be pointed out that the orientation of crystallites in the nanofibers is not axial, but random. Stretching before and during the stabilization of nanofiber bundles of PAN copolymer (PAN/itaconic acid) has been applied, with the expectation of producing a well-oriented nanotexture and high mechanical properties [20,21], but, regrettably, no detailed experimental data were presented.

Loading of nanoparticles of various metals and metal oxides has been performed via the electrospinning process. Magnetic $CoFe_2O_4$ nanoparticles were embedded in PAN-based carbon nanofibers via electrospinning of a PAN/DMF solution dispersed with oleic-acid-modified $CoFe_2O_4$ nanoparticles of 5 nm in size [22]. $CoFe_2O_4$-embedded nanofibers were superparamagnetic because of the nano-sized magnetic particles, and saturation magnetization increased from 45 to 63 Am^2/kg by carbonization. SiO_2-embedded carbon nanofibers were prepared by electrospinning of PAN/DMF solutions containing different amounts of SiO_2 [23]. SiO_2 particles embedded in carbon nanofibers were washed out by HF, but S_{BET} and V_{total} increased only to 340 m^2/g and 0.472 cm^3/g, respectively. Vanadium-embedded carbon nanofibers were prepared by electrospinning of PAN/DMF solutions containing different amounts of V_2O_5 [24]. After activation using KOH at 750 °C, nanoporous nanofibers were obtained, S_{BET} reaching 2780 m^2/g, V_{total} 2.67 cm^3/g, and V_{micro} 1.52 cm^3/g, which gave a hydrogen-storage capacity of 2.41 mass% at 303 K under 10 MPa. Mn-loaded carbon nanofibers, which were activated by steam at 850 °C and had V_{micro} of 0.42 cm^3/g, showed a relatively high adsorption capacity for toluene at 289 K [25].

8.1.2 Pitch

A DMF-insoluble fraction of petroleum-derived isotropic pitch was successfully electrospun from tetrahydrofuran (THF) solution to form webs of carbon fibers of 2–6 μm in diameter [26,27]. However, the low boiling point (65–67 °C) of

THF led to difficulty in preparing thinner fiber, because the viscosity of the jet increased owing to the volatilization of THF during electrospinning. After activation, the webs were microporous, showing a very high S_{BET} of 2200 m^2/g. By mixing PAN with a pitch in a binary solvent, DMF + THF, spinnability was improved, resulting in fibers with a diameter of 750 nm [28]. After activation using steam at 900 °C, S_{BET} of 1877 m^2/g and V_{total} 1.11 cm^3/g were obtained, owing to both micropores and mesopores. The THF-soluble component of the pitch, with a low molecular weight of 556, gave better spinnability in THF solution, but the carbon nanofibers prepared from a high-molecular-weight pitch of 2380 showed higher development of micropores, giving S_{BET} of 2053 m^2/g, after carbonization at 1000 °C and activation at 900 °C in steam/N$_2$ flow [29]. Highly porous carbon nanofibers were obtained by electrospinning from THF solution of polycarbosilane, followed by pyrolysis at different temperatures and chlorination to extract Si [30]. The nanofibers pyrolyzed at 900 °C and chlorinated at 850 °C had a very high S_{BET} of 3116 m^2/g and V_{total} of 1.66 cm^3/g, and were reported to have a high storage capacity for hydrogen, 3.86 mass% at 77 K under 1.7 MPa.

8.1.3 Polyimides

Polyimide (PI) is often electrospun to form nanofibers, but there are not many reports on the conversion to carbon nanofibers [31–35]. Carbon nanofibers with a diameter less than 2–3 μm, which were prepared from a PI of polymellitic dianhydride and 2,4′-oxydianiline (PMDA/ODA), gave a relatively high tensile strength of 74 MPa and electrical conductivity of 5.3 S/cm after heat treatment at 2200 °C [31]. A thermotropic PI (Matimid* 5218) dissolved in dimethyl acetamide (DMAc) with the addition of 0.3–3.0 mass% Fe(III) acetylacetonate (AAI) was spun to nanofibers in an atmosphere of 24% humidity and then carbonized at 400–1200 °C [33]. During the carbonization process, AAI decomposed to α-Fe and Fe$_3$O$_4$. By the addition of 3 mass% AAI and heat treatment at 1200 °C, d_{002} decreased from 0.37 to 0.34 nm, $Lc(002)$ increased from 1.0 to 4.2 nm, and I_D/I_G decreased from 3.4 to 1.8. The addition of PAN to the PI solution improved spinnability and decreased the diameter of the resultant carbon nanofibers [35].

8.1.4 Poly(vinylidene fluoride)

Poly(vinylidene fluoride) (PVDF) nanofibers have been electrospun from DMF solutions with the addition of poly(ethylene oxide) (PEO) and water, of which the webs were dehydrofluorized using 1,8-diazabicyelo[5,4,0]undec-7-ene at 90 °C, followed by carbonization at 1000 °C [36]. The resultant webs of carbon nanofibers contained three kinds of pores: the largest pores were the interstices of fibers, intermediate-sized pores of 100–300 nm were formed on the fiber surface due to liquid-liquid phase separation, and micropores were formed by the decomposition of PEO during carbonization. The dehydrofluorination process was found to be key to retaining the pore morphology in as-spun fibers during carbonization.

8.1.5 Phenolic resins

For electrospinning of phenolic resin (novolac type), the concentration was critical; a solution more than 65 mass% was difficult to spin and one less than 50 mass% gave fibers including beads [37]. Spinnability of phenolic resin solution was improved by the addition of a small amount of poly(vinyl butyral) (PVB) with a molecular weight (M_w) of 110,000. The resultant carbon nanofiber fabrics prepared at 900 °C were flexible with a relatively high S_{BET} of c. 500 m^2/g. Addition of a high M_w PVB (340,000) markedly improved the spinnability of phenolic resin solution owing to decreasing solution viscosity, and an additional electrolyte (pyridine or Na_2CO_3) allowed production of thinner fibers because of increasing electrical conductivity of the precursor solution; 0.1 mass% Na_2CO_3 resulting in carbon nanofibers with an average diameter of 110 nm and S_{BET} of 790 m^2/g [38]. Microporous carbon nanofibers with V_{micro} of 0.9 cm^3/g were prepared from novolac-type phenol-formaldehyde (PF) by adding PVB and Na_2CO_3, followed by carbonization at 800 °C without activation [39]. Electrical conductivity of these carbon nanofiber fabrics was 5.29 S/cm. Carbon nanofibers with a narrow pore-size distribution of 0.4–0.7 nm were prepared from novolac-type phenolic resin via electrospinning of its methanol solution, curing in a formaldehyde/HCl solution, and carbonization at 800 °C. Even though no activation was applied, the nanofibers had S_{BET} of 812 m^2/g and V_{total} of 0.91 cm^3/g.

8.2 Applications

8.2.1 Electrode materials for electrochemical capacitors

In commercially available electrochemical capacitors, activated carbons are commonly used as electrodes. For the improvement of capacitive performance, various carbon materials have been proposed, including activated carbon fibers, templated carbons, and carbon nanotubes [40]. Pore structure of electrospun carbon nanofibers was controlled for the application as the electrode for electric double-layer capacitors (EDLCs), and fine metal particles were loaded to the nanofibers via electrospinning in order to add some pseudocapacitance. An advantage of electrospinning is the ability to produce webs; the web shape makes activation and electrode formation easy.

Webs of PAN-based carbon nanofibers, which were prepared by electrospinning, stabilization, carbonization, and activation in a flow of N_2 containing 30 vol% steam, were used as electrodes of an EDLC in 30 mass% KOH aqueous solution [41]. The 700 °C-activated webs gave the highest capacitance of 173 F/g at low discharge current density of 10 mA/g, but at a high current density of 1000 mA/g, the 800 °C-activated webs gave the highest capacitance, 120 F/g. The former was micropore-rich and had S_{BET} of 1230 m^2/g, but the latter was mesopore-rich and had S_{BET} of 850 m^2/g. Similar results have been reported [42]. The importance of the presence of mesopores has been reported in many papers, using different carbon materials [40]. PAN-based carbon nanofibers prepared by mixing PAN with 15 mass% cellulose acetate gave a microporous surface area, S_{micro}, of 919 m^2/g and mesoporous surface area, S_{meso}, of 241 m^2/g, and consequently showed a capacitance of 245 F/g in 6 mol/L KOH aqueous solution at 1 mA/cm^2 [43]. Electrospun PAN-based carbon nanofibers, which were prepared from a PAN/DMF solution containing PVP followed by carbonization at 970 °C in N_2 and activation at 800 °C in CO_2, were tested in 1 mol/L H_2SO_4 aqueous solution [15]. Side-by-side bicomponent carbon nanofibers prepared by using 8 mass% PVP gave a capacitance of 221 F/g, higher than the capacitance of carbon nanofibers prepared from a blended solution of PVP with PAN.

An MWCNT-dispersed PAN/DMF solution was successfully spun and carbonized to get webs consisting of carbon nanofibers that gave improved EDLC performance in aqueous electrolytes [44,45]. The addition of 3 mass% MWCNTs to the precursor PAN increased S_{BET} to 1170 m^2/g, electrical conductivity to 0.98 S/cm, and consequently increased EDLC capacitance to 180 F/g in 6 mol/L KOH, and fiber diameter decreased to about 230 nm [44]. Coating of polypyrole (PPy) on these MWCNT-embedded nanofibers led to further increase in capacitance to 333 F/g [44]. Addition of MWCNTs in carbon nanofibers prepared by electrospinning, carbonization at 700 °C, and activation at 650 °C in H_2O_2 vapor was shown to be effective for increasing electrical conductivity and consequently increasing capacitance: electrical conductivity increased from 0.86 to 5.32 S/cm and capacitance in 1 mol/L H_2SO_4 from 170 to 310 F/g [45]. Control of pore structure during carbonization of pitch/PAN nanofibers has been tried by mixing tetraethoxy orthosilicate and $AgNO_3$ into PAN [46–48].

EDLC performance has been studied with the carbon nanofiber webs prepared by electrospinning of a DMAc solution of PBI [49–51]. Dependence of capacitance in 1 mol/L H_2SO_4 aqueous electrolyte on discharge current density is shown for webs activated at different temperatures in Figure 8.5 [51]. A capacitance of about 202 F/g at discharge current density of 1 mA/cm^2 was measured on the web, which was activated at 800 °C and had S_{BET} of 1220 m^2/g, V_{micro} of 0.71 cm^3/g, and V_{meso} of 0.20 cm^3/g.

PI-derived carbon nanofiber webs have also been tested as electrode materials in EDLCs using 30 mass% KOH aqueous solution, and showed a relatively high capacitance of 175 F/g at a high current density of 1000 mA/cm^2 [32]. The precursor for spinning was PMDA/ODA dissolved in a mixed solution of THF and methanol (8/2 in mass) and the webs obtained after spinning were thermally imidized by heating to 350 °C. PI webs were carbonized at 1000 °C and then activated in air flow containing 40 vol% steam in the temperature range of 650–800 °C. All activated webs gave peculiar behavior: capacitance increased with increasing current density up to 10 mA/cm^2 and then decreased.

Addition of inorganic salts into the precursor PAN influenced the pore structure of the resultant carbon nanofibers [52–55]. Electrospinning from PAN/DMF solutions containing 1–15 mass% $ZnCl_2$, followed by stabilization in air and carbonization at 700 or 800 °C, resulted in porous carbon nanofibers without any activation process after removing Zn compounds by HCl [52,55]. The nanofibers became microporous and showed S_{BET} of 550 m^2/g and, as a consequence, EDLC capacitance in 6 mol/L

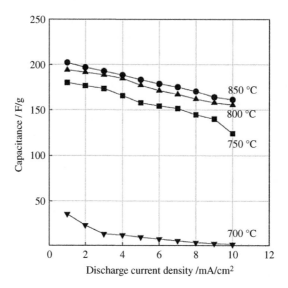

FIGURE 8.5 Dependence of Electric Double-layer Capacitance on Discharge Current Density for Polybenzimidazol-derived Carbon Nanofiber Webs Heat-treated at Different Temperatures

From [51]

KOH electrolyte was improved [52]. Addition of 5 mass% $Ni(NO_3)_24H_2O$ into a PAN/DMF solution resulted in an increase in S_{BET}, mainly owing to the increase in mesopores, without any activation process after carbonization [54].

The principle of EDLCs, i.e. physical adsorption/desorption of ions in solution, has been applied to water purification (capacitive deionization) by using different carbon materials [56]. Electrochemical adsorption capacity of Na^+ of electrospun PAN-based carbon fiber webs, which was measured in NaCl aqueous solution, is compared with various carbon materials in Figure 8.6 [57]. They showed a capacity of 3.2 mg/g, which was comparable with those for carbon nanotube films and carbon aerogels.

In order to gain pseudo-capacitance in addition to electric double-layer capacitance, loading of Ru on electrospun carbon nanofibers was carried out [58]. By the addition of Ru acetylacetonate into PAN/DMF solution, metallic Ru particles with 2–15 nm size were embedded in the nanofibers. Nanofibers loaded by 7.31 mass% Ru showed a capacitance of 391 F/g in 6 mol/L KOH aqueous solution, although the nanofibers without Ru loading gave 140 F/g. Carbon nanofibers containing nanoparticles of metallic Ni were prepared via electrospinning of a PAN/DMF solution with addition of 35, 100, or 135 mass% Ni acetate and carbonization at 1000 °C [59]. The fibers obtained had a diameter of about 150 nm and fine Ni particles were deposited on the surface. The nanofiber webs, with 22.4 mass% Ni loading, showed a capacitance of 164 F/g in 6 mol/L KOH electrolyte at a current density of 250 mA/g, which was much larger than the ~50 F/g for nanofiber webs without Ni loading. Embedding of

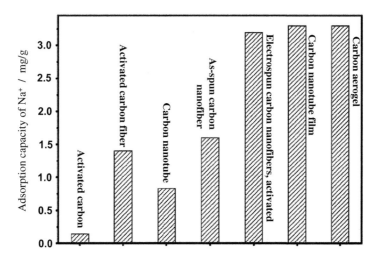

FIGURE 8.6 Electrochemical Adsorption Capacity of Na Ions in NaCl Solution

From [57]

metallic V into PAN-based carbon nanofibers was performed by electrospinning of a PAN/DMF solution containing V_2O_5 [60]. After activation using KOH at 800 °C, S_{BET} reached 2800 m^2/g, owing to micropores of 0.7–1.2 nm and mesopores of 2–4 nm. These nanofibers showed a capacitance of 107 F/g at a current density of 20 A/g in 1.0 mol/L $(C_2H_5)_4NBF_4/PC$ electrolyte.

8.2.2 Anode materials for lithium-ion rechargeable batteries

Carbon nanofiber webs prepared via electrospinning have an advantage in the application as an anode of lithium-ion rechargeable batteries (LIBs), because there is no necessity to use an electrically conductive additive, such as acetylene black, or organic binder, such as polytetrafluoroethylene (PTFE). However, low irreversible capacity like that of natural graphite, which is currently used in commercially available batteries, cannot be expected in carbon nanofibers, because most of them are not graphitized sufficiently. A few papers have reported the results of application to anodes of LIBs, but high irreversible capacities could not be avoided.

PAN-based carbon nanofiber webs have been prepared by electrospinning of a PAN/DMF solution [61]. On 1000 °C-treated webs, a reversible discharge capacity, C_{dis}, of 450 mAh/g was obtained at a current density of 30 mA/g, a little higher than that of natural graphite, but irreversible capacity, C_{irr}, was as high as 500 mAh/g. High-temperature treatment of the webs could not improve LIB performance. Pores were introduced into carbon nanofibers by adding poly-L-lactic acid (PLLA) to a PAN/DMF solution [62,63]. Addition of PLLA created micropores without an activation process, and increased C_{dis} to 435 mAh/g after the fiftieth cycle at 50 mA/g [63]. Addition of minute fumed-SiO_2 particles to the precursor PAN/DMF solution

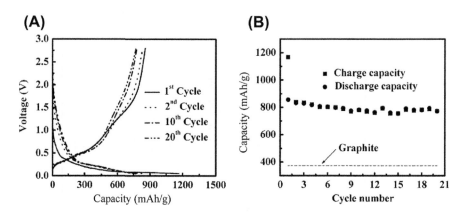

FIGURE 8.7 Lithium-ion Rechargeable Battery Performance of Carbon Nanofiber Webs Prepared From 15 mass% Si-dispersed PAN/DMF Solution

(A) Charge-discharge curves, and (B) change in capacities with cycle

From [65]

was carried out in order to create pores after leaching out of SiO_2, but it showed a low S_{BET} of 92 m^2/g and a very high C_{irr}, more than 1000 mAh/g, at the first cycle [64].

Much work applying electrospinning is aimed at loading electrochemically active metallic particles to carbon nanofibers in order to improve LIB performance. Si-loaded carbon nanofibers were prepared from PAN/DMF solutions with Si nanoparticles (c. 70 nm) [63,65–67] and also from a PVA/H2O solution containing Si nanoparticles (c. 40 nm) dispersed with a surfactant [68]. Galvanostatic charge-discharge curves and cycle performance at a current density of 100 mA/g in 1 mol/L $LiPF_6$/ (EC + EMC) (ethylene carbonate + ethyl methyl carbonate) solution are shown in Figure 8.7 for webs prepared from a 15 mass% Si-dispersed PAN/DMF solution by carbonization at 700 °C [65]. C_{dis} for the first cycle is very high at 855 mAh/g, with a relatively large C_{irr} of 312 mAh/g. C_{dis} decreases slowly with cycling, 781 mAh/g after the tenth cycle and 773 mAh/g after the twentieth cycle, C_{irr} becoming small after the second cycle. The C_{dis} obtained after the second cycle is much higher than the theoretical capacity of graphite and also the capacity of carbon nanofiber webs prepared without Si addition.

Sn-loading via electrospinning has been performed by adding a Sn compound into the precursor solution [69–72]. Sn-containing carbon nanofiber webs were prepared from $SnCl_2$ dissolved in PVA/H_2O solution by electrospinning and carbonization in Ar/H_2 (95/5 v/v) [71,72]. The diameter of the resultant nanofibers was about 4 µm and they contained Sn/SnO_x particles of about 20–40 nm in size, most of which were located in pores of the carbon nanofibers, as shown in Figure 8.8A [72]. The webs show a relatively high C_{dis} of 735 mAh/g and relatively large C_{irr} for the first cycle (457 mAh/g), which becomes smaller after the second cycle (Figure 8.8B). Sn-encapsulated carbon nanofibers were prepared from electrospun PAN nanofibers containing tributyltin (TBT), most of which located at the core of the fiber, by

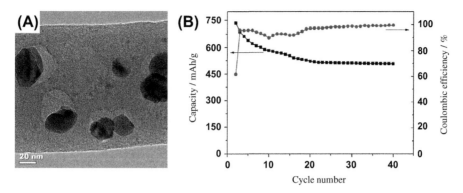

FIGURE 8.8 Sn/SnO$_x$-containing Carbon Nanofibers

(A) TEM image of the fiber, and (B) cycle performance in 1 mol/L LiPF$_6$/(EC + DMC) electrolyte with 30 mA/g

From [72]

heating at 1000 °C in an Ar/H$_2$ atmosphere to carbonize the outer PAN sheath and to decompose the TBT core to metallic Sn [69]. The resultant nanofibers showed high C_{dis} of 737 mAh/g, even after 200 cycles. Sn-encapsulated carbon nanofibers were also prepared from PAN/DMF containing PMMA and tin octoate [70]. The nanofibers had a diameter of about 2 μm and consisted of hollow channels of about 100 nm diameter containing metallic Sn nanoparticles. Particles of Sn were covered by a carbon layer with a thickness of about 5 nm. The nanofibers showed a high C_{dis} of 648 mAh/g with a current density of 100 mA/g, even after 140 cycles.

Loading of Co, Fe, Mn, Ni, and Cu has been carried out, mostly by adding their acetates into PAN/DMF solution [59,73–79]. Carbon nanofibers containing metallic Co, Ni, or Cu nanoparticles delivered relatively high C_{dis} in 1 mol/L LiPF$_6$/(EC + DEC) (ethylene carbonate + diethyl carbonate) electrolyte [59,73,79]. When Mn acetate (Mn(OAc)$_2$) was used, carbon nanofibers containing crystalline MnO and Mn$_3$O$_4$ nanoparticles were obtained after carbonization at 700 °C, which showed steady cycle performance [75,78]. The carbon nanofibers include MnO$_x$ nanoparticles in the matrix, as shown in Figure 8.9A, and give high capacity with good cycle performance; the nanofibers prepared by adding 50 mass% Mn(OAc)$_2$ giving a steady capacity value of about 600 mAh/g, much higher than the nanofibers prepared without Mn(OAc)$_2$, as shown in Figure 8.9B [78]. When Fe(III) acetylacetonate, Fe(acac)$_3$, was added into PAN/DMF solution, Fe$_3$O$_4$-loaded carbon nanofibers were obtained [74]. After carbonization at 600 °C, the resultant carbon nanofibers contained crystalline Fe$_3$O$_4$ particles of about 20 nm in size, of which the content was calculated to be 31 mass%. C_{dis} of the nanofibers decreased rapidly during the first few cycles and then tended to increase with increasing cycle number, giving a higher discharge capacity than that without Fe$_3$O$_4$.

Nanofibers of LiFePO$_4$/C composite were prepared via electrospinning of a PAN/DMF solution containing equimolar LiCO$_2$CH$_3$, Fe(CO$_2$CH$_3$)$_2$, and H$_3$PO$_4$,

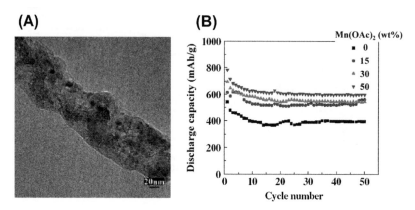

FIGURE 8.9 MnO$_x$-loaded Carbon Nanofibers Prepared by Adding 0, 10, 30, and 50 mass% Mn(OAc)$_2$ into PAN/DMF Solution

(A) TEM image of the fiber, and (B) cycle performance in 1 mol/L LiPF$_6$/(EC + DEC) electrolyte

From [78]

followed by stabilization at 280 °C in air and carbonization at 600–800 °C for different times in Ar [80]. LiFePO$_4$/C composite was obtained as nanofibers with a wide range of diameters of 10–200 nm, together with beads. The nanofibers carbonized at 700 °C gave the highest C_{dis} of 160 mAh/g in 1 mol/L LiPF$_6$/(EC + EMC) solution as the cathode. Cathode material Li$_2$ZnTi$_3$O$_8$ was synthesized in fibrous form by electrospinning of an ethanol solution of tetrabutyl titanate, Zn acetate, and Li acetate with PVP, followed by calcination at 750 °C in air [81].

8.2.3 Catalyst support

Carbon nanofiber webs loaded with nanoparticles of catalyst Pt have been prepared via electrospinning [82–87]. Carbon nanofiber webs prepared from PAN/DMF solution by carbonization at 1200 °C were loaded with Pt by either electrochemical deposition in H$_2$PtCl$_6$ aqueous solution [82] or coating with commercially available Pt/C catalyst by using a water/Nafion mixed solution [83]. The webs consisted of nanofibers with relatively uniform diameters of 130–170 nm with a high electrical conductivity of about 50 S/cm, and a low S_{BET} of about 7 m^2/g. After being loaded with Pt, the webs showed a high catalytic peak current for methanol oxidation, about 420 mA/mg-Pt, which was significantly higher than the 185 mA/mg-Pt of commercially available Pt/C electrodes for fuel cells [82]. Electrochemical deposition of Pt$_x$Au$_{100-x}$ on the electrospun PAN-based carbon nanofibers was done using H$_2$SO$_4$ solution of H$_2$PtCl$_6$ and HAuCl$_4$ [85]. In Figure 8.10, SEM images of Pt$_x$Au$_{100-x}$-loaded carbon nanofibers are shown. Loaded metal particles have uniform size and distribute on the surface homogeneously. The particles deposited from Pt solution have an average size of 225 nm, but those from Pt$_{50}$Au$_{50}$ solution have a slightly smaller

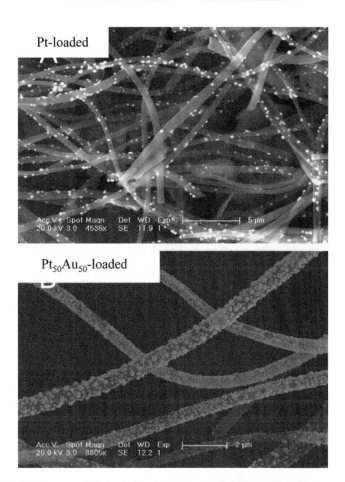

Pt-loaded

Pt₅₀Au₅₀-loaded

FIGURE 8.10 SEM Images of Pt-loaded and Pt$_{50}$Au$_{50}$-loaded Carbon Nanofibers

From [85]

particle size of 118 nm. Efficiency for formic acid oxidation was improved by alloying Au to Pt, Pt$_{50}$Au$_{50}$-loading giving the highest efficiency. Pt deposition was also performed for PAN-based carbon nanofibers after 1-aminopyrene (1-AP) functionalization [87]. Non-covalent functionalization using 1-AP was shown to be effective to deposit small Pt particles homogeneously on the surface of carbon nanofibers; c. 3.2 nm on 1-AP-treated nanofibers without noticeable aggregation, but c. 4.2 nm on oxidized nanofibers with some aggregation and c. 21 nm on as-prepared nanofibers with severe aggregation. The resultant Pt-loaded carbon nanofiber webs using 1-AP funtionalization exhibited higher active surface area and better performance for methanol oxidation than carbon nanofiber webs oxidized using H$_2$SO$_4$/HNO$_3$.

Loading of Pt has been performed on electrospun PAN-based carbon nanofibers in ethylene glycol solution of H$_2$PtCl$_6$ [86]. Pt particles with a size of about 2.6 nm were dispersed uniformly on the nanofiber surface, suggesting high catalytic activity in a polymer electrolyte membrane fuel cell (PEMFC). Pt-loading was also performed on electrospun PI nanofibers after hydrolysis in KOH using acetone solution of $Pt(acac)_2$, followed by carbonization at 900 °C in Ar [84]. Hydrolysis of the surface of PI nanofibers was experimentally shown to be effective for immobilization of the Pt nanoparticles, which were uniformly distributed on the surface and had a size of about 10 nm. By electrospinning of lignin/ethanol solutions containing 0.2 and 0.4 mass% $Pt(acac)_2$, followed by stabilization in air and carbonization at 600–1000 °C, Pt-loaded microporous carbon nanofibers were obtained [88]. Pd-loaded carbon nanofibers were prepared from a PAN/DMF solution containing 4.8 mass% $Pd(OAc)_2$ by electrospinning, stabilization by heating in steps from 230 to 300 °C, and carbonization at 1100 °C [89]. The resultant nanofibers showed high electrocatalytic activity for the reduction of H_2O_2.

8.2.4 Composite with carbon nanotubes

Various CNTs have been included in carbon nanofibers via electrospinning for the reinforcement of their mechanical properties [8,90,91]. MWCNTs were successfully introduced into electrospun carbon nanofibers prepared from a PAN/DMF solution in which 2–35 mass% MWCNTs were dispersed [91]. In as-spun nanofibers, most MWCNTs are well aligned along the fiber axis, but slightly curved MWCNTs resulted in curved nanofibers, and winding or helical MWCNTs could not be embedded in the nanofibers. PAN nanofibers containing 5 mass% MWCNTs showed a tensile strength of 80 MPa and tensile modulus of 3.1 GPa. The fibrous morphology was maintained after carbonization at 850 °C, even though the PAN shrunk a great deal (but MWCNTs did not). Carbon nanofibers were prepared via electrospinning from a PAN/DMF solution in which 1–4 mass% SWCNTs were dispersed [90]. SWCNTs were distributed in parallel to the fiber axis, maintaining their straight shape, in PAN nanofibers with a diameter of 50–200 nm, suggesting that reinforcement of carbon nanofibers by adding SWCNTs is possible.

Carbon nanotubes have been grown on the surface of electrospun carbon nanofibers [92]. Fe-loaded carbon nanofibers were prepared from PAN/DMF solutions containing 3.3 and 6.7 mass% $Fe(acac)_3$ by electrospinning, stabilization in air, reduction of Fe(III) to metallic Fe at 500–550 °C in H_2, and carbonization at 1100 °C. Metallic Fe particles of 10–20 nm in size were dispersed in the nanofibers. On the surface of the carbon nanofibers thus prepared, MWCNTs were grown by catalytic CVD of hexane, as shown in Figure 8.11.

8.3 Concluding remarks

8.3.1 Carbon precursors

The diameter of electrospun carbon nanofibers is 100 to 1000 nm, which might have some influence on the formation of nanotexture during their carbonization process and further structural modification at high temperatures. Size of carbon layers, evaluated

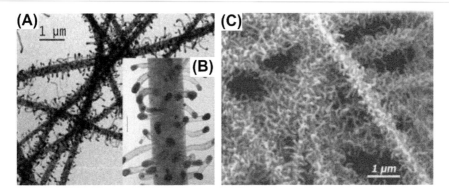

FIGURE 8.11 Electrospun Carbon Nanofibers with Multi-walled Carbon Nanotubes Grown on the Surface

(A) TEM images, and (B) SEM image

From [92]

as *L*a by X-ray diffraction (XRD), was experimentally shown to depend strongly on the particle size of spherical carbon materials [93]. PAN-derived carbon fibers prepared via melt-spinning, stabilization, and carbonization have a random nanotexture and are non-graphitizing (called general-purpose grade). In order to improve the nanotexture to axial orientation, stretching of the as-spun fibers is known to be necessary during further treatment, and the resultant PAN-based carbon fibers have greater strength and higher Young's modulus (high-performance grade). Carbon prepared in the gallery of layer-structured montmorillonite (template carbonization) is easily converted to graphite at temperatures as high as 2800 °C, because the pyrolysis of PAN is performed in a monolayer of carbon atoms between the template layers [94]. Therefore, the formation of nanotexture in nano-sized fibrous carbons (carbon nanofibers) and their change in structure and properties with high-temperature treatment might be a little different from our knowledge based on large-sized particles, larger than micrometer size, on carbons with different nanotextures. PAN-based continuous nanofibers have been prepared under tension during stabilization and carbonization with the expectation of achieving superior mechanical properties [20]. However, no structural studies have been reported yet and the heat treatment has been done only up to 1500 °C.

Stabilization of PAN fibers with micrometer-sized diameters is known to be the most important process in the production of carbon fibers. In most of the literature on PAN-based carbon nanofibers, however, as-electrospun PAN fibers were stabilized at 280 °C in air and carbonized in a temperature range of 600–1100 °C in an inert atmosphere. It may be necessary to explore the optimum conditions for stabilization and carbonization from the structural view point, parameters such as the size and orientation of BSUs of carbon layers in nanofibers; carbonization yield; the content of foreign atoms, such as hydrogen, nitrogen, and oxygen; and changes in diameter distribution.

Aromatic PIs are interesting carbon precursors because they give a wide variety in structure, from well-crystallized graphite film to amorphous carbon with random nanotexture, after high-temperature treatment, according to the selected combination of starting dianhydride and diamine [95]. However, one kind of PI, PMDA/ODA, has been predominantly used for electrospinning [32,34,84]. The thickness of the PMDA/ODA film is known to govern the graphitizability: thin films of less than 25 μm can be converted to highly crystalline graphite by high-temperature treatment, but thicker films cannot, which is explained by the orientation of precursor molecules in the film [95]. Studies using different PIs coupled with heat treatment at high temperatures and detailed structural analyses may give a wider variety of nanotextures in the resultant carbon nanofibers and, consequently, may open a wider range of applications, because of the possibility to achieve a controlled structure, from graphite to amorphous carbon. For applications requiring high electrical conductivity, PIs are worthwhile trying as precursors.

Pitch is less costly than other carbon precursors, such as PAN and PI, but its use for electrospinning is less reported in the literature. Since pitch is a mixture of various hydrocarbons, certain additional components might be necessary for electrospinning according to the solvent selected, as in the case for the DMF-insoluble part of pitch with THF [27]. Mesophase pitch, which is used as a precursor for commercial carbon fibers, is worthwhile using for electrospinning, even though a stabilization process is needed to get pitch-based nanofibers, which may lead to lower graphitizability. Although PVA has a very low carbonization yield (a few mass%), it might be an interesting precursor because it is water soluble [68,71,72]. In 2012, an aqueous solution of PVA-dispersed mesophase pitch was used for electrospinning to prepare carbon nanofibers [96]. Phenolic resins, which give non-graphitizing carbon in the conventional carbonization process, are often used as carbon precursors owing to an advantage of relatively high carbonization yield. However, they are very rarely used in studies on carbon nanofibers via electrospinning [37–39].

8.3.2 Pore-structure control

In most cases, electrospun carbon nanofibers have been prepared in the form of mats (webs), which is an advantage in various applications, such as electrodes for EDLCs and LIBs, as mentioned above. In the webs consisting of carbon nanofibers, different kinds of pores are formed: large pores (they may be better called spaces) including macropores (>50 nm in size), in addition to mesopores (2–50 nm), and micropores (<2 nm) in nanofibers.

Large spaces mainly due to entanglement of carbon nanofibers are very difficult to control and characterize because they are flexible (flexible interparticle pores). Bulk density or porosity of the webs is the parameter that is easy to determine and effective in characterization of these large spaces. However, information on the bulk density or porosity has not been presented in most of the literature reporting on electrospun carbon nanofiber webs. It has to be emphasized that the spaces in the webs are very important for their applications: they are pathways in energy storage devices for electrolytes to diffuse into smaller pores to be adsorbed, but too large an amount of pores makes the volumetric storage capacity small.

Most micropores and some mesopores are intrinsically formed in fibrous carbon particles during the carbonization process owing to evolution of different gaseous species, such as hydrocarbons with various sizes, CO, CO_2, etc., depending strongly on the carbon precursor used and carbonization conditions. To increase and control the amount of micropores, the conventional activation process has been applied by using either steam, KOH, or H_3PO_4 as an activation reagent for electrospun nanofibers. The addition of $ZnCl_2$, which is one of the reagents for conventional activation, into a PAN/DMF solution before electrospinning has also been adopted, but no pronounced improvement of S_{BET} was obtained [52]. In order to prepare the electrodes of electrochemical capacitors, an activation process has been applied on electrospun carbon nanofibers derived from PAN [41,42,57], PBI [49–51], and PI [32]. By the activation of PAN-based carbon nanofibers in a flow of N_2 containing 30 mass% steam at 800 °C, S_{BET} increased to 1160 m^2/g, V_{total} to 0.64 cm^3/g, and capacitance to 134 F/g in 6 mol/L KOH aqueous electrolyte at 1 mA/cm^2 [42]. By the activation of PBI-derived carbon nanofibers at 800 °C, capacitance of 202 F/g in 1 mol/L H_2SO_4 aqueous electrolyte was obtained with good rate performance [51]. Also, the addition of cellulose acetate into PAN has been tried to control pore structure without an activation process, giving increases in S_{BET} from 740 to 1160 m^2/g and in capacitance from 141 to 245 F/g in 6 mol/L KOH at 1 mA/cm^2 [43]. By the activation of carbon nanofibers, micropores are thought to be formed on the surface, and they are expected to be advantageous for adsorption/desorption. With commercially available activated carbon fibers (ACFs), it is well known that micropores on the surface are advantageous for gas adsorption/desorption. Reported good rate performance of electrochemical capacitors using electrospun carbon nanofiber webs [42,51] is probably due to the formation of micropores on the nanofiber surface, but the experimental evidence has not yet been reported.

Mesopores can be increased by applying a conventional activation process to the electrospun carbon nanofibers, but this is possible only by sacrificing micropores. It might be interesting to add a template, either block copolymer surfactants or MgO precursors, into the electrospinning precursor solution, that could leave mesopores after carbonization [97].

Most macropores and some mesopores have to be introduced into fibrous particles intentionally, by using either an additional carbon precursor that gives pores after carbonization because of its low carbonization yield, such as PMMA in PAN to introduce tubular macropores [13], or an additional solvent for the electrospinning solution, such as water in DMF [36].

In EDLCs, carbon materials composing electrodes are the most important component for accepting electrolyte ions by physical adsorption, forming electric double-layers to store electric energy [21]. Since micro- and mesopores are known to work effectively for adsorption of electrolyte ions, this is the reason why the carbon materials are used as an electrode material; it was considered that high S_{BET} is advantageous to give high capacitance. But the diffusion of ions to micropore and/or mesopore surfaces to be adsorbed is also known to have a strong influence on capacitive performance, particularly the charge/discharge process at high rates, which is reasonably supposed to be governed by pore structure, i.e. not only pore sizes but also connection of the pores. The presence of mesopores, in addition to micropores, has been pointed out to be important for achieving good rate performance; in other words, for having high retention in capacitance at high-rate charging/discharging. One of the merits of electrospinning is the direct formation of webs consisting of carbon nanofibers, because no binder or electroconductive additives (commonly carbon blacks) are needed to form electrode sheets, and the presence of large spaces between nanofibers may be advantageous for the diffusion of electrolyte ions to the micropores and mesopores in nanofibers. However, it should be mentioned that the webs usually have low bulk density, which may lead to low volumetric capacitance.

8.3.3 Improvement of electrical conductivity

High electrical conductivity of electrospun carbon nanofibers is required in various applications, such as in electrode materials for EDLCs, LIBs, and catalyst supports.

As the anode of LIBs, carbon nanofiber webs are required to have both high electrical conductivity and well-developed graphite structure. High-temperature treatment is known to be essential to improve these two factors. However, graphitizability of a carbon depends strongly on the carbon precursor used; for example, PAN-based carbon fibers cannot reach a high degree of graphitization but mesophase-pitch-based carbon fibers can be graphitized. Since the stabilization process is essential to keep fibrous morphology after carbonization for the preparation of carbon nanofibers from PAN and pitch, the possibility of forming highly graphitized nanofibers is lower. PAN-derived carbon nanofibers were not graphitized even by heat treatment at 2800 °C, giving d_{002} of 0.341 nm and electrical conductivity of 20 S/cm, and C_{dis} was as low as about 120 mAh/g in 1 mol/L LiClO$_4$/(EC + DEC) electrolyte [61]. Since the nanofibers have a low graphitization degree, large C_{irr} could not be avoided; 1000 °C-treated nanofibers showed C_{dis} of c. 450 mAh/g and C_{irr} of c. 500 mAh/g in the first charge/discharge cycle. On the other hand, PIs in very thin films are known to show very high graphitizability after high-temperature treatment without stabilization and to become thin graphite film [95]. Carbon nanofibers prepared from PMDA/ODA and heat-treated at a high temperature of 3000 °C are worthwhile testing as the anode of LIBs.

High-temperature treatment is known to improve conductivity not only by the development of graphitic structure but also by exclusion of foreign atoms in carbon materials; however, it is also known to reduce mesopores and micropores [97,98], which is not desirable for EDLC applications. Embedding of MWCNTs was performed in order to increase electric conductivity of electrospun carbon nanofibers [44,45]. Embedding of 0.8 mass% MWCNTs led to an increase in electrical conductivity of the webs from 0.86 to 5.32 S/cm, which resulted in an increase in EDLC capacitance from 170 to 310 F/g in 1 mol/L H$_2$SO$_4$ aqueous electrolyte [45].

In order to control the electrical conductivity of carbon nanofibers prepared via electrospinning, more fundamental study on the changes in structure and nanotexture with heat treatment over a wide range of temperatures is needed, for a wide range of diameters of nanofibers prepared from various carbon precursors, including PAN, pitch, and PIs.

8.3.4 Loading of metallic species

One of the requirements for electrospun nanofibers to be applied to energy storage devices is uniform loading of metallic nanoparticles. There are two loading processes: one is the addition of metal precursors to carbon precursors before electrospinning, and the other is the modification of the electrospun fiber surface with metal precursors before or after carbonization. By the former process, metallic nanoparticles are embedded and immobilized in carbon nanofibers, but some proportion of nanoparticles might be lost from the fiber surface during carbonization, and particles placed in the center of fibers might be inactive. By the latter process, they are preferentially deposited on the fiber surface, but immobilization and grain-growth inhibition of the deposited nanoparticles have to be taken into consideration.

For LIBs, loading of various metallic nanoparticles that are electrochemically active for lithium storage is carried out by mixing the precursors, with or without organic additives, into the carbon precursor solution before electrospinning, as described in the previous section. Most of the Sn/SnO_x nanoparticles formed in carbon nanofibers, which were electrospun from PVA/H_2O solution containing $SnCl_2$, were located in pores, as shown in Figure 8.8A. Encapsulation of Sn/SnO_x nanoparticles is the reason why the nanofiber webs show stable cycle performance (Figure 8.8B): pore space neighboring Sn/SnO_x particles may absorb large volume expansion due to alloying with lithium [72]. Carbon nanofibers with encapsulated Sn nanoparticles (c. 100 nm in size) via electrospinning using a co-axial spinneret showed a steady cycle performance at high C_{dis} of 737 mAh/g (91% of theoretical capacity) [69]. Electrospinning of a carbon precursor with a Sn precursor can be done easily and effectively compared with the Sn-deposition after carbonization, because Sn and/or SnO_x nanoparticles included in the carbon matrix can be active in LIBs; in other words, lithium ions in LIBs can penetrate into carbon nanofibers to react with encapsulated metal particles. A cathode material for LIBs, $LiFePO_4$, was loaded into carbon nanofibers via electrospinning [80], but it did not show any advantage of the electrospinning process, probably because appropriate spinning conditions had not been selected.

In order to gain pseudo-capacitance due to redox reaction for EDLC of carbon nanofiber webs, loading of metallic species to carbon nanofibers via electrospinning was carried out using $Ru(acac)_3$ [58] and V_2O_5 [60]. By the addition of 20 mass% $Ru(acac)_3$ into a PAN/DMF solution, nano-sized Ru particles (2–15 nm) were embedded, and the capacitance in 6 mol/L KOH electrolyte increased from 140 to 391 F/g by embedding 7.31 mass% Ru, about 250 F/g being pseudo-capacitance [58]. In contrast to LIBs, metallic species have to be located on the surface and the species included in the carbon matrix are difficult work effectively, because most of the electrolyte ions in capacitors cannot penetrate into the carbon matrix, which may reduce the utilization efficiency of the loaded metal.

For the application of electrospun carbon nanofibers as catalyst supports, particularly for fuel cells, it has been attempted to load catalyst particles on the nanofiber surface. In most cases, immobilization of catalyst nanoparticles is crucial to success. When PI was selected as carbon precursor, hydrolysis of as-spun PI nanofibers in KOH prior to Pt deposition from Pt(acac)$_2$, followed by carbonization, was very effective for homogeneously loading Pt particles as small as 100 nm [84]. When PAN was used, surface treatment with 1-aminopyrene (1-AP) after carbonization of nanofibers was reported to be effective to deposit fine particles of Pt in H$_3$PtCl$_6$/ethylene glycol solution [87].

References

[1] Oberlin A, Endo M, Koyama T. J Cryst Growth 1976;32:335–49.
[2] Iijima S. Nature 1991;354:56–8.
[3] Iijima S, Ichihashi T. Nature 1993;363:603–5.
[4] Bethune DS, Kiang CH, Devries MS, et al. Nature 1993;363:605–7.
[5] Endo M, Muramatsu H, Hayashi T, et al. Nature 2005;433:476.
[6] Inagaki M, Yang Y, Kang F. Adv Mater 2012;50:3247–66.
[7] Zussman E, Chen X, Ding W, et al. Carbon 2005;43:2175–85.
[8] Zhou Z, Lai C, Zhang L, et al. Polymer 2009;50:2999–3006.
[9] Prilutsky S, Zussman E, Cohen Y. Nanotechnology 2008;19:165603.
[10] Shim WG, Kim C, Lee JW, et al. J Appl Polym Sci 2006;102:2454–62.
[11] Nataraj SK, Yang KS, Aminabhavi TM. Prog Polym Sci 2012;37:487–513.
[12] Zussman E, Yarin AL, Bazilevsky AV, et al. Adv Mater 2006;18:348–53.
[13] Kim C, Jeong Y, Ngoc B, et al. Small 2007;3:91–5.
[14] Zhang ZY, Li XH, Wang CH, et al. Macromol Mater Eng 2009;294:673–8.
[15] Niu H, Zhang J, Xie Z, et al. Carbon 2011;49:2380–8.
[16] Wang Y, Serrano S, Santiago-Aviles JJ. J Mater Sci Lett 2002;21:1055–7.
[17] Wang Y, Santiago-Aviles JJ. J Appl Phys 2003;94:1721–7.
[18] Kang SC, Im JS, Lee Y-S. Carbon Lett 2011;12:21–5.
[19] Arshad SN, Naraghi M, Chasiotis I. Carbon 2011;49:1710–9.
[20] Liu J, Yue Z, Fong H. Small 2009;5:536–42.
[21] Zhang W, Liu J, Wu G. Carbon 2003;14:2805–12.
[22] Chen IH, Wang CC, Chen C Y. Carbon 2010;48:604–11.
[23] Im JS, Park SJ, Lee Y S. J Coll Interf Sci 2007;314:32–7.
[24] Im JS, Kwon O, Kim YH, et al. Microp Mesop Mater 2008;115:514–21.
[25] Oh GY, Ju YW, Jung HR, et al. J Anal Appl Pyrolysis 2008;81:211–7.
[26] Park SH, Kim C, Choi YO, et al. Carbon 2003;41:2655–7.
[27] Park SH, Kim C, Yang KS. Synth Met 2004;143:175–9.
[28] Bui N-N, Kim B-H, Yang KS, et al. Carbon 2009;47:2538–9.
[29] Kim B-H, Wazir AH, Yang KS, et al. Carbon Lett 2011;12:70–80.
[30] Rose M, Kockrick E, Senkovska I, et al. Carbon 2010;48:403–7.
[31] Yang KS, Edie DD, Lim DY, et al. Carbon 2003;41:2039–46.

[32] Kim C, Choi YO, Lee WJ, et al. Electrochim Acta 2004;50:883–7.

[33] Chung GS, Jo SM, Kim BC. J Appl Polym Sci 2005;97:165–70.

[34] Xuyen NT, Ra EJ, Geng HZ, et al. J Phys Chem B 2007;111:11350–3.

[35] Kim C, Cho YJ, Yun WY, et al. Solid State Commun 2007;142:20–3.

[36] Yang Y, Centrone A, Chen L, et al. Carbon 2011;49:3395–403.

[37] Suzuki K, Matsumoto H, Minagawa M, et al. Polymer J 2007;39:1128–34.

[38] Imaizumi S, Matsumoto H, Suzuki K, et al. Polymer J 2009;41:1124–8.

[39] Wang MX, Huang ZH, Kang F, et al. Mater Lett 2011;65:1875–7.

[40] Inagaki M, Konno H, Tanaike O. J Power Sources 2010;195:7880–903.

[41] Kim C, Yang KS. Appl Phys Lett 2003;83:1216–8.

[42] Kim C, Yang KS, Lee WJ. Electrochem Solid State Lett 2004;7:A397–9.

[43] Ju YW, Park SH, Jung HR, et al. J Electrochem Soc 2009;156:A489–94.

[44] Ju YW, Choi GR, Jung HR, et al. Electrochim Acta 2008;53:5796–803.

[45] Guo QH, Zhou XP, Li XY, et al. J Mater Chem 2009;19:2810–6.

[46] Kim BH, Yang KS, Woo H G. Electrochem Commun 2011;13:1042–6.

[47] Kim BH, Yang KS, Woo H G. J Nanosci Nanotech 2011;11:7193–7.

[48] Kim BH, Yang KS, Kim YA, et al. J Power Sources 2011;196:10496–501.

[49] Kim C, Park SH, Lee WJ, et al. Electrochim Acta 2004;50:877–81.

[50] Kim C, Kim JS, Kim SJ, et al. J Electrochem Soc 2004;151:A769–73.

[51] Kim C. J Power Sources 2005;142:382–8.

[52] Kim C, Ngoc BTN, Yang KS, et al. Adv Mater 2007;19:2341–6.

[53] Li J, Liu EH, Li W, et al. J Alloys Compd 2009;478:371–4.

[54] Nataraj SK, Kim BH, Yun JH, et al. Mater Sci Eng B 2009;162:75–81.

[55] Ji LW, Zhang XW. Electrochem Commun 2009;11:684–7.

[56] Oren Y. Desalination 2008;228:10–29.

[57] Wang M, Huang ZH, Wang L, et al. New J Chem 2010;34:1843–5.

[58] Ju YW, Choi GR, Jung HR, et al. J Electrochem Soc 2007;154:A192–7.

[59] Ji LW, Lin Z, Medford AJ, et al. Chem-A Eur J 2009;15:10718–22.

[60] Im JS, Woo SW, Jung MJ, et al. J Coll Interf Sci 2008;327:115–9.

[61] Kim C, Yang KS, Kojima M, et al. Adv Func Mater 2006;16:2393–7.

[62] Ji LW, Zhang XW. Electrochem Commun 2009;11:1146–9.

[63] Ji LW, Zhang XW. Nanotechnology 2009;20:155705.

[64] Ji LW, Lin Z, Medford AJ, et al. Carbon 2009;47:3346–54.

[65] Ji LW, Jung KH, Medford AJ, et al. J Mater Chem 2009;19:4992–7.

[66] Ji L, Zhang X. Carbon 2009;47:3219–26.

[67] Choi HS, Lee JK, Lee HY, et al. Electrochim Acta 2010;56:790–6.

[68] Fan X, Zou L, Zheng YP, et al. Electrochem Solid State Lett 2009;12:A199–201.

[69] Yu Y, Gu L, Wang CL, et al. Angew Chem Int Ed 2009;48:6485–9.

[70] Yu Y, Gu L, Zhu CB, et al. J Am Chem Soc 2009;131:15984–5.

[71] Zou L, Gan L, Kang FY, et al. J Power Sources 2010;195:1216–20.

[72] Zou L, Gan L, Lv R, et al. Carbon 2011;49:89–95.

[73] Wang L, Yu Y, Chen PC, et al. Scripta Mater 2008;58:405–8.

[74] Wang L, Yu Y, Chen PC, et al. J Power Sources 2008;183:717–23.

[75] Ji LW, Medford AJ, Zhang XW. J Mater Chem 2009;19:5593–601.

[76] Nataraj SK, Kim BH, Dela Cruz M, et al. Mater Lett 2009;63:218–20.

[77] Nataraj SK, Kim BH, Yun JH, et al. Synth Met 2009;159:1496–504.

[78] Ji LW, Zhang XW. Electrochem Commun 2009;11:795–8.

[79] Ji LW, Lin Z, Zhou R, et al. Electrochim Acta 2010;55:1605–11.

[80] Toprakci O, Ji L, Lin Z, et al. J Power Sources 2011;196:7692–9.
[81] Wang L, Wu L, Li Z, et al. Electrochim Acta 2011;56:5343–6.
[82] Li M, Han B, Yang B. Electrochem Commun 2008;10:880–3.
[83] Li MY, Zhao SZ, Han GY, et al. J Power Sources 2009;191:351–6.
[84] Xuyen NT, Jeong HK, Kim G, et al. J Mater Chem 2009;19:1283–8.
[85] Huang JS, Hou HQ, You TY. Electrochem Commun 2009;11:1281–4.
[86] Park JH, Ju YW, Park SH, et al. J Appl Electrochem 2009;39:1229–36.
[87] Lin Z, Ji LW, Woodroof MD, et al. J Phys Chem C 2010;114:3791–7.
[88] Ruiz-Rosas R, Bedia J, Lallave M, et al. Carbon 2010;48:696–705.
[89] Huang J, Wang D, Hou H, et al. Adv Funct Mater 2008;18:441–8.
[90] Ko F, Gogotsi Y, Ali A, et al. Adv Mater 2003;15:1161–4.
[91] Hou HQ, Ge JJ, Zeng J, et al. Chem Mater 2005;17:967–73.
[92] Hou H, Reneker DH. Adv Mater 2004;16:69–73.
[93] Inagaki M. TANSO 1985; No. 122: 114–121 [in Japanese].
[94] Kyotani T, Sonobe N, Tomita A. Nature 1988;331:331–3.
[95] Inagaki M, Takeichi T, Hishiyama Y, et al. Chem Phys Carbon 1999;26:245–333.
[96] Kitano T, Iwata A, Okino F. TANSO 2012; No. 254: 160–164 [in Japanese].
[97] Inagaki M, Orikasa H, Morishita T. RSC Adv 2011;1:1620–40.
[98] Li Z, Jaroniec M, Lee YJ, et al. Chem Commun 2002:1346–7.

Carbon Foams

One of characteristics of carbon materials is the possibility to have high porosity. High specific surface area and high concentration of micropores in carbon materials has attracted the attention of scientists and engineers, as can be seen in activated carbons [1,2]. More recently, mesopores as well as micropores have been pointed out to be important in the development of specific functionalities, such as in electrode materials for electric double-layer capacitors, as described in Chapter 11 and also in review articles [3–5]. Also, macropores in carbon materials have been reported to have important roles for some applications, such as for heavy oil recovery and sorption of biomedical liquids, as described in Chapter 14.

As a unique carbon material having a variety of spaces and pores, the carbon foam has been prepared by different processes and attempted to be applied to various fields. Carbon foams may be classified into the so-called porous carbons, represented by activated carbon, but these two materials must be differentiated distinctly. Carbon foams are characterized by large pores of micrometer sizes, whereas activated carbons are characterized by nanometer-sized pores, micropores and mesopores. In Figure 9.1, scanning electron microscopy (SEM) images of carbon foam are shown, to differentiate carbon foam from conventional activated carbons and to show the terminology used to characterize carbon foams. Two different types of pores are usually seen in the foam: one is surrounded by a carbon wall, and is called a "cell," and the other exists in the carbon wall connecting two neighboring cells, called a "window" or "connecting pore." The walls of cells consist of two parts, the "ligament" and the "joint." In the carbon wall, there are micropores and mesopores, even macropores in some cases, but these cannot be seen under SEM and are identified usually by N_2-adsorption analysis.

In this chapter, the preparation processes for carbon foam are reviewed, focusing on how cell structure is controlled, and some of their applications are described. Mats (webs) consisting of fibrous carbon materials, which also have an open cell structure, are not included here, but are briefly described in Chapter 10, on nanoporous carbon membranes.

Prerequisite for readers: Chapter 2.7 (Pore development in carbon materials) in *Carbon Materials Science and Engineering: From Fundamentals to Applications*, Tsinghua University Press.

Advanced Materials Science and Engineering of Carbon.

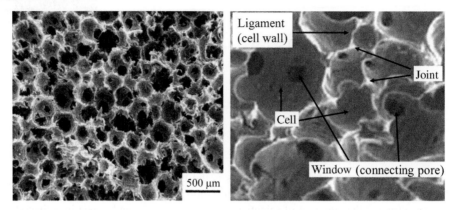

FIGURE 9.1 SEM Images of Carbon Foams, and the Terminology Employed to Characterize them

9.1 Preparation of carbon foams

Carbon materials possessing foam morphology have been prepared via three routes: exfoliation and compaction of graphite, blowing of carbon precursors (mostly pitches) with subsequent carbonization, and template carbonization of carbon precursors.

9.1.1 Exfoliation and compaction of graphite

By compression in a mold, exfoliated graphite (EG) becomes self-standing compact exfoliated graphite foam (EG-foam). EG is now produced on an industrial scale and used, by compressing into flexible graphite sheets, for sealing and packing for gases and liquids [6–9]. Natural graphite flakes are converted to graphite intercalation compounds at room temperature in concentric sulfuric acid, in which a small amount of nitric acid is added as an oxidant. By rapid heating of this intercalation compound to a high temperature, around 1000 °C, EG can be easily obtained. Upon this rapid heating, however, a large amount of corrosive gases, such as SO_x and NO_x, are evolved. Therefore, the intercalation compounds with sulfuric acid are usually changed to so-called residue compounds by washing in water. Residue compounds do not exhibit a regular stage structure, unlike pristine intercalation compounds, and contain much less sulfuric acid intercalates. These residue compounds keep the flaky morphology of the starting graphite and, when they are exposed to a high temperature, they decompose suddenly and exfoliate along the normal to the flakes by changing their morphology to worm-like, as shown in Figures 9.2A and 9.2B. In Figure 9.2C, a cross-sectional view of the worm-like particle is shown, exhibiting the formation of ellipsoidal pores.

In Figure 9.3, exfoliation volume, bulk density, and mass loss of a residue compound are plotted against exfoliation temperature, at which it was kept for 60 s [10,11]. Exfoliation volume and mass loss increase and bulk density decreases markedly with increasing temperature. The change in bulk density covers so large a range that it has

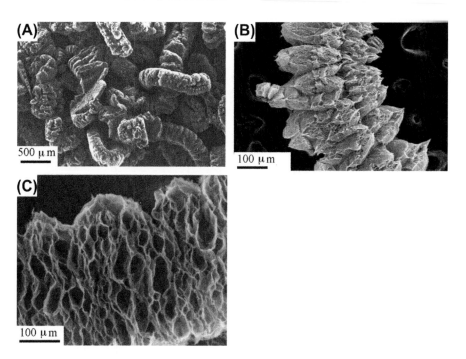

FIGURE 9.2 SEM Images of Exfoliated Graphite (EG)

(A) Aggregates, (B) appearance, and (C) the cross-section of the worm-like particle

to be plotted in logarithmic scale. Abrupt change in these parameters is due to the exfoliation of graphite flakes preferentially along the c-axis to form worm-like particles, which contain a number of ellipsoidal pores inside, as shown in Figure 9.2C, and are entangled with each other, as shown in Figure 9.2A. Changes in these parameters in Figure 9.3 suggest that exfoliation tends to saturate above 800 °C but is not completed even at 1000 °C after 60-s holding.

Pores in EG can be classified into two, those inside worm-like particles and those among the particles, both of which have been quantitatively evaluated by the image-processing technique [9–13]. Pores inside the worm-like particles were characterized by their cross-sectional area, lengths of major and minor axes, aspect ratio, and fractal dimension, on the basis of a large number of pores, more than 2000. Averaged values of these parameters are listed for EG exfoliated at different temperatures in Table 9.1 [9]. With increasing exfoliation temperature, pores inside the worm-like particles grow, increasing the cross-sectional area and lengths of both axes, keeping the aspect ratio and fractal dimension constant. From the distribution histograms of these parameters, pores were found to grow by sacrificing small pores. The pores formed among the worm-like particles became larger with increasing exfoliation temperature, but the volume ratio between the pores among and inside the worm-like particles was almost constant, about 75/25 [13].

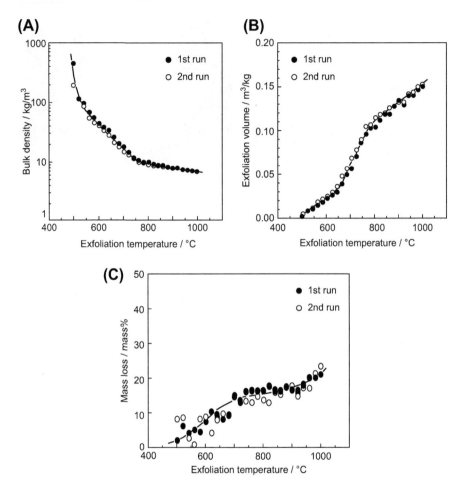

FIGURE 9.3 Changes in Parameters of Exfoliated Graphite as a Function of Exfoliation Temperature

(A) Bulk density, (B) exfoliation volume, and (C) mass loss. Holding time at the temperature was 60 s

Exfoliation behavior of residue compounds depends strongly on the host graphite (crystallinity and size of flakes), the amount of residual sulfuric acid, and holding time at the exfoliation temperature, as well as on the exfoliation temperature itself. Residue compounds with sulfuric acid that can be used for the preparation of EG are commercially available under the name "expandable graphite."

EG can be prepared via various intercalation compounds: compounds with $FeCl_3$ [14], formic acid [15], Na tetrahydrofuran (Na-THF) [16], K-THF [16], and $MoCl_5$-$CHCl_3$ [17]. In EG formed via intercalation compounds with sulfuric acid, a small amount of sulfur remains even after exfoliation, which is not desirable for some

Table 9.1 Average Values of Pore Parameters of Exfoliated Graphite (EG) Prepared at Different Temperatures

Pore Parameters	Exfoliation Temperature (°C)		
	600	800	1000
Bulk density (kg/m³)	40.3	8.8	6.6
Cross-sectional pore area (μm²)	193	217	321
Major axis (μm)	24.4	26.0	31.2
Minor axis (μm)	8.8	9.7	112
Aspect ratio	0.412	0.424	0.412
Fractal dimension	1.09	1.10	1.09
Number of pores used	2583	2161	2059

applications. Therefore, intercalation compounds with formic acid were proposed to be used for EG production, to avoid the presence of sulfur in the resultant EG [15]. Intercalation compounds with Na-THF and K-THF can be synthesized at room temperature and the exfoliation occurs at low temperatures, in addition they do not contain any sulfur [16].

Carbon foams can be prepared from EG by compression in a mold. Bulk density and porosity of the EG-foam are controlled by changing the pressure; bulk densities can be achieved of 0.04 to 0.3 g/cm³, or larger. Compression has been performed on EG powder, prepared using flakes with a size of 420–500 μm from its residue compound of H_2SO_4 and exfoliated at a temperature of 500–800 °C for 2 min in air, to get a constant bulk density of 0.1–0.4 g/cm³ [18]. In Figure 9.4, porosity and thermal conductivity are plotted against bulk density of the EG-foams thus prepared. Porosity of the foam depends strongly on exfoliation temperature; exfoliation at a lower temperature results in lower porosity due to incomplete exfoliation (cf. Figure 9.3). However, thermal conductivity of the foam depends strongly on the bulk density with marked anisotropy, and not on the exfoliation temperature.

9.1.2 Blowing of carbon precursors

Blowing procedures of carbon precursor polymers to prepare carbon foams are divided into two: the first under pressure and the second by using chemical additives. In the former process, the precursors are decomposed to evolve gases and saturate in a closed vessel, and then by rapid reduction of pressure a number of bubbles are nucleated (called the "micro-cellular foaming technique"); this has been applied mainly to various pitches. The latter has been applied on a variety of polymers and blowing chemicals.

Carbon foams have been prepared from pitches via blowing in a pressure chamber, followed by carbonization. For example, a pitch was heated in an autoclave up to a temperature of slightly above the softening point, T_s, at a constant rate, and kept

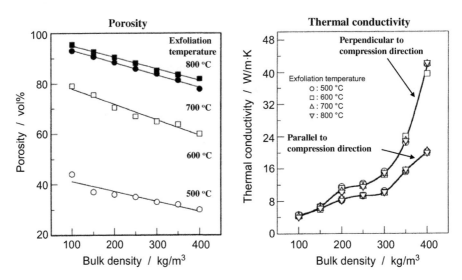

FIGURE 9.4 Porosity and Thermal Conductivity Against bulk Density for EG-foams

From [18]

for a relatively long time. During this heating, the pressure was controlled either by built-up pressure due to volatilized matters from the precursor pitch or by adding an inert gas up to a few MPa. After maintenance under pressure at a predetermined temperature for the required time, the product was cooled down to room temperature and then the pressure was released rapidly. The product was carbonized at around 1000 °C in an inert atmosphere and then graphitized above 2500 °C, if necessary. The so-called stabilization process before carbonization to maintain foam morphology was not necessary when heat treatment was long enough under pressure.

When high thermal conductivity and high electrical conductivity were required for the resulting carbon foams, mesophase pitches were selected as precursors and heat treatment at high temperatures was applied [19–36]. In Figure 9.5, SEM images of the carbon foam prepared from a mesophase pitch are shown. The foam morphology was maintained after the treatment at high temperature. Effects of foaming temperature, pressure, and pressure-releasing time on the properties of carbon foams were studied using a commercially available mesophase pitch with T_s of 283 °C [26]. When foaming temperature was increased from 280 to 300 °C, bulk density and compressive strength of the carbon foams increased from 0.38 to 0.56 g/cm^3 and from 1.47 to 3.31 MPa, respectively, at a constant pressure of 6.8 MPa and releasing time of 5 s. Increase in foaming pressure from 3.8 to 7.8 MPa resulted in increases in bulk density and compressive strength, but increase in releasing time from 5 to 600 s decreased both density and strength. Mesophase-pitch-derived carbon foams, which were prepared under different pressures and had a bulk density of 0.2–0.6 g/cm^3 and cell sizes of around 300 μm, showed high thermal conductivity and high degree of graphitization after 2800 °C-treatment, thermal conductivity of 40–150 W/m·K, d_{002}

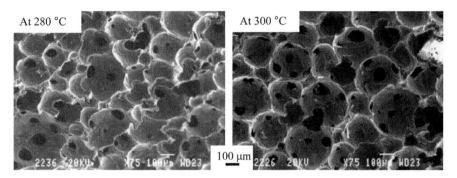

FIGURE 9.5 SEM Images of the Foams Prepared From a Mesophase Pitch at 280 and 300 °C Under 6.8 MPa and a Releasing Time of 5 s

From [26]

spacing of 0.3354 nm, and crystallite size, La, of about 40 nm [19]. Low heating rate for graphitization of foams was reported to be preferable to gain high thermal conductivity and high degree of graphitization [21]. For two mesophase pitches with different T_s of 280 and 330 °C, the bulk density of resultant foams was demonstrated to depend on the T_s of the pitches, as well as pressure during foaming, as shown in Figure 9.6 [31]. The foams prepared from the pitch of low T_s showed better orientation and fewer microcracks in both ligaments and junctions, and less shrinkage during graphitization than those from the pitch of high T_s [36].

Anisotropy in pore morphology was observed in the carbon foams prepared from a mesophase pitch of $T_s = 285$ °C at 500 °C under 3–10 MPa, followed by carbonization at 700–1000 °C, and graphitization at 2300–2800 °C: pores extended parallel to the direction of gravity and led to higher compressive strength and modulus [27]. The detailed nanotexture of the carbon layer alignment in the ligament and junction was analyzed by polarized-light microscopy [28].

Since the bubbles are caused by decomposition gases from the pitches, foaming conditions and properties of the resultant foam depend strongly on the precursor pitch, particularly on its composition. Various pitches, including coal tar pitches with or without a quinoline-insoluble fraction, and petroleum pitches, were used for the preparation of carbon foams at 500 °C under 3 MPa [25]. Foaming was shown to be possible in a mold with a heavy enough lid, the pressure in the mold being controlled by changing the amount of pitch supplied [22]. Extraction of a toluene-soluble fraction from the mesophase pitch was shown to influence pore structure in the foam; the reduction in the toluene-soluble fraction resulted in a marked increase in bulk density, as shown in Figure 9.7 [30]. Thermo-oxidative pretreatment of a commercial pitch using a mineral acid was effective to give high mechanical strength to the foams [35]. The pretreatment of a coal tar pitch was performed by heating in either H_2SO_4 or HNO_3 at 120 °C, followed by heating at 350 °C at atmospheric pressure, and this modified pitch was then heated at 580 °C under a pressure up to 1 MPa to

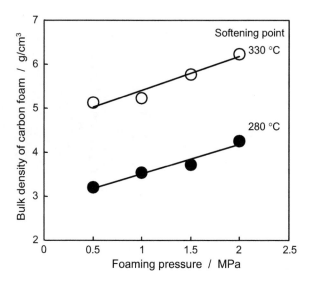

FIGURE 9.6 Dependence of Bulk Density of the Carbon Foams From Pitches of Different Softening Points on Foaming Pressure After Carbonization

Plotted from data reported in [31]

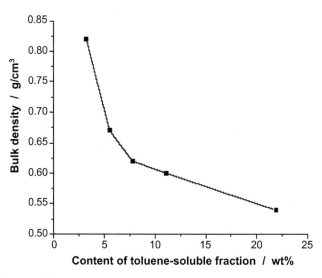

FIGURE 9.7 Change in Bulk Density of Carbon Foam with Content of the Toluene-soluble Fraction in the Precursor Mesophase Pitch

From [30]

prepare foam. A bituminous coal has also been used as the precursor for carbon foam preparation at 450–500 °C under 3–7 MPa [37].

The formation mechanism of carbon foams from mesophase pitch has been discussed in relation to various foaming conditions [36]. With increasing foaming temperature, small bubbles are formed owing to the release of decomposition gases from the pitch, and then the coalescence with neighboring bubbles occurs to form cells in the resultant foam, for which the driving force is thought to be surface tension of the molten pitch. The shear stress originating from bubble growth makes pitch molecules orient their aromatic planes preferentially along the bubble wall. This oriented nanotexture is preserved in the ligaments and joints, because of the high viscosity of the partially decomposed pitch. Therefore, the viscosity and surface tension of molten pitch are key factors to control not only cell size of the foam but also nanotexture in ligaments and junctions of the foam. The nanotexture in ligaments and junctions strongly influences graphitizability of the foam.

In order to improve the mechanical properties of carbon foams, different additives to the precursor pitches have been used. Graphite particles (<30 μm in size) were mixed into a mesophase pitch, followed by blowing under 8 MPa, carbonizing at 1200 °C, and graphitizing at 3000 °C [38]. By the addition of 5 mass% graphite, compressive strength improved from 3.7 to 12.5 MPa and thermal conductivity from 70.2 to 107.4 W/m·K. Addition of 4 vol% vapor-grown carbon fibers heat-treated at 3000 °C was very effective for improving the crush strength of the foam [39]. Starting from homogeneous dispersions of multi-walled carbon nanotubes by using surfactants in a pitch, carbon foams have been prepared of which bulk density was 0.15–0.22 g/cm^3 [40]. In an alternative method, mesocarbon microbeads (MCMBs) were added into a mesophase pitch up to 55 mass%; bulk density and compressive strength increased markedly, but thermal conductivity decreased with increasing MCMB content [41]. Na-montmorillonite has also been used as an additive to mesophase pitch, but the effect on the properties was not significant [42]. Addition of pitch fluoride by 3 mass% to a mesophase pitch markedly enhanced thermal conductivity of the carbon foam heat-treated at 3000 °C, although bulk density decreased a little, from 0.55 to 0.51 g/cm^3 [43].

Carbon foams have been prepared from a commercial mesophase pitch by using the barium salt of 5-phenyltetrazote as the foaming agent (decomposition temperature 375 °C) [20]. The resultant carbon foam consisted of large cells, more than a few hundred micrometers, much larger than that prepared by blowing under pressure (c. 10 μm), and was graphitized to a d_{002} of 0.336 nm after 2400 °C-treatment. Foaming of mesophase pitch has also been carried out by mixing with toluene at 320 °C and 12 MPa (the supercritical foaming method) [33]. Pore structure and properties of mesophase-pitch-derived carbon foams have been reported to depend strongly on the residence time at foaming temperature [34].

Carbon foams have been prepared from phenol-formaldehyde resin (PF) by blowing its ethanol solution of different concentrations at 170 °C under 4 MPa, followed by carbonization [44]. PF concentration had a strong influence on bulk density and properties of the resultant carbon foams after carbonization at 800 °C; the concentration change from 0.13 to 0.36 g/cm^3 resulted in an increase in bulk density from 0.24 to 0.73 g/cm^3 and in thermal conductivity from 0.06 to 0.24 W/m·K. The cell structure in the carbon foams also changed markedly with the change in PF concentration. From an emulsion of resorcinol (water phase), formaldehyde, and liquid paraffin (oil phase) with surfactants (Span 80 and Tween 80), carbon foams were prepared after carbonization at 1000 °C (the oil-in-water emulsion method) [45]. Carbon foams have also been prepared from the mixture of hollow phenolic resin and carbon spheres by using either furfuryl alcohol [46,47] or phenolic resin [48] as binder. Phenolic resin containing 3–15 mass% aluminosilicate (particle size of 100–200 μm) was converted to foam at 80 °C by using sodium bicarbonate as foaming agent and then carbonized at 800 °C to get carbon foam [49]. Bulk density and compressive strength of the foam increased from 0.261 to 0.316 g/cm^3 and from 3.78 to 4.71 MPa with increasing content of aluminosilicate from 0 to 15 mass%. Oxidation mass loss of the foams tended to decrease with increasing aluminosilicate content.

Via chemical blowing, a carbon foam with bulk density of 0.6 g/cm^3 and compressive strength of 25.8 MPa was prepared from poly(arylacetylene) (PAA) with high carbonization yield [50,51]. PAA prepolymer mixed with pentane (foaming agent), Tween 80 (bubble stabilizer), and sulfuric acid (catalyst) was heated slowly to 100 °C to blow pentane, and maintained at 100 °C. The PAA foam thus prepared was heated up to 350 °C stepwise to complete polymerization and then carbonized at 1000 °C. The ligament thickness of carbon foams depended on the concentration of sulfuric acid; changing from 10 μm at 70 mass% to more than 100 μm at 90 mass%. A mimosa tannin extract, which contained 84 mass% phenolic material, was mixed with formaldehyde (cross-linking agent), furfuryl alcohol (strengthening agent), diethylether (foaming agent), and 65 mass% toluene-4-sulfonic acid (catalyst) in water, to be transformed at 40 °C to a foam with bulk density 0.05–0.08 g/cm^3, which could be converted to carbon foam at 900 °C [52]. The resultant carbon foams were composed of glass-like carbon, of which the bulk density was 0.067 g/cm^3 and porosity was 96.4%. Sucrose has also been used to prepare carbon foam by concentrating its acidic aqueous solution at 110 °C in air, followed by carbonization at 300–950 °C in Ar [53]. Carbon foams have been prepared from olive stone by heating to 500 °C under 1.0 MPa; bulk densities were 0.2–0.3 g/cm^3 and cells were larger than 1 mm [54].

9.1.3 Template carbonization

Carbon foam has been prepared by impregnating phenol into sintered NaCl (template), followed by carbonization at 700 °C and dissolution of NaCl. The carbon foam was composed of cells of approximately 8 mm diameter and had a bulk density in the range of 0.035 to 0.075 g/cm^3 [55]. Mesocellular carbon foams with cell size larger than 20 nm were prepared by selecting mesocellular aluminosilicates with various cell and window sizes as the template, and phenol/formaldehyde as the carbon precursor [56].

A carbon material with honeycomb-type open cell structure, as shown in Figure 9.8, was developed by carbonization of a polyurethane (PU) foam

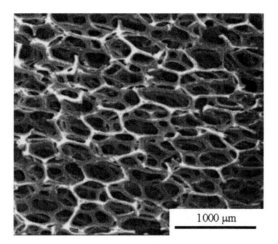

1000 μm

FIGURE 9.8 Carbon Foam Prepared from Polyurethane Foam, Reticulated Vitreous Carbon (RVC)

impregnated with phenol, furfuryl alcohol, or epoxy resin. It was commercialized around the 1980s, under the name of Reticulated Vitreous Carbon (RVC) [57,58]. The cell size and the fraction of open cells in the template PU foam was controlled by the addition of montmorillonite clay. The addition of clay by 4 parts per 100 parts of polyol increased open cells of the template foam from 56 to 81%, and after carbonization at 900 °C, porosity and density of the foam increased from 26 to 78 cells per inch and from 0.49 to 0.87 g/cm^3, respectively. The addition of small particles of clay to furfuryl alcohol for the impregnation has been applied to control the cell structure of carbon foam [59]. The resultant carbon foam was composed of glass-like carbon cell walls.

To improve the thermal and mechanical properties, copper was coated on the RVC by electroplating [60–62]. PU foams, which were prepared from precursors containing 30 mass% charcoal powder and then heated at 400 °C in air, were impregnated with furfuryl alcohol, followed by carbonization at 1000 °C in N$_2$ [63].

In an alternative method, PU or melamine foam was impregnated with poly(amic acid), followed by imidization and carbonization [64,65]. Ester-type PU foams were immersed into a N,N'-dimethylacetamide (DMAc) solution of poly(amic acid) composed of pyromellitic dianhydride (PMDA) and 4,4-oxydianiline (ODA) at room temperature, followed by heating at 60 °C to evaporate DMAc, imidization of poly(amic acid) to polyimide (PI) at 200 °C, and then carbonization at 1000 °C [64]. In Figure 9.9, SEM images of cross-sections are shown for the foam of the template PU, after PI impregnation, and after carbonization at 1000 °C. Although some cells of PI-impregnated foam are collapsed, probably because of insufficient impregnation of PI into the PU foam, foam morphology is well preserved after carbonization. Bulk density of the foam increased from 0.06 g/cm^3 of the original PU foam to 0.29 g/cm^3 after PI impregnation (PU/PI composite foam) and 0.18 g/cm^3 after carbonization. The template PU is pyrolyzed at relatively low temperature around

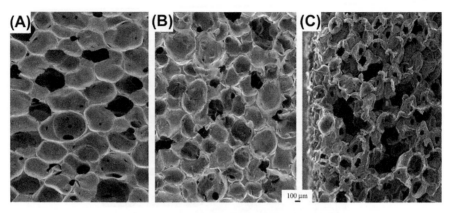

FIGURE 9.9 SEM Images of Carbon Foam Prepared From PU Foam and Polyimide

Cross-sections of (A) the template polyurethane foam, (B) the foam after polyimide impregnation, and (C) that after carbonization at 1000 °C

From [64]

300–400 °C and gives only a negligibly small amount of carbon residues after carbonization. Since PI is stable up to 550 °C, it inherits the cell structure of the PU even after the decomposition of template PU, and is carbonized above 600 °C. Carbon foams prepared from PI (POMD/ODA) were able to be converted to graphite foam by heat treatment at 3000 °C [64]. The foam morphology was maintained, d_{002} was reduced to 0.3367 nm, and the crystallite size, $Lc(002)$, became 30 nm.

By using fluorinated polyimide (2,2-bis(3,4-anhydrocarboxiphenyl hexafluoropropane (6FDA) and 2,2-bis(trifluoromethyl)benzidine (TFMB)) as carbon precursors, carbon foams with microporous cell walls, S_{BET} of 1540 m^2/g, and micropore volume, V_{micro}, of 0.63 cm^3/g, were prepared by a one-step carbonization process [65]. By impregnating a mixture of PI with T-type zeolite powder into the PU foam, followed by carbonization at 900 °C and dissolving out of zeolite by HF, carbon foams were prepared [66]. The walls of the foams were composed of microporous carbon having pore size less than 0.45 nm.

Template PU foam has been impregnated with a petroleum pitch of high softening point from water slurry, followed by drying, heating at 450 °C in N$_2$ to form "interparticulate bonded" pitch foam, stabilizing at 350 °C in air, and then carbonizing at 1000 °C [67]. The carbon foam after the heat treatment at 2800 °C had d_{002} of 0.338 nm and $Lc(002)$ of 40 nm. A mesophase pitch has also been used to impregnate into PU foam as water slurry, followed by stabilization at 350 °C in air, carbonization at 1000 and 1400 °C, and graphitization at 2400 °C [68]. Bulk density of the carbon foams after 2400 °C-treatment was in the range of 0.23–0.58 g/cm^3, and depended mainly on the pitch concentration of the water slurry and density of template PU foam. For a foam with bulk density of 0.58 g/cm^3, compressive strength of 5.0 MPa and thermal conductivity of 60 W/m·K were obtained.

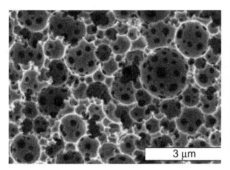 

FIGURE 9.10 Carbon Foam Prepared From an Emulsion of Resorcinol-formaldehyde in Oil Phase at Different Magnifications

From [72]

By carbonization of commercially available melamine foam at 600–1200 °C in an inert atmosphere, carbon foams were prepared with a yield of about 7 mass% after 1200 °C-carbonization [69]. The appearance of the resultant foam was very similar to RVC in Figure 9.8, but the carbon walls contained a relatively large amount of nitrogen, about 20 and 7 mass% N after carbonization at 800 and 1000 °C. Carbon foams with pore sizes larger than 20 nm have been prepared by selecting mesocellular aluminosilicates as the templates and phenol-formaldehyde as the carbon precursor [70]. Carbon foams composed of ordered and three-dimensionally interconnected macropores with uniform mesoporous walls were synthesized from divinylbenzene by using polystyrene spheres for the macropore template and silica particles for the mesopore template [71]. From an emulsion of resorcinol-formaldehyde in silicone oil, carbon foam was obtained after heat treatment above 800 °C [72]. The sizes of cells and windows (Figure 9.10) were controlled by the content of oil phase, and the size distribution of mesopores was governed by the concentration of carbon precursors in the emulsion.

Carbon foams with microporous walls have been prepared from poly(urethane-imide) films at 900 °C, with average macropore size changing from 0.6 to 10 µm with increasing urethane content from 10 to 70 mass% [73]. Graphite foams have been prepared from porous PI films with 50% porosity by heat treatment up to 3000 °C [74]. The foamy films had high graphitizability, d_{002} of 0.3363 nm, maximum magnetoresistance, $(\Delta\rho/\rho)_{max}$, of 18%, and a high degree of preferred orientation of graphite layers along the film surface.

9.2 Applications of carbon foams

Among the various morphologies of carbon materials, such as fibrous, tubular, spherical, lamellar, and granular, carbon foams have a specific porous, open cell structure. They have novel features owing to this cell structure, such as a large geometric surface area, in addition to the characteristics of carbon materials, such as light weight, high thermal stability, high thermal and electrical conductivities, etc. In bulky carbon materials, thermal and electrical conductivities can be changed in a relatively wide range by controlling the structure from graphitic to glass-like, and the nanotexture from oriented to random. The variation range of properties in carbon foams seems to be wider than in conventional carbon materials, by controlling their cell structure in addition to control of the structure and nanotexture in the carbon ligaments and joints. Carbon foams have been studied for thermal energy storage, as a container having high thermal conductivity; for various electrochemical measurements as electrodes; for adsorption/desorption of gaseous species as molecular sieves; and for absorption of electromagnetic waves.

9.2.1 Thermal energy storage

It has been proposed to use the latent heat associated with the phase change for thermal energy storage (latent heat thermal storage, LHTS). The use of a liquid-solid phase change is more practical, because volume changes associated with phase change are much smaller than those associated with gas-liquid and gas-solid phase changes. The materials used for this thermal energy storage are called phase-change materials (PCMs), and are generally divided into two groups, organic and inorganic compounds. Organic PCMs have several advantages, including the ability to melt congruently, self-nucleation, and non-corrosive behavior. The properties of PCMs have been reviewed [75]. Paraffin waxes, mainly consisting of mixed normal alkanes, have been used as PCMs by coupling with carbon foam as a container, in order to offset the low thermal conductivity of paraffin (about 0.24 W/m·K) [76–84].

EG-foams with various bulk densities were impregnated by soaking with either molten paraffins (melting point, T_m, of 73–80 °C and -9°C) or hexadecane (T_m of 18.1 °C) to prepare PCM/EG-foam composites [76]. The saturated amount of PCM impregnated depends principally on the bulk density of the EG-foam, as shown in Figure 9.11A [78]. Latent heat of the composite, i.e. thermal energy stored in the composite, decreases with increasing bulk density of the EG-foam, as a consequence of the decrease in the amount of PCM loaded [78]. Thermal conductivity of the composite shows a marked anisotropy mainly due to the preferred orientation of graphite layers, as shown in Figure 9.11B; conductivity along the direction perpendicular to compression force, i.e. parallel to the graphite layers oriented in the compact, is higher and strongly depends on bulk density of the EG-foam. In Figure 9.12, thermal conductivity and latent heat of the paraffin/EG-foam composite are plotted against bulk density of the matrix EG-foam [81]. With increasing bulk density of EG-foam, thermal conductivity of the composite increases markedly, to 28–180 times that of the original paraffin itself, but latent heat of the composite decreases because of the decrease in the amount of paraffin loaded.

High thermal conductivity makes heat exchange faster: rapid heat storage from and release to water are demonstrated using the composite, which is prepared from paraffin with T_m of 48–50 °C and EG-foam, as shown in Figure 9.13 [79]. The

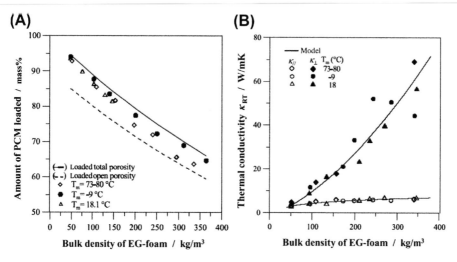

FIGURE 9.11 Characteristics of Paraffin/EG-foam Composites

(A) Change in amount of PCM loaded with bulk density of EG-foam (PCM: paraffin with different melting points), and (B) changes in thermal conductivity, κ_{RT}, of the composite along the directions parallel and perpendicular to compression force, $\kappa_{//}$ and $\kappa_{\perp}$, with bulk density

From [76]

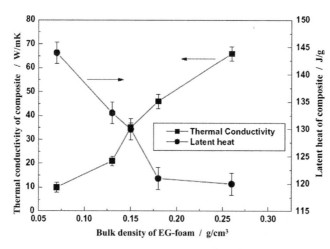

FIGURE 9.12 Dependence of Thermal Conductivity and Latent Heat of the Paraffin/EG-foam Composite on Bulk Density of the Matrix EG-foam

From [81]

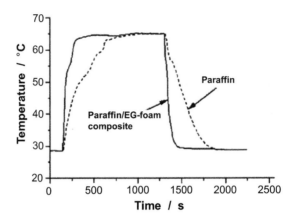

FIGURE 9.13 Heat Storage and Release Curves of Paraffin/EG-foam Composite, in Comparison with Paraffin

From [79]

composite can be heated from 29 to 65 °C and cooled down to 29 °C quickly, faster than the paraffin. A paraffin/EG-foam composite (80 mass% paraffin) showed faster heat exchange with air than the paraffin, although stored energy reduced a little [77].

The proper mass fraction of EG in the paraffin/EG-foam composite was determined to be 10 mass%, and the thermal conductivity was enhanced to 0.82 W/m·K, by using paraffin (*n*-docosane) with T_m of 42–44 °C and EG prepared at 900 °C [82]. Duration for complete melting of paraffin shortened noticeably and latent heat capacity was approximately equivalent to the value calculated from the mass content of paraffin. The paraffin/EG-foam composite allowed no leakage of melted paraffin during solid-liquid phase change owing to the capillary and surface tension forces of EG. A shift of phase-change temperature (melting/freezing) was noticed on a paraffin/EG-foam composite with 10 mass% EG by using a cylinder-shaped composite with a size of $120\phi \times 18\phi \times 400$ mm^3 [83]. EG-foam with a bulk density of 0.044 g/cm^3 was converted to a composite with a bulk density of 0.83 g/cm^3 by paraffin impregnation. Anomaly of the latent heat for the composites was observed at small EG content of about 1 mass% [84].

Mesophase-pitch-derived carbon foams with different cell structures have been used to fabricate composites with paraffin [85]. Four carbon foams with bulk density of 0.20–0.57 g/cm^3 and cell diameter of 600–200 μm, and paraffin with T_m of 58 °C and latent heat of 143.7 J/g were selected. Latent heat of the composites was proportional to mass fraction of paraffin in the composite, as shown in Figure 9.14A, but thermal diffusivity seemed to depend on cell size and wall thickness of the carbon foam: the foam with the smallest cell size and the highest bulk density gave marked improvement in thermal diffusivity, as shown in Figure 9.14B.

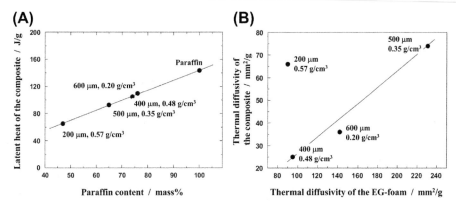

FIGURE 9.14　Characteristics of Mesophase-pitch-derived Carbon Foams with Different Cell Structures

Relationships of (A) latent heat to paraffin content of the paraffin/EG-foam composite, and (B) thermal diffusivity of the composite to pristine EG-foam

From [85]

Paraffin/carbon composites were prepared from four commercially available carbon foams [86]. Experimental results, as well as numerical calculation, demonstrated that high thermal conductivity of the carbon foam matrix increases the heat-exchange rate of the composite. The use of carbon foam to improve thermal conductivity of the composite was expected to increase the output power greatly; the composite prepared from carbon foam with 97% porosity gave five times larger energy storage output power than pure PCM [87].

A composite prepared by impregnation of Wood's alloy (50Bi/27Pb/13Sn/10Cd) (T_m 71 °C) into EG-foam (bulk density 0.35 g/cm³) resulted in a high thermal conductivity of 193 W/m·K, higher than the 58.9 W/m·K of Wood's alloy, and a small thermal expansion coefficient of 7.8 ppm/K, smaller than the 24.8 ppm/K of Wood's alloy, but the latent heat of 29.2 J/g was almost the same as Wood's alloy [88].

Graphite foam works well as a thermal enhancer and a container of PCMs, particularly for LHTS systems at low temperature, mainly owing to its high thermal conductivity, low density, low thermal expansion coefficient, and chemical inertness, although other materials have been proposed, such as metal [89], porous silica [90], and expanded perlite [91]. For LHTS at high temperatures, inorganic salts, such as $NaNO_3$ and KNO_3, are currently employed, but their low thermal conductivity is a more serious problem compared with LHTS at low temperatures. The composites of $NaNO_3$ with metal foams (copper and copper/steel alloys) were compared to the composite with EG-foam, and the overall performance of metal foams is superior to that of EG-foam, although both enhanced heat transfer significantly [92].

Addition of highly conductive materials, including natural graphite (NG), EG, carbon fiber (CF), and carbon nanofiber (CNF), into PCMs was effective to improve the thermal performance of the PCMs [93–107]. Addition of EG nanoplatelets (<10 nm thick) prepared by ultrasonic pulverization of EG was reported to be effective [93,94]. In order to improve the thermal conductivity and the flame retardant property of shape-stabilized PCM, EG was added to the mixture of paraffin and high-density polyethylene (HDPE) together with an intumescent flame retardant [95]. Thermal conductivity enhancement with the mixture of paraffin and HDPE was achieved by mixing with EG and/or NG [96]. Addition of CNFs and carbon nanotubes (CNTs) at less than 10 mass% was effective to improve thermal performance of paraffin [97,98]; CNFs were more effective than CNTs probably because of their better dispersion in the matrix PCM [98]. By adding 10 mass% EG or chopped CFs, thermal conductivity of stearic acid increased more than 200%, mainly owing to high thermal conductivity of EG (4–100 W/m·K) and CF (190 W/m·K) in comparison with that of stearic acid (0.3 W/m·K) [99]. Mixtures of stearic acid with EG in different mass ratios up to 5:1 showed no leakage of the stearic acid even after melting [100]. Also, the addition of 10 mass% EG into acetamide PCM was effective to improve thermal conductivity [101]. Mixing of NG and EG into different salt PCMs, KNO_3, KOH, etc., was reported to enhance their thermal conductivity [102–104]. Addition of stainless steel and copper pieces ($16\phi \times 18\phi \times 25$ mm^3) was compared with a commercial PCM/graphite composite in an ice/water (PCM), reporting the effectiveness of the latter [105]. Dispersion of high-conducting materials [106] in a PCM and microcapsulation of PCM [107] were reported. When EG was used as a conductive additive, it worked both as a thermal enhancer and a container, where PCMs were incorporated in large pores in worm-like particles of EG. Consequently, shape stability of the composite and no leakage of melted PCMs were achieved, the same situation as in the case using EG-foams.

Activated carbons have also been used as a container for a PCM, poly(ethylene glycol) (PEG) [108]. Impregnation of 30–70 mass% PEG into a mesoporous activated carbon was performed through its ethanol solution, followed by evaporation of ethanol, and the composites showed shape stability and no leakage of melted PEG even at 80 °C (above the melting point of PEG). In this case, confinement of PCM molecules to walls of micropores and mesopores of activated carbons has to be taken into consideration because of their physical adsorption, which may affect PCM performance by shifting the melting and solidification temperatures and by modifying latent heat capacity.

LHTS using solid-liquid phase changes of materials has attracted attention in relation to energy saving, and many reviews have been published from various viewpoints, focusing on the phase-change materials and their applications [109,110], on enhancement of thermal conductivity of PCMs [111], and on the applications of PCMs in building parts, such as in wall, roof, and floor materials [112,113]. Paraffin/EG-foam composites were studied for their application to thermal management of Li-ion batteries; thermal conductivity, tensile strength, compression strength, and burst performance under a gas pressure up to 1.2 MPa were tested as a function of paraffin content in the composites [84].

9.2.2 Electrodes

Reticulated vitreous carbon (RVC) has been used as an electrode since 1977, just after its industrial development, because it has honeycomb-like structure, as shown in Figure 9.8, exceptionally high pore volume of 97%, high geometrical surface area, rigid structure, low resistance to fluid flow and high heat resistance in a non-oxidizing environment, and high corrosion resistance. RVC was shown to be useful as an optically transparent electrode for simultaneous spectroscopic and electrochemical measurements in corrosive solutions at elevated temperatures [114]. Various applications as a flow-through electrode have been proposed: flow injection analysis with the detection limits of a few tenths of a nanogram [115]; flowing potential coulometry for microgram and submicrogram amounts of ferricyanide ions, catechol, hydroquinone, ascorbic acid, and epinephrine [116]; monitoring the acidity of solutions, owing to its sensitive response to pH change even in flowing streams [117]; flow-through electrolysis with a detection limit around 1×10^{-9} mol/L ferrocyanide [118]; and as a rotating-disk electrode for voltammetry to detect submicromolar concentrations of electroactive species [119]. Performance of an RVC electrode has been compared with that of other electrodes, including carbon paste, pyrolytic graphite, and platinum [120]. RVC has also been tested as an indicator electrode in solid-state sensors and it showed high sensitivity for gaseous NO_2 in air in a range of 0.2–2.2 ppm (v/v) [121]. To produce H_2O_2 for the removal of toxic organics in industrial wastes, an electrochemical cell with an RVC cathode was proposed, and the removal of formaldehyde was achieved at a relatively high rate in a flow system [122]. For the production of H_2O_2, a three-dimensional electrode consisting of RVC was shown to be efficient for effluent treatment [123].

RVC has also been used as an electrode material in various electrochemical energy storage cells. RVC modified with a polyaniline-poly(p-styrenesulfonic acid) composite was tested as an anode of a lithium-ion rechargeable battery in 1 mol/L $LiClO_4$/PC (propylene carbonate) electrolyte solution [124]. Carbon foams prepared from commercial melamine foam at 800 °C, which had a very similar honeycomb structure to RVC, gave high capacitive performance in 1 mol/L H_2SO_4 aqueous electrolyte, as shown by high gravimetric capacitance and good rate performance in galvanostatic charge/discharge in Figure 9.15 [69]. High content of nitrogen in the carbon matrix, about 20 mass% after carbonization at 800 °C, was thought to give high pseudo-capacitance. Redox couples for redox rechargeable batteries were studied on RVC electrodes; in H_2SO_4 solution of V^{5+}/V^{4+} couple using static and rotating-disk electrodes by cyclic voltammetry [125], and in an aqueous system of Br_2/Br^- using a rotating-disk electrode [126]. Deposition of zinc on an RVC electrode was studied in zinc-bromine and zinc-zinc cells by circulating electrolytes of $ZnBr_2$, KBr, and NaCl aqueous solutions [127]. RVC has been used as a current collector for lead-acid batteries after the deposition of Pb-Sn (1 mass%) by electroplating [128].

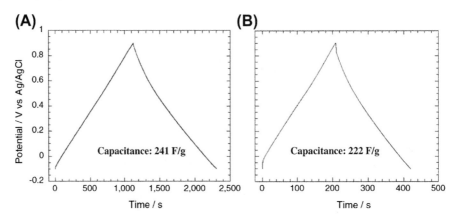

FIGURE 9.15 Galvanostatic Charge/Discharge Curves of N-enriched Carbon Foam in 1 mol/L H₂SO₄ Aqueous Electrolyte at Different Current Densities

(A) 200 mA/g, and (B) 1000 mA/g

Courtesy of Dr M. Kodama of AIST, Japan

Various applications of RVC in electrochemical devices have been reviewed in comparison with solid glassy carbon [57], and by emphasizing the relationships to its properties [129].

9.2.3 Adsorption

Carbon foam prepared from sucrose and RVC has been used for the trapping (adsorption) of radioactive ^{137}Cs [53]. Trapping efficiency of ^{137}Cs and distribution coefficient of ^{137}Cs between carbon foam and sodium were determined to be 73–77% and $4.9–6.0\times10^2$ for the carbon foam from sucrose, and 83–93% and $1.4–3.5\times10^3$ for RVC.

Because of their macropores, some carbon foams can easily be activated by using gaseous reagents to increase the micropores in their cell walls, which is a simpler process than that using solid reagents, such as $ZnCl_2$ and KOH. Adsorption/desorption behavior of water vapor in the atmosphere has been studied on carbon foams prepared from PI via templating in PU foams [64,65]. An example of cyclic adsorption/desorption behavior measured in thermogravimetry apparatus is shown in Figure 9.16A; adsorption was performed by a flow of N_2 gas bubbled through water at 40 °C, and desorption in a flow of dry N_2. After degassing by heating at 200 °C (first cycle), adsorption and desorption of water vapor are highly reproducible and relatively fast. In Figure 9.16B, the reproducible adsorptivity for water vapor is plotted against V_{micro} for carbon foams prepared at different temperatures, revealing a good linear relationship. Carbon foams with microporous cell walls prepared via template carbonization using PU foam and fluorinated PI showed reproducible adsorption/desorption cycles of atmospheric water vapor [65].

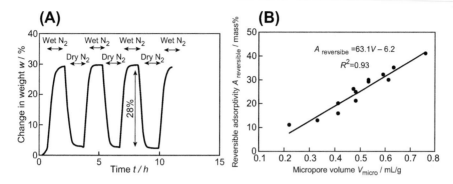

FIGURE 9.16 Characteristics of Carbon Foams Prepared Using PU Foam as Template

(A) Adsorption/desorption behavior of water vapor, and (B) relationship between reversible adsorptivity for water vapor and V_{micro} of the foam

From [64]

Carbon foams prepared by PU-templating have been used as a support of photocatalytic TiO_2 particles [64]. Fine anatase-type TiO_2 particles less than 100 nm were successfully loaded using 0.1 mol/L aqueous solution of $TiOSO_4$ under hydrothermal conditions at 180 °C. This foam had very low bulk density and floated on the water surface, giving high efficiency for the use of UV rays after photocatalyst loading.

9.2.4 Other applications

Carbon foam prepared from mesophase pitch under pressure (bulk density of 0.25–0.65 g/cm³, thermal conductivity of 0.3–180 W/m·K) was tested as a heat-sink material for cooling of power electronics, because of its light weight, high thermal conductivity, and high geometrical surface area to contact with a coolant gas or liquid [130]. Cooling efficiency of the carbon foam by air and water was assessed to evaluate the heat transfer coefficient. A very high coefficient of c. 2600 W/m²K was obtained for the carbon foam by using water coolant, in comparison with c. 250 W/m²K for a conventional Al-foam heat sink. The response time for the carbon foam was much shorter than that of the Al-foam.

Electromagnetic response of a carbon foam, which was prepared from a mesophase pitch (softening point of 272 °C) at 450 °C under 2.0 MPa and with a cell size of 400–600 μm, was studied [131]. Microwave reflectivity for the carbon foam is shown in Figure 9.17 as a function of carbonization temperature. The carbon foam carbonized at 700 °C shows excellent microwave absorption, reflection loss reaching more than −10 dB (90% absorption) at around 7 GHz. Cells of carbon foam were thought to act as a darkroom for microwaves. With RVC, microwave absorption performance has been confirmed [132]. Carbon foams prepared via template carbonization using PU had smaller dielectric constants but several times larger dielectric loss in comparison with corresponding pulverized ones [133]. Moreover, carbon foams showed magnetic loss, although this could not be observed with corresponding

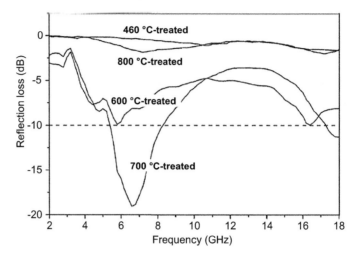

FIGURE 9.17 Microwave Reflectivity of Carbon Foams Carbonized at Different Temperatures

From [131]

pulverized ones. Carbon foams prepared from mesophase pitch showed anisotropic microwave absorption along the gravitational force, as well as cell structure and mechanical properties [27].

9.3 Concluding remarks

Fabrication of carbon foams is accomplished by three routes: exfoliation and compaction of natural graphite, blowing of carbon precursors, and template carbonization. The second route, blowing, may be divided into two: foaming by rapid release of pressure built up by the pyrolysis of carbon precursors, such as pitches, and by the use of a foaming agent. In the latter, the selection of thermosetting precursors, such as phenol resins, makes the fabrication process simple, but high graphitization degree and, as a consequence, high electrical and thermal conductivity cannot be expected for the resultant foams. Foaming of thermoplastic precursors, such as pitch and PVC, using additives requires stabilization to keep the foam morphology during further heat treatment at high temperatures. Since the stabilization is an oxidation process, graphitizability of the resultant foams is lowered.

Control of cell size in the carbon foam is important for all applications and is carefully done in each fabrication route. In the first route, the optimum degree of exfoliation of natural graphite and optimum compression conditions have to be determined. In the second route, the carbon precursor and foaming agent, in addition to the foaming conditions, have to be selected. In the third route, a carbon precursor appropriate for impregnation to the template has to be selected. In addition, non-uniformity in cell size in the foam must be avoided by careful processing, particularly in the first and third routes, as explained in the corresponding sections and discussed below.

For the application of carbon foams as containers for PCM, both high electrical and thermal conductivities are required, in addition to some mechanical strength to resist volume change of PCM due to thermal energy storage/release cycles. To satisfy the requirements for electrical and thermal conductivities, a highly crystalline graphite structure has to be developed in the foam and consequently the carbon precursor for foam fabrication has to be selected for this purpose. Flaky natural graphite has been used as one of precursors via exfoliation and compaction to obtain the desired porosity. However, homogeneity of pores was very difficult to attain in a block of foam during compaction; in most cases, the outside of the block near the wall of the mold being denser than the central part. Homogeneous foaming can be expected by blowing mesophase pitch, although heat treatment at a high temperature is necessary to realize high electrical and thermal conductivities. If the stabilization of pitch is needed after foaming to maintain foam morphology during further heat treatment at high temperatures, the development of graphitic structure is disturbed. Therefore, a stabilization process after foaming is crucial to obtain high electrical and thermal conductivities. When PU or melamine foam is used as a template, polyimide is an interesting precursor to prepare carbon foam of high electrical and thermal conductivity, because of the simple process of impregnation and carbonization, although heat treatment at high temperature is essential. In this route, the impregnation process is the most important to obtain a carbon foam with homogeneous porosity and structure. In order to impregnate poly(amic acid) homogeneously into the center of the template PU foam, porosity and connectivity of cells in the template have to be optimized.

For the application as electrodes, RVC has been used, particularly for optically transparent electrodes for simultaneous spectroscopic and electrochemical measurements, and also as flow-through electrodes. Carbon foams may be used as electrodes containing electrochemically active materials, requiring high electrical and thermal conductivities of the carbon foams. To have high conductivity both electrically and thermally, carbon foams prepared via the template route using PI or mesophase pitch as the carbon precursor are interesting, rather than RVC, because these foams can have high conductivity after high-temperature treatment and be self-standing.

For the application as adsorbents, micropores and mesopores in the ligaments must be developed by activation. Gas activation can be applied at relatively low temperatures because of easy diffusion through the large cells in the carbon foam. Foaming of thermosetting resins, such as phenol resin, can be carried out using the agent at low temperatures, the resultant foams being carbonized without any stabilization and able to have microporous cell walls without activation.

For any application, size of cells and connectivity between neighboring cells seem to be very important, but not many studies concerning this topic have been reported. Cell-size distribution should be statistically clarified by applying image analysis techniques, and cell connectivity may be evaluated by measuring the sizes of windows. These parameters have to be determined corresponding to the precursor used and foaming conditions, although no systematic studies have yet been carried out.

References

[1] Inagaki M. New Carbon Mater 2009;24:193–222.

[2] Bandosz TJ, editor. Activated Carbon Surfaces in Environmental Remediation. Elsevier; 2006.

[3] Tascon JMD, editor. Adsorption by Carbons. Elsevier; 2008.

[4] Béguin F, Frackowiak E, editors. Carbons for Electrochemical Energy Storage and Conversion Systems. CRC Press; 2010.

[5] Inagaki M, Konno H, Tanaike O. J Power Sources 2010;195:7880–903.

[6] Chung DDL. J Mater Sci 1987;22:4190–8.

[7] Furdin G. Fuel 1998;77:479–85.

[8] Inagaki M, Kang F, Toyoda M. Chem Phys Carbon 2004;29:1–69.

[9] Inagaki M, Toyoda M, Kang F, et al. New Carbon Mater 2003;18:241–9.

[10] Inagaki M, Tashiro R, Washino Y, et al. J Phys Chem Solids 2004;65:133–7.

[11] Inagaki M, Suwa T. Carbon 2001;39:915–20.

[12] Inagaki M, Tashiro R, Toyoda M, et al. J Ceram Soc Japan, Suppl 2004;112:S1513–6.

[13] Zheng YP, Wang HN, Kang F, et al. Carbon 2004;42:2603–7.

[14] Berger D, Maire J. J Mater Sci Eng 1977;31:335–9.

[15] Kang F, Leng Y, Zhang TY. Carbon 1997;35:1089–96.

[16] Inagaki M, Muramatsu K, Maekawa K, et al. TANSO 1985; No.123: 160–165 [in Japanese].

[17] Soneda Y, Inagaki M. Solid State Ionics 1993;63–65:523–7.

[18] Han JH, Cho KW, Lee KH, et al. Carbon 1998;36:1801–998.

[19] Klett J, Hardy R, Romine E, et al. Carbon 2000;38:953–73.

[20] Mehta R, Anderson DP, Hager JW. Carbon 2003;41(11):2174–6.

[21] Klett J, McMillan AD, Gallego NC, et al. Carbon 2004;42:1849–52.

[22] Li TQ, Wang CY, An BX, et al. Carbon 2005;43:2030–2.

[23] Beechem T, Lafdi K, Elgafy A. Carbon 2005;43:1055–64.

[24] Rosebrock G, Elgafy A, Beechem T, et al. Carbon 2005;43:3075–87.

[25] Chen C, Kennel EB, Stiller AH, et al. Carbon 2006;44:1535–43.

[26] Eksilioglu A, Gencay N, Yardim MF, et al. J Mater Sci 2006;41(10):2743–8.

[27] Ge M, Shen ZM, Chi WD, et al. Carbon 2007;45:141–5.

[28] Fethollahi B, Zimmer J. Carbon 2007;45:3057–9.

[29] Wang MX, Wang CY, Li YL, et al. Carbon 2007;45:687–9.

[30] Li S, Guo Q, Ya Song, et al. Carbon 2007;45(14):2843–5.

[31] Wang M, Wang CY, Li TQ, et al. Carbon 2008;46:84–91.

[32] Wang M, Wang C, Chen M, et al. New Carbon Mater 2009;24:61–6.

[33] Li J, Wang C, Zhan L, et al. Carbon 2009;47:1204–6.

[34] Wang Y, Min Z, Cao M, et al. New Carbon Mater 2009;24:321–6.

[35] Tsyntsarski B, Petrova B, Budinova T, et al. Carbon 2010;48:3523–30.

[36] Li S, Tian Y, Zhong Y, et al. Carbon 2011;49:618–24.

[37] Calvo M, Garcia R, Arenillas A, et al. Fuel 2005;84:2184–9.

[38] Zhu J, Wang X, Guo L, et al. Carbon 2007;45:2547–50.

[39] Fawcett W, Shetty DK. Carbon 2010;48:68–80.

[40] Leroy CM, Carn F, Backov R, et al. Carbon 2007;45:2307–20.

[41] Li SZ, Song YZ, Song Y, et al. Carbon 2007;45:2092–7.

[42] Wang XY, Zhong JM, Wang YM, et al. Carbon 2006;44:1560–4.

[43] Li S, Guo Q, Song Y, et al. Carbon 2010;48:1312–20.

[44] Lei S, Guo Q, Shi J, et al. Carbon 2010;48:2644–6.

[45] Liu M, Gan L, Zhao F, et al. Carbon 2007;45:2710–2.

[46] Benton ST, Schmitt CR. Carbon 1972;12:185–90.

[47] Nicholson J, Thomas CR. Carbon 1973;11:65–6.

[48] Zhang L, Ma J. Carbon 2009;47:1451–6.

[49] Wu X, Liu Y, Fang M, et al. Carbon 2011;49:1782–6.

[50] Liu M, Gan L, Zhao F, et al. Carbon 2007;45:3055–7.

[51] Zhang S, Liu M, Gan L, et al. New Carbon Mater 2010;25:9–14.

[52] Tondi G, Fierro V, Pizzi A. Carbon 2009;47:1480–92; Erratum: Carbon 2009; 47: 2761.

[53] Jana P, Ganesan V. Carbon 2009;47:3001–9.

[54] Rios RVRA, Martinez-Escandell M, Molina-Sabio M, et al. Carbon 2006;44: 1448–54.

[55] Pekala R, Hopper RW. J Mater Sci 1987;22:1840–4.

[56] Lee J, Sohn K, Hyeon T. J Am Chem Soc 2001;123:5146–7.

[57] Wang J. Electrochim Acta 1981;26:1721–6.

[58] Chakovskoi AG, Hunt CE, Forceberg G, et al. J Vac Sci Technol B 2003;21:571–5.

[59] Harikrishnan G, Patro TU, Khakhar DV. Carbon 2007;45:531–5.

[60] Lafdi K, Almajali M, Huzayyin O. Carbon 2009;47:2620–6.

[61] Almajali M, Lafdi K, Prodhomme PH, et al. Carbon 2010;48:1604–8.

[62] Almajali M, Lafdi K. Carbon 2010;48:4238–47.

[63] Amini N, Aguey-Zinsou KF, Guo ZX. Carbon 2011;49:3857–64.

[64] Inagaki M, Morishita T, Kuno A, et al. Carbon 2004;42:497–502.

[65] Ohta N, Nishi Y, Morishita T, et al. New Carbon Mater 2008;23:216–20.

[66] Xiaon N, Ling Z, Wang L, et al. New Carbon Mater 2010;25:321–4 [in Chinese].

[67] Ya Chen, Chen B, Shi X, et al. Carbon 2007;45:2132–4.

[68] Yadav A, Kumar R, Bhatia G, et al. Carbon 2011;49:3622–30.

[69] Kodama M, Yamashita J, Soneda Y, et al. Carbon 2007;45:1105–7.

[70] Lee J, Sohn K, Hyeon T. J Am Chem Soc 2001;123:5146–7.

[71] Chai GS, Shin IS, Yu J S. Adv Mater 2004;16:2057–61.

[72] Gross AF, Nowak AP. Langmuir 2010;26:11378–83.

[73] Takeichi T, Yamazaki Y, Zuo M, et al. Carbon 2001;39:257–65.

[74] Kaburagi Y, Aoki H, Yoshida A. TANSO 2012; No. 253: 95–99.

[75] Himran S, Suwono A. Energy Sources 1994;16:117–28.

[76] Py X, Olives R, Mauran S. Int J Heat Mass Transfer 2001;44:2727–37.

[77] Marin JM, Zalba B, Cabeza LF, et al. Int J Heat Mass Transfer 2005;48:2561–70.

[78] Mills A, Farid M, Selman JR, et al. Appl Thermal Eng 2006;26:1652–61.

[79] Zhang Z, Fang X. Energy Convers Manag 2006;47:303–10.

[80] Lafdi K, Mesalhy O, Shaikh S. Carbon 2007;45:2188–94.

[81] Zhong Y, Li S, Wei X, et al. Carbon 2010;48:300–4.

[82] Sart A, Karaipekli A. Appl Therm Eng 2007;27:1271–7.

[83] Xia L, Zhang P, Wang RZ. Carbon 2010;48:2538–48.

[84] Alrashdan A, Mayyas AT, Al-Hallaj S. J Mater Process Technol 2010;210:174–9.

[85] Zhong Y, Guo Q, Li S, et al. Solar Energy Mat Solar Cells 2010;94:1011–4.

[86] Mesalhy O, Lafdi K, Elgafy A. Carbon 2006;44:2080–8.

[87] Lafdi K, Mesalhy O, Elgafy A. Carbon 2008;46:159–68.

[88] Zhong Y, Guo Q, Li S, et al. Carbon 2010;48:1689–92.

[89] Hoogendoorn CJ, Bart GC. Solar Energy 1992;48:53–8.

[90] Erk H, Dudukovic M. AIChE J 1996;42:791–808.

[91] Nomura T, Okinaka N, Akiyama T. Mater Chem Phys 2009;115:846–50.

[92] Zhao CY, Wu ZG. Solar Energy Mater Solar Cells 2011;95:636–43.

[93] Kim S, Drzal LT. Solar Energy Mater Solar Cells 2009;93:136–42.

[94] Xiang J, Drzal LT. Solar Energy Mater Solar Cells 2011;95:1811–8.

[95] Zhang P, Hu Y, Song L, et al. Solar Energy Mater Solar Cells 2010;94:360–5.

[96] Cheng W-I, Zhang R, Xie K, et al. Solar Energy Mater Solar Cells 2010;94:1636–42.

[97] Sanusi O, Warzoha R, Fleischer AS. Int J Heat Mass Transfer 2011;54:4429–36.

[98] Cui Y, Liu C, Hu S, et al. Solar Energy Mater Solar Cells 2011;95:1208–12.

[99] Karaipekli A, Sari A, Kaygusuz K. Renew Energy 2007;32:2201–10.

[100] Fang G, Li H, Chen Z, et al. Energy 2010;35:4622–6.

[101] Xia L, Zhang P. Solar Energy Mater Solar Cells 2011;95:2246–54.

[102] Pincemin S, Olives R, Py X, et al. Solar Energy Mater Solar Cells 2008;92:603–13.

[103] Acem Z, Lopez J, Del Barrio EP. Appl Therm Eng 2010;30:1580–5.

[104] Lopez J, Acem Z, Del Barrio EP. Appl Therm Eng 2010;30:1586–93.

[105] Cabeza LF, Mehling H, Hiebler S, et al. Appl Therm Eng 2002;22:1141–51.

[106] Siegel R. Int J Heat Mass Transfer 1977;20:1087–9.

[107] Bedecarrats JP, Strub F, Falcon B, et al. Int J Refrig 1996;19:187–96.

[108] Feng L, Zheng J, Yang H, et al. Solar Energy Mater Solar Cells 2011;95:644–50.

[109] Zalba B, Marin JM, Cabeza LF, et al. Appl Therm Eng 2003;23:251–83.

[110] Farid MM, Khudhair AM, Siddique KR, et al. Energy Convers Manag 2004;45: 1597–615.

[111] Fan L, Khodadadi JM. Renew Sustain Energy Rev 2011;15:24–46.

[112] Kuznik F, David D, Johannes K, et al. Renew Sustain Energy Rev 2011;15:379–91.

[113] Tyagi VV, Kaushik SC, Tyagi SK, et al. Renew Sustain Energy Rev 2011;15:1373–91.

[114] Norvell VE, Mamantov G. Anal Chem 1977;49:1470–2.

[115] Strohl AN, Curran DJ. Anal Chem 1979;51:1045–9.

[116] Strohl AN, Curran DJ. Anal Chem 1979;51:1050–3.

[117] Strohl AN, Curran DJ. Anal Chim Acta 1979;108:379–83.

[118] Blaedel WJ, Wang J. Anal Chem 1979;51:799–802.

[119] Blaedel WJ, Wang J. Anal Chem 1980;52:76–80.

[120] Saidman SB, Bessone JB. Electrochim Acta 2000;45:3151–6.

[121] Hrncirova P, Opekar F, Stulik K. Sens Actuators B 2000;69:199–204.

[122] Ponce de Leon C, Pletcher D. J Appl Electrochem 1995;25:307–14.

[123] Alvarez-Gallegos A, Pletcher D. Electrochim Acta 1998;44:853–61.

[124] Tsutsumi H, Yamashita S, Oishi T. J Appl Electrochem 1997;27:477–81.

[125] Zhong S, Skyllas-Kazacos M. J Power Sources 1992;39:1–9.

[126] Mastragostino M, Gramellini C. Electrochim Acta 1985;30:373–80.

[127] Iacovangelo CD, Will FG. J Electrochem Soc 1985;132:851–7.

[128] Gyenge E, Jung J, Mahato B. J Power Sources 2003;113:388–95.

[129] Friedrich JM, Ponce-de-Leon C, Reade GW, et al. J Electroanal Chem 2004;561: 203–17.

[130] Gallego NC, Klett JW. Carbon 2003;41:1461–6.

[131] Yang J, Shen ZM, Hao ZB. Carbon 2004;42:1882–5.

[132] Fang ZG, Cao XM, Li CS, et al. Carbon 2006;44:3368–70.

[133] Fang Z, Li C, Sun J, et al. Carbon 2007;45:2873–9.

Nanoporous Carbon Membranes and Webs

10

Pores commonly existing in carbon materials are schematically illustrated in Figure 10.1 using a carbon grain. The pores are sorted by whether a specific gaseous molecule, usually nitrogen, can be adsorbed or not, into open and closed pores. Open pores are classified by their widths: micropores less than 2 nm wide, mesopores 2–50 nm, and macropores more than 50 nm, according to the recommendations of the International Union of Pure and Applied Chemistry. Micropores can accommodate various molecules in their spaces and so they are very important for applications as adsorbents, as has been pointed out for activated carbons. Mesopores work to adsorb large molecules, but they also function as the paths for small molecules to be adsorbed into micropores, i.e. diffusion paths. Often, these micropores and mesopores are called "nanopores," because their widths are on the scale of nanometers, and this nomenclature has been adopted in this chapter. If carbon is in the form of either fiber or thin film, most of the micropores tend to be directly open to the atmosphere, so that gas molecules can be adsorbed and desorbed directly to and from the micropores, resulting in faster adsorption/desorption.

Nanoporous carbon membranes, thin films and webs, can be synthesized by pyrolysis of organic polymer precursors, chemical vapor deposition (CVD), physical vapor deposition (PVD), template carbonization, and electrochemical methods. The membranes synthesized by CVD and PVD mostly contain disordered micropores that render them suitable for separation and adsorption of small gas molecules, but their applicability to large molecules is limited, including biomolecules and bulky organics. Membranes or webs of carbon nanotubes (CNTs) and carbon nanofibers (CNFs) have been developed, which work as nanoporous carbon membranes. Carbon membranes containing mesopores are synthesized by means of templates, such as mesoporous silicas or block copolymers, which have been reviewed by several authors [1–4] and also in Chapter 7. The webs consisting of CNTs and CNFs discussed in this chapter have a characteristic texture of macropores and mesopores formed between fibrous carbons, in addition to micropores formed within fibrous carbon.

Here, the fabrication methods of nanoporous carbon membranes are reviewed, surveying their applications in order to understand the advantages of nanoporosity of the membranes. Since nanoporous carbon membranes have properties of high

---

Prerequisite for readers: Chapter 3.5 (Porous carbons) in *Carbon Materials Science and Engineering: From Fundamentals to Applications*, Tsinghua University Press.

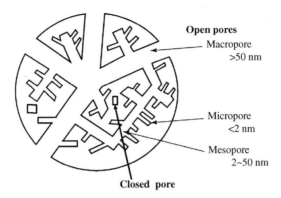

FIGURE 10.1 Schematic Illustration of Pores Commonly Existing in Carbon Materials

nanoporosity, relatively good electrical and thermal conductivity, low density, and chemical and mechanical stability, they are attractive for energy-saving environmentally friendly processes and have potentially wide applications in industrial processes as adsorbents, gas separators, electrodes, and sensors.

10.1 Synthesis
10.1.1 Pyrolysis and carbonization of organic precursors

As carbon precursors for the synthesis of nanoporous carbon films, various polymers, such as polyimides, phenol-formaldehyde resins, poly(vinylidene fluoride), poly(furfuryl alcohol), and hyper-cross-linked polystyrenes, have been employed. Porous films of thermosetting resins are easily transferred to carbon membranes by heat treatment at high temperatures, although heating conditions have to be selected carefully to avoid the formation of cracks during the heating and cooling processes. In controlling the sizes of pores, thermal shrinkage during carbonization from porous organic polymers to carbon must also be taken into consideration.

Self-standing carbon membranes have been prepared from poly(furfuryl alcohol) and their gas permeability was measured [5], although it was difficult to obtain a large area. A nanoporous carbon membrane (carbon xerogel film) with tailored thickness was prepared at 600 °C from resorcinol-formaldehyde resin without a supercritical drying process [6]. The resultant carbon membrane has bimodal pore structure consisting of uniform micropores and mesopores; its pore-size distribution is shown together with adsorption/desorption isotherms of N_2 in Figure 10.2. The membrane has a Brunauer-Emmett-Teller (BET) surface area, S_{BET}, of 915 m^2/g and a total pore volume, V_{total}, of about 1.1 cm^3/g.

Nanoporous carbon membranes were prepared from hot-pressed poly(vinylidene fluoride) (PVDF) films by dehydrofluorination with strong organic bases, 1,8-diazabicyclo[5,4,0]undec-7-ene (DBU), followed by carbonization at 1300 °C

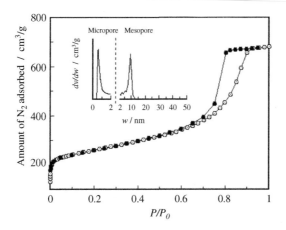

FIGURE 10.2 Adsorption/Desorption Isotherms of N_2 at 77 K and Pore-size Distribution (Inserted Figures) for the Nanoporous Carbon Membrane Prepared from Resorcinol-formaldehyde Resin

From [6]

and activation in CO_2 at 850 °C [7]. The resultant membrane had high adsorptivity at 538 mg/g for methylene blue. Macroporous honeycomb-patterned carbon membranes of high thermal and chemical stability were prepared from fluorescent hyperbranched poly(phenylene vinylene) (hypPPV) at 600 °C [8]. They showed that hypPPV formed honeycomb membranes by a simple casting from organic solvents of $CHCl_3$ or CS_2 under a flow of humid air. Carbonization of this polymer film above 873 K produced a carbon membrane of hexagonally packed macropores.

Carbon thin membranes are prepared from films of aromatic polyimides via a simple heat treatment at high temperatures, of which the structure can be controlled from amorphous to graphitic and also the texture from nanoporous to dense by selecting the molecular structure of the precursor polyimide, i.e. various combinations of dianhydrides with aromatic diamines [9]. By carbonization of polyimide PMDA/ODA (pyromellitic dianhydride/4,4′-oxydianiline), a commercially available one (Kapton) and a laboratory-made one, nanoporous carbon membranes with molecular sieving characteristics were prepared, which had a high selectivity for hydrogen [10,11]. The molecular structure of polyimides has been experimentally shown to govern the pore structure in the resultant carbon membranes [12–14].

From the polyimides with repeating units containing pendant groups of $-CH_3$ and $-CF_3$, microporous carbon membranes consisting of ultramicropores with pore width of c. 0.5 nm were obtained, of which the microporous surface area, S_{micro}, depended strongly on the content of pendant groups, particularly $-CF_3$ [12]. In Figure 10.3, S_{micro} determined from an α_s plot of fluorinated aromatic polyimides (Figure 10.3A) is plotted against fluorine content in the repeating unit of the polyimides [13]. The precursor polyimide, containing 31.3 mass% F, gives S_{micro} of about 1340 m²/g and V_{micro} of 0.44 cm³/g without any activation treatment (Figure 10.3B). This microporous

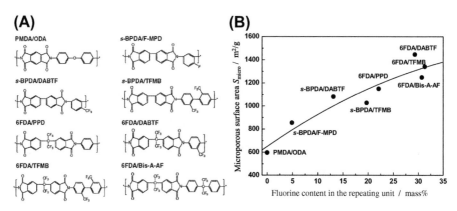

FIGURE 10.3 Microporous Carbon Membranes Prepared From Polyimides

(A) Repeating unit of polyimides used, and (B) relationship between microporous surface area, S_{micro}, and fluorine content in the repeating unit

From [13]

carbon membrane had a high adsorption capacity for water vapor, about 0.46 g/g, of which 90% was adsorbed at the relative pressure of 0.46 [14]. Carbonized nonporous films prepared from PMDA/ODA could be activated under mild conditions, 375 °C in a flow of air, to give cyclic adsorption/desorption of water vapor in air [15]. Carbon membranes were prepared by the carbonization of a polyimide deposited on macroporous carbon disks (35 mmϕ and 2.2–2.5 mm thick) applying a phase inversion technique [16]. The resultant membranes were confirmed to be suitable for gas separation of O_2/N_2, CO_2/N_2, and CO_2/CH_4 mixtures. Glass-like carbon membranes have been prepared from polycarbodiimide by stabilization and carbonization [17], but no data related to pore structure were reported.

Mesoporous carbon membranes have been successfully obtained from blended polymers consisting of poly(ethylene glycol) and polyimide, in which the pore size tends to increase with an increase of the poly(ethylene glycol)/polyimide ratio [18]. Macroporous carbon membranes having micropores in the macropore walls were prepared from poly(urethane-imide) films at 900 °C, the average macropore size increasing from 0.6 to 10 μm with increasing urethane content from 10 to 70 mass% [19]. Macroporous carbon membranes (carbon foams) have also been prepared by applying the template method: impregnation of poly(amic acid) into a template of either polyurethane or melamine foam, followed by carbonization at 700–1000 °C [20]. Membranes composed of a continuous carbon matrix and a dispersed silica domain, C/SiO_2 membranes, were prepared from the mixture of polyimide and tetraethoxysilane at 600 °C [21]. Macroporous graphite films have been prepared from porous aromatic polyimide films with 50% porosity by heat treatment up to 3000 °C, and had high graphitizability, expressed by the interlayer spacing, d_{002}, of 0.3363 nm and maximum magnetoresistance, $(\Delta\rho/\rho)_{max}$, of 18%, and high degree of preferred orientation of graphite layers along the film surface [22].

Superhydrophilic carbon membrane was prepared from superhydroforbic natural lotus leaf by carbonization at 600 °C in Ar, followed by washing with HCl to remove inorganic impurities [23,24]. On the carbonized lotus leaf, a water droplet spread quickly and thoroughly within 256 ms, the contact angle against water being about 0°.

Membranes of microporous carbon were formed by the deposition of novolac-type phenolic resin on the inner surface of an Al_2O_3 tube, followed by carbonization at 700 °C and air-activation at 300–400 °C [25,26]. These carbon membranes showed high selectivity for hydrocarbon mixtures; 2–11 for an ethylene/ethane mixture, 10–50 for propylene/propane, and 10–40 for n-butane/iso-butane [26].

Carbon membranes were deposited on a Si (100) substrate by electrolysis of methanol under a DC voltage of 1200 V with a current density of about 52 mA/cm^2 at 50–55 °C [27]. The membranes consisted of amorphous carbon and high-quality single crystalline diamond particles of 200–300 nm in size.

CNF webs, prepared via electrospinning of various carbon precursors followed by stabilization and carbonization, could have high functionality for various applications, similar to nanoporous carbon membranes after activation [28–35]. Various carbon precursors have been employed to prepare CNF webs with well-developed nanopores, such as polyimide (PMDA/ODA) [28], polybenzimidazol (PBI) [29], polyacrylonitrile (PAN) [34], pitch with PAN [35], PAN with multi-walled CNTs (MWCNTs) [30,31], PAN with Si nanoparticles [32], and polycarbosilane coupled with chlorination after carbonization [33]. The resultant CNF webs were tested mainly as electrodes for electrochemical capacitors. By changing the concentration of pitch in tetrahydrofuran (THF) solution, microporosity in electrospun CNFs was controlled in the range of 0.27–0.38 cm^3/g by a simple heat treatment at 1000 °C [35].

10.1.2 Templating

Nanoporous carbon membranes have been fabricated by film-type replication (templating) of colloidal silica and ordered mesoporous silica SBA-15 using resorcinol-crotonaldehyde as a carbon precursor [36,37]. The membranes had a very high porosity of 84–95% and a total pore volume of 5–9 cm^3/g calculated from the N_2 adsorption isotherm: the high porosity is due to the formed bimodal mesopores, consisting of pores reflecting the unfilled space of hollow colloidal silica templates and of pores created by the dissolution of silica pore walls [37]. In order to obtain porous thin carbon membranes of tunable thickness by using colloidal silica as a template, chemical vapor infiltration of methane at 1050 °C was performed [38]. In Figure 10.4, scanning electron microscopy SEM images (side views) are shown for the template of silica spheres and the resulting carbon membrane. The nanoporous structure of the carbon membranes can be tailored by selecting the thickness and particle size of silica colloid template. Sucrose/silica composite films prepared by spin-coating on Si wafers were converted to mesoporous carbon membranes by carbonization at 900 °C, and had S_{BET} of 2603 m^2/g and V_{total} of 1.39 cm^3/g [39].

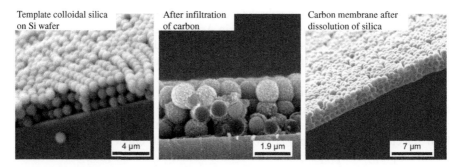

Template colloidal silica on Si wafer

After infiltration of carbon

Carbon membrane after dissolution of silica

4 μm   1.9 μm   7 μm

FIGURE 10.4 Colloidal Silica Template Consisting of 1.1 μm Particles on a Si Wafer, and Free-standing Carbon Membrane After Infiltration of Carbon and Dissolution of Silica

From [38]

Nanoporous carbon membranes have also been synthesized by using self-assembly of block copolymers as soft templates [40–49]. Highly cross-linked resorcinol-formaldehyde resin was prepared with the assistance of solvent-induced self-assembly from polystyrene-block-poly(4-vinylpyridine) (PS-P4VP) and converted to carbon membrane with highly ordered and well-oriented nanochannels of 2–20 nm in diameter [40]. Ordered mesoporous carbon membranes, powders, and fibers have been synthesized from a (resorcinol + phloroglucinol)-formaldehyde polymer with a triblock copolymer template in an ethanol/water solution on a silicon substrate by dip-coating [47]. After carbonization at 800 °C with slow heating, continuous and flat membranes were obtained with designed thickness, controlled by both the ethanol/water molar ratio and the withdrawal rate during dip-coating. Free-standing nanoporous carbon membranes with uniform mesopore sizes and controlled thickness have been fabricated from the composite films of self-assembled phenolic resin and block copolymer [49]. In these mesoporous carbon films, micropores could be created by KOH activation without noticeable change in the mesopore network [46].

Carbon membranes with nanochannels (cylindrical mesopores) have been fabricated from the polymer films of self-assembled phenol-formaldehyde oligomers with amphiphilic triblock copolymers (Pluronic P123) by pyrolysis at 350–600 °C [48]. Synthesized polymer films after pyrolysis at 350 °C were characterized by highly aligned cylindrical mesopores along the [110] direction with quasi-hexagonal symmetry along the (001) plane, as shown in Figure 10.5, but challenges have remained in obtaining the ordered cylindrical mesopores in carbon membranes after high-temperature treatment. For example, wall thickness has been insufficient to withstand the uniaxial stress experienced during template removal. Mesoporous carbon membranes prepared by using triblock copolymer template were further treated in a NH_3 atmosphere at 850 °C to improve the performance in redox flow batteries by nitrogen-doping [50].

A vapor-based soft-templating method has been developed: vapor of benzyl alcohol was penetrated into triblock copolymer films prepared on a silicon substrate,

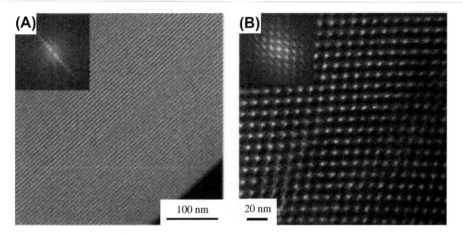

FIGURE 10.5 Cross-section Transmission Electron Microscopy Images of Mesoporous Polymer Films After Pyrolysis at 350 °C in Nitrogen

(A) (110) and (B) (001) planes

From [48]

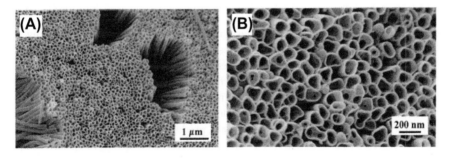

FIGURE 10.6 SEM Images of a Carbon Nanofiber Array Prepared Using an AAO Template

(B) is magnified image of (A)

From [52]

resulting in ultrathin carbon membranes (c. 15 nm thick) with ordered cylindrical mesopores after carbonization at 800 °C [51]. Benzyl alcohol thickened the pore wall of the triblock copolymer film, thus preventing structural shrinkage.

Well-aligned CNF arrays have been prepared by using epoxy-toluene solution as the carbon precursor and anodic aluminum oxide (AAO) film as the template [52]. Open-ended tubular carbons are aligned vertically in the membrane, as shown in Figure 10.6, and micropores are formed in the walls of tubes. Rapid mono-layer adsorption of *n*-hexane is attained due to small diffusion resistance during adsorption.

10.1.3 **Chemical and physical vapor deposition**

Application of CVD and PVD to produce nanoporous carbon membranes has been rare, but the films formed from carbon nanotubes and nanofibers that were synthesized by the catalytic CVD method have often been prepared, and their characteristics have been studied in many papers.

Porous carbon membranes have been prepared with a microwave-plasma-enhanced CVD technique [53]. These membranes showed superhydrophobicity with a contact angle of 150°, which was explained by the hydrogen-terminated edges of carbon layers. Nanoporous carbon membranes were deposited on porous Al_2O_3 disks using hexamethyldisiloxane as the carbon precursor by remote inductively coupled–plasma CVD, the resultant membranes having high performance for gas separation [54,55]. Porous carbon membranes have been deposited on an Al substrate with Co-Fe catalyst layer by means of pulsed discharge plasma CVD in a H_2/C_2H_2 gas mixture [56].

The PVD technique has been applied to synthesize nanoporous carbon membranes [57–59]. The laser-produced carbon plasma and vapor plume were deposited onto a rotating Si substrate with a highly oblique incident angle by pulse laser deposition to produce porous carbon membranes [57]. A graphite target was exposed to a laser pulse, which produced an expansion plume consisting of carbon ions, electrons, neutral molecules, clusters, and particulate ejecta of nanometer-size, to get porous carbon membranes. Electrochemical characterization was performed using these carbon membranes as an electrode, showing high faradaic and background currents, higher than for a polished glass-like carbon electrode, probably owing to its large substantial surface area in the electrolyte (63 m^2/m^2) [58]. Porous carbon membranes have been produced by low-energy cluster beam deposition, as shown in SEM and atomic force microscopy (AFM) images in Figure 10.7 [59]. The structure and properties of the carbon membrane were controlled by varying the cluster mass distribution prior to deposition and by co-depositing metallic nanoparticles together with carbon clusters. The carbon membranes produced by PVD contain almost no

FIGURE 10.7 Carbon Membranes Produced by the Low-energy Cluster Beam Deposition Technique

(A) SEM image, and (B) AFM image

From [59]

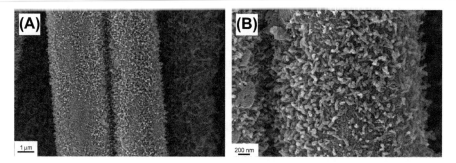

FIGURE 10.8 Activated Carbon Fibers Modified by Carbon Nanofiber Deposition

(B) is magnified image of (A)

From [60]

hydrogen but a large proportion of sp³ bonds, and are therefore very hard and show excellent wear resistance, in comparison with those produced by CVD.

Surface modification of carbon materials by the deposition of CNTs was reported to be effective to improve adsorption performance [60,61]. In Figure 10.8, activated carbon fibers modified by the deposition of CNFs are shown as an example [60].

10.1.4 Formation of carbon nanotubes and nanofibers

Membranes can be fabricated from CNTs and CNFs (CNT and CNF webs) by dispersing in solvents, in most cases oxidative solvents, for their de-bundling, and then filtering [62–64]; these have been called "buckypaper" [62]. Buckypaper has been successfully prepared from double-walled carbon nanotubes (DWCNTs), which produced a thin, flexible, and tough membrane. Because of the densely packed bundle texture, DWCNT membranes are highly porous, with S_{BET} of 510 m²/g, V_{total} of 2.05 cm³/g, and V_{micro} of 0.11 cm³/g, much larger than for membranes prepared from single-walled carbon nanotubes (SWCNTs) by the same procedure [63]. Heat treatment at 2000 °C modified the pore structure and hydrophilicity of DWCNT membranes [65]. Ultramicropores existing in the DWCNT membrane could adsorb water vapor, although the membrane itself was hydrophobic. Enrichment in ultramicropores in SWCNT membranes was possible by treatment in a mixed acid of HNO_3/H_2SO_4; S_{micro} increasing from 115 to 880 m²/g and V_{micro} from 0.10 to 0.38 cm³/g [66]. Hydrogen storage has been studied in the membranes of SWCNTs and DWCNTs [67]. DWCNT membranes having a thickness of a few tens of nanometers were obtained by flotation on the water surface after treatment in H_2O_2 and HCl [68]. SWCNT membranes as electrodes of electrochemical capacitors were prepared by the spray-drying of a water suspension of SWCNTs [69]. Thin membranes consisting of randomly or horizontally aligned SWCNTs were formed into multilayer films by transferring using Au and polyimide film as a stamp [70].

An electrophoretic process has also been employed to deposit a CNT membrane on a metallic substrate from a CNT-dispersed solution [71,72]. Since the deposition kinetics of CNTs are controlled by the applied electric field and deposition time, CNT membranes can be fabricated with designed thickness and excellent macroscopic homogeneity. From an as-grown CNT forest, CNT membranes have been formed either by spinning [73–75] or by mechanical compression with shear forces [76], as described in Chapter 2.

CNF webs have been fabricated from the webs of electrospun polymer fibers by carbonization and activation, with polyimides, polyacrylonitrile, or phenols used as carbon precursors, as described in Chapter 8.

10.2 Applications

10.2.1 Adsorbents

Nanoporous carbon membranes with S_{BET} of 1400 m^2/g (woven activated carbon fibers) were applied to purify laundry waste water, containing organic surfactants, by using dodecyl-benzene sulfate as a model pollutant [77]. Capacity and rate performances in cyclic adsorption have been studied by the regeneration at 800 °C under high vacuum. Microporous carbon membranes were experimentally shown to adsorb water vapor efficiently, suggesting their application for environmental control [14,15]. Nanoporous carbon membranes, prepared from fluorine-containing polyimide film and with micropores about 0.6 nm wide, adsorbed a large amount of water vapor, as large as 465 mg/g at room temperature; the adsorption commenced at the relative pressure, P/P_0, of about 0.2 and saturated above 0.5 [14]. Macroporous carbon membranes prepared via template carbonization of polyimides using polyurethane or melamine foam were easily converted to nanoporous carbon membranes by mild activation in air, and gave a rapid adsorption/desorption of water vapor in ambient air [78]. Adsorption/desorption isotherms are shown for the membranes (foams), before and after activation, in Figure 10.9A, showing that the adsorption begins at P/P_0 below 0.1. Repeated adsorption and desorption by alternating flow of wet and dry air was possible, revealing that desorption is much faster than adsorption (Figure 10.9B), and that reversible adsorption capacity is related to the micropore volume of the membranes (Figure 10.9C).

Ordered mesoporous carbons, which were prepared via template carbonization of phenol-formaldehyde using F127 at 900 °C and with S_{BET} of 2580 m^2/g, V_{total} of 2.16 cm^3/g, and bimodal pore-size distribution (population maxima at 6.4 and 1.7 nm), were found to have high adsorption capacity, twice that of commercial activated carbon, for bulky dyes, such as methylthionine chloride, fuchsin basic, and rhodamine B [79].

CNF webs, prepared by different techniques (electrospinning, catalytic CVD, and the template method) and thinner than commercially available activated carbon fibers (ACFs), were found to have highly developed microporosity after activation [80,81] and have been studied as the adsorbent for volatile organic compounds (VOCs), as described in Chapter 15.

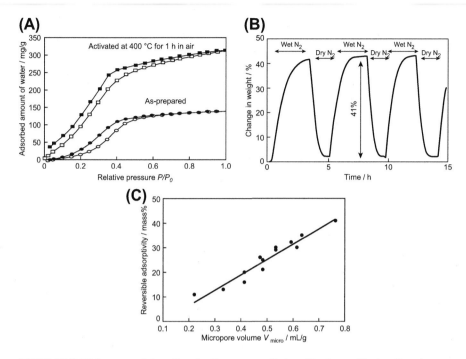

FIGURE 10.9 Water-vapor Adsorption for Nanoporous Carbon Membrane (Carbon Foam) Prepared From Polyimides

(A) Adsorption/desorption isotherms, (B) the repetition of water adsorption/desorption of the membrane, and (C) the dependence of the reversible adsorptivity on micropore volume, V_{micro}, of the membrane

From [78]

10.2.2 Separation membranes

Carbon membranes with micropores and mesopores are, perhaps, the most important material for gas separation owing to their high separation efficiency, high permeability, and relatively high thermal and chemical stability [82]. Carbon membranes prepared from different precursors using different supports gave a high permselectivity for O_2/N_2 up to 14, with a high O_2 permeance of about 2×10^{-7} mol/m^{2}sPa at 25 °C [54,55,83–85]. Carbon membranes prepared from polyimides have shown high permselectivity for different gas mixtures, including H_2/N_2, He/N_2, CO_2/N_2, O_2/N_2, H_2/N_2, CO_2/CH_4, and C_2H_4/C_2H_6 [10,11,86–94]. The permselectivities attained by a Kapton-derived carbon membrane were 4700 for H_2/N_2, 2800 for He/N_2, 122 for CO_2/N_2, and 36 for O_2/N_2 at 308 K [90]. Molecules with sizes larger than C_2H_6 (0.40 nm) can penetrate the polyimide-derived carbon membrane at 700 °C, but only a small amount of CO_2 (0.33 nm) penetrates the membrane after carbonization up to 1000 °C [10,11]. For carbon membranes prepared from the commercial polyimide Kapton with 0.125 mm in thickness, gas permeability of hydrogen and selectivity against CO and CO_2 are listed in Table 10.1, as a function of

Table 10.1 Hydrogen Permeability and Selectivity Against CO and CO_2 at 50 °C

Carbonization Temperature (°C)	H_2 permeability, P_{H2} (mol/msPa)	Selectivity	
		P_{H2}/PCO	P_{H2}/PCO_2
900	2.80×10^{-14}	200	17
1000	7.87×10^{-15}	1770	50
1000	2.40×10^{-15}	5900	161
1100	1.06×10^{-16}	$-^*$	343

*Permeation of CO could not be detected
From [11]

carbonization temperature [11]. Both permeability and permselectivity depend strongly on carbonization temperature. The carbon membrane carbonized at 1000 °C has very high selectivity for H_2, because the micropores formed in this membrane are mostly smaller than the minimum dimensions of CO and CO_2. Carbon membranes prepared from commercial polyimides via the phase inversion technique with asymmetric and symmetric pore structures had molecular sieve properties that are suitable for gas separation of O_2/N_2, CO_2/CH_4, and CO_2/N_2 [16]. Carbon membranes composed of hollow fibers of polyimide after carbonization at 750 °C had high separation performance for organic vapors of benzene and cyclohexane, and also worked as a dehydrogenation reactor from cyclohexane to benzene [95].

Carbon membranes prepared from polyimide (Figure 10.10A) gave high permselectivity for O_2/N_2 (air) separation [96]. In Figure 10.10B, permselectivities for carbon membranes prepared at different temperatures are plotted, as a function of permeability for O_2, together with those for polymer membranes. Permeability for O_2 increases by carbonization (pyrolysis) up to 535 °C and then permselectivity increases markedly with increasing carbonization temperature up to 800 °C, though permeability decreases slightly. The separation properties of carbonized membranes for O_2/N_2 are much better than most polymer membranes and well above the so-called "upper-bound trade-off curve."

Zeolite/carbon composite membranes have exhibited high gas permeability together with high selectivity for small molecules (H_2, CO_2, O_2, N_2) [97]. A composite membrane was prepared from the mixture of poly(amic acid) (PMDA/ODA) and nano-sized zeolite ZMS-5 (20–50 nm size) with zeolite content of 4.76–16.7 mass% by pyrolysis at 600–800 °C. High permeability for H_2 (kinetic diameter of 0.289 nm), but low permeability for N_2 (0.364 nm) was obtained. Carbon membranes were prepared from a polyimide (BPDA/ODA) on the surface of a porous α-Al_2O_3 tube, which gave high permselectivities for CO_2/CH_4, C_2H_4/C_2H_6, and C_3H_6/C_3H_8 gas mixtures [98–101]. In Figure 10.11, the micropore volume of the carbon membrane determined with different probe gases is shown as a function of carbonization temperature and oxidation conditions [101]. Volume of micropores with a width of less than 0.4 nm reaches a maximum under carbonization at 800 °C, and increases by additional oxidation at 300 °C. Carbon membranes, which were deposited on porous Al_2O_3 by a remote inductively coupled–plasma CVD of hexamethyldisiloxane and by combining the surface treatment with high-energy-ion bombardment, exhibited

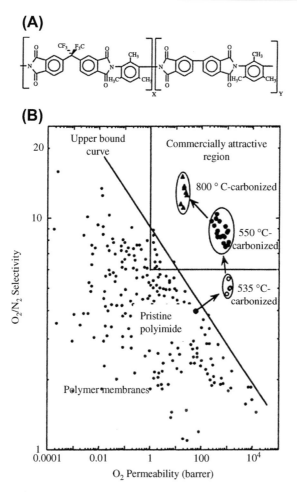

FIGURE 10.10 Carbon Membranes Prepared From Polyimide for O₂/N₂ Separation

(A) Repeating unit of pristine polyimide, and (B) permselectivity vs permeability for O_2, together with that for polymer membranes

From [96]

H_2/N_2 selectivity as high as 30–45 with an extremely high permeance of H_2 of around 1.5×10^{-6} mol/m²sPa at 150 °C [55].

The hydrophilic characteristics of carbon membranes are also useful for water separation from water/organic solvent mixtures, and their use in this application is highly desired in industrial processes for energy-efficient purification and separation, in contrast to the energy-consuming distillation widely used today. The high stability of carbon membranes under various severe conditions, such as in concentrated acid or alkaline solutions, makes them very suitable for pervaporation separation, which is a potentially useful technique for separating water from organic solvents or organic

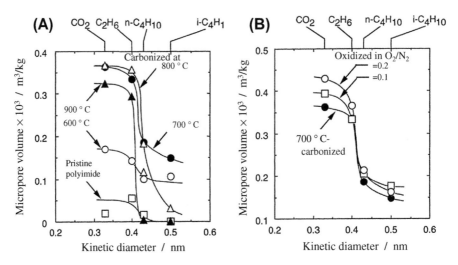

FIGURE 10.11 Micropore Volume Determined with Probe Gases of Different Kinetic Diameters for Nanoporous Carbon Membranes Prepared From Polyimide

(A) Effects of carbonization temperature, and (B) of oxidation conditions

From [101]

mixtures [83]. Porous carbon membranes have been shown to be useful for collecting fine aerosol particles in the atmosphere [102], although the work was aiming to apply this to transmission electron microscopy, suggesting a possible application in environment cleaning. The pore structure in carbons prepared from furfuryl alcohol using various templates has been discussed, with a focus on the application for gas separation and storage [103].

Modification of the separation properties (permeance and permselectivity) with storage period in different environments (air, nitrogen, and propylene) has been studied on carbon membranes prepared on the inner surface of porous alumina tubes from phenol resin at 700 °C [104]. The membranes stored in air showed a marked decrease in permeance for various gases, but permselectivity increased slightly. Storage of carbon membranes in propylene had a positive effect on permeance, whereas permselectivities decreased slightly. The results suggest that the performance of carbon membranes for the separation of gas mixtures containing oxygen, such as air, is expected to depend strongly on storage atmosphere, whereas the separation of oxygen-free gas mixtures, such as olefin/paraffin, n-butane/i-butane, and hydrogen/hydrocarbons, is a more reliable option.

10.2.3 Chemical sensors and biosensors

Nanoporous carbon membranes have been used in chemical sensors and biosensors that were based on the electrochemical behavior of the target substances. Owing to their high S_{BET} and nanopore volume, carbon membranes have the advantage of high enzyme loading in these applications, such as amperometric transducers for bio-catalytic sensors. Carbon membranes that were prepared by spin-casting of diluted PAN solution in humid air and carbonizing at 1100 °C have worked as electrocatalytic sensors with high sensitivity and rapid response in redox systems, such as nicotinamide adenine dinucleotide, uric acid, ascorbic acid, and acetaminophen [105]. The membranes, coupled with the efficient catalytic action of dispersed Ru and Pt particles, are attractive for use in low-potential electrocatalytic sensors [106].

CNT-based nanoporous carbon membrane sensors have been fabricated for the selective detection of various chemical and biological molecules and ions. MWCNT/polyvanillin composite membranes prepared via electropolymerization of vanillin aqueous solutions containing suspended MWCNTs on a carbon-fiber microelectrode were used as chemical sensors for nitrite [107]. Their sensitive response to NO_2 made them quite suitable for the determination of NO_2 in water. CNT-based sensors were also efficient for measurement of hydrogen molecules, carbon dioxide, nitrogen dioxide, proteins, DNA, organic chemicals, and ammonia [108–111]. The characteristics of sensors can be improved by treatment of CNTs with catalytic materials.

A pyrolytic graphite electrode, of which the surface was modified with either mesoporous nanofibers or ordered mesoporous carbon particles by deposition from suspension, showed high electrochemical response for low concentrations of dopamine, uric acid, and ascorbic acid at characteristic potentials [112].

10.2.4 Electrodes

Carbon membranes with different nanostructural and electrochemical characteristics have been tested as electrode materials in electrochemical capacitors and fuel cells. Compared with traditional electrodes made of carbon particles, nanoporous carbon films much reduce interparticle empty volume and increase the effective surface area, increasing the volumetric capacitance substantially.

A nanoporous carbon membrane was prepared from resorcinol-formaldehyde xerogel by carbonization at 1050 °C, and its performance as the electrode of electric double-layer capacitors (EDLCs) was studied [6]. In Figure 10.12, its cyclic voltammogram in 2 mol/L H_2SO_4 solution at a scan rate of 10 mV/s is shown, giving a high volumetric capacitance. The membranes had a rapid current response at each potential end and high values for both gravimetric and volumetric capacitances, facilitated by an S_{BET} of 915 m^2/g, V_{micro} of 0.37 cm^3/g, and V_{meso} of 0.73 cm^3/g, and a bimodal pore-size distribution with maxima at 0.6 and 9 nm. Micro-supercapacitors were fabricated by applying a unique method to synthesize monolithic-carbide-derived carbon membranes, by removing Ti metals from TiC with chlorine at elevated temperatures [113,114]. Microporous carbon membranes on a conductive TiC substrate could lead to a volumetric capacity of 180 F/cm^3 in organic 1 mol/L TEABF$_4$ electrolyte, and 160 F/cm^3 in 1 mol/L H_2SO_4 electrolyte.

Activated carbon fibers (ACFs) have been tested as electrode materials of EDLCs in aqueous and non-aqueous electrolyte solution by many research groups, as

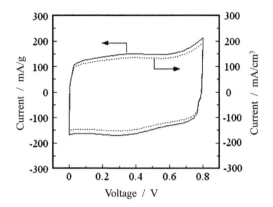

FIGURE 10.12 Cyclic Voltammogram of Nanoporous Carbon Membrane Prepared From Carbon-based Xerogel Membrane in 2 mol/L H$_2$SO$_4$ at a Scanning Rate of 10 mV/s

From [6]

described in Chapter 11. In most cases, ACF mats (webs) were used. CNF webs with developed pore structures prepared via electrospinning and carbonization have also been applied to the electrodes of electrochemical capacitors [28–33]. PAN-based CNF webs were tested as electrodes for capacitive desalination of Na$^+$ [34].

Porous carbon membranes reinforced by CNTs have been produced by the carbonization and KOH activation of PAN/CNT films cast on a glass substrate, and tested as an electrode of EDLCs [115]. A high capacitance of 300 F/g was achieved in 6 mol/L KOH electrolyte for the membranes activated at 800 °C. Energy density of 22 Wh/kg was obtained in an ionic liquid/organic electrolyte.

Carbon membranes produced by glancing angle deposition (PVD) exhibited good performance in redox systems [58]. The membrane electrodes exhibited peak-shaped voltammograms for all the redox systems investigated, as shown in Figure 10.13, in comparison with glass-like carbon in two redox systems. The membrane electrode demonstrates high faradaic and background currents on its surface, higher than the polished glass-like carbon electrode. CNF webs with high nitrogen content of about 16 mass% were prepared by carbonization-activation of polypyrrole nanofibers at 650 °C after mixing with KOH [116]. The resultant porous CNF webs gave a high reversible capacity of 943 mAh/g at 2 A/g after 600 cycles as an anode of lithium-ion batteries.

Nanoporous carbon membranes have also been applied as supports of metallic catalysts in fuel cells [117–119]. Onto the mesoporous carbon membranes prepared via soft-template carbonization of phloroglucinol-formaldehyde, Pt nanoparticles were loaded after heat treatment at 2600 °C to give a good electric conductivity of the support, resulting in stable catalyst membranes [117]. Pt-loading was also performed on membranes prepared by a similar procedure, after the pore walls of the membranes were non-covalently functionalized with poly(diallyldimethylammonium chloride) to enhance the activity and durability of the supports [119]. Pt particles

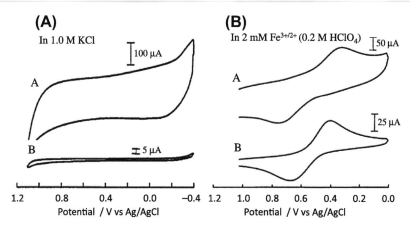

(A)

In 1.0 M KCl

I 100 μA

A

B $\mathbb{I}$ 5 μA

1.2   0.8   0.4   0.0   −0.4

Potential / V vs Ag/AgCl

(B)

In 2 mM $Fe^{3+/2+}$ (0.2 M $HClO_4$) I 50 μA

A

I 25 μA

B

1.2   1.0   0.8   0.6   0.4   0.2   0.0

Potential / V vs Ag/AgCl

FIGURE 10.13 Cyclic Voltammograms of Carbon Membranes in Two Redox Systems

(A) Nanoporous carbon membrane, and (B) glass-like carbon plate

From [58]

were uniformly dispersed in the nanopores of the carbon membranes, resulting in stable fuel-cell operation.

10.2.5 Other applications

Thin and flexible transparent films of CNT networks or CNT nanocomposites have been fabricated with excellent electrical conductivity and mechanical strength. The discharge capacity and performance of lithium-ion batteries were comparable to other flexible energy-storage devices or thick CNT-based devices [69,119–122]. High-quality SWCNTs were coated on the flexible polyethylene terephthalate substrate by spraying to form crisscross networks with uniformly distributed channels [122]. The film had low sheet resistance of 472 Ω/square at 85% optical transmittance. These thin films were developed for applications in flexible display devices [122], as well as high-performance supercapacitors and batteries [69,119–121]. Such CNT transparent conductive films have been fabricated for various devices, such as solar cells, secondary cells, and organic light-emitting diodes [122–127].

10.3 Concluding remarks

Nanoporous carbon membranes have been synthesized by various methods, as explained above. Most of the prepared membranes have randomly distributed nanopores, usually classified into micropores and mesopores. The number of open pores and pore sizes are important factors for applications such as adsorption, separation, sensing, and as electrodes.

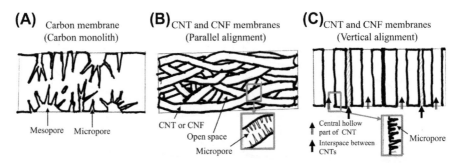

FIGURE 10.14 Illustrations of Pore Structure of Carbon Film and Carbon Fiber Webs with Different Orientations

Pyrolysis of organic films is the simplest way to prepare carbon membranes and has been used for a long time. However, most organic precursors give a large amount of volatile decomposition gases and, as a consequence, show a large shrinkage during pyrolysis and carbonization. This carbonization behavior of organic precursors has a high risk of causing cracks in the membrane, so that carbonization conditions have to be controlled precisely. Among various organic materials, aromatic polyimide is one of the precursors that can give carbon membranes easily, without any cracks. Different combinations of dianhydrides with aromatic diamines in the repeating unit give different polyimides, some of them being commercially available. The molecular structure in their repeating units is known to govern the structure, nanotexture, and properties of the resultant carbon membranes [9]. By selecting the precursor polyimide, microporous carbon membranes have been synthesized by one-step carbonization without any activation treatment.

CVD and PVD methods form carbon membranes directly on the substrates, and are attractive for fundamental studies of materials chemistry and engineering because of their potential applications in nanodevices and sensors.

The soft-template carbonization process allows the fabrication of carbon membranes with ordered nanopores of uniform size. Carbon membranes with ordered channels of mesopore-size diameters are of particular interest because of the variety in potential applications, such as nanometer-scale patterning, nanodevices, and size- and/or shape-selective reaction control. This soft-templating is based on the patterning of precursor polymers by using block copolymers, followed by direct carbonization, so that it does not require any removal procedure of templates.

Membranes consisting of CNTs and/or CNFs are attractive for applications related to nanoscience and technology. CNTs and CNFs are formed into membranes by different processes, such as filtration from their dispersing solutions, in situ formation, and spinning from their vertically aligned forests. These membranes may be classified according to the orientation of CNTs and/or CNFs, the axes of fibrous carbons being preferentially parallel or perpendicular. In Figure 10.14, the difference in pore structure is illustrated, and compared with carbon films (monolith). In nanoporous carbon films (Figure 10.14A), some of the micropores, which are important pores for most applications, are formed directly on the film surface, but some of them on the surface of mesopores. So, adsorbates, gaseous molecules, or ions, have to diffuse through the mesopores before arriving at the micropore entrance, although diffusion resistance is much smaller than that in granular activated carbons, where the adsorbate has to pass through macropores and then mesopores (Figure 10.1). In the case of webs with parallel alignment of CNTs and CNFs (Figure 10.14B), most micropores are formed directly on the surface of fibrous carbons and so adsorbates can arrive at the micropore entrances through the open spaces formed by the entangled fibrous carbons, most of the open spaces being macropores. In webs, the adsorbate is reasonably supposed to diffuse more easily than in films, but still passes through the open spaces. Since these open spaces have irregular shape and flexibly change their size, quantitative characterization of the spaces is very difficult. This is the reason why there has been no experimental investigation into controlling these spaces, even though it seems to be important. In the case of webs with perpendicular alignment (Figure 10.14C), such as vertically aligned CNT arrays (Figure 10.6), adsorbate molecules have to diffuse through the central hollow part of CNTs and the interspaces between CNTs, both being straight channels if alignment of CNTs is perfect. The sizes of these central hollow parts and interspaces between CNTs can be changed by changing the diameter and wall thickness of CNTs, although there has been no report on the effect of CNTs for gas permeability. In the case of hollow CNFs aligned vertically, micropores can be formed in the walls by activation and this may give interesting adsorptive performance, although no experiments have been reported.

Nanoporous carbon membranes are known to have superior properties over polymeric and inorganic membranes for different applications. Since pore structure and porosities of the membranes are determined mostly by the precursor used and its carbonization conditions, much effort is required to control the pore structure (size, shape, and number of pores) and surface nature for different applications. For practical applications, however, low-cost preparation methods have to be developed. Moreover, further potential for applications may be opened by developments in the modifications of carbon membranes, such as nitrogen-doping and addition of surface functionalities.

Nanoporous carbon membranes have high potential for various fields, and one of the most promising applications is environmental cleaning by adsorption and separation of organic pollutants. For example, the use of carbon membranes in water separation is a low-energy consumptive process, in comparison with water separation via distillation. Further study should be carried out on water adsorption and separation from organic solvent/water mixtures.

References

[1] Lee J, Kim J, Hyeon T. Adv Mater 2006;18:2073–94.
[2] Su F, Zhou Z, Cuo W, et al. Chem Phys Carbon 2008;30:63–128.
[3] Stein A, Wang Z, Fierke MA. Adv Mater 2009;21:265–93.

[4] Inagaki M, Orikasa H, Morishita T. RSC Adv 2011;1:1620–40.

[5] Bird AJ, Trimm DL. Carbon 1983;21:177–80.

[6] Tao Y, Endo M, Ohsawa R, et al. Appl Phys Lett 2008;93:193112.

[7] Yamashita J, Shioya M, Kikutani T, et al. Carbon 2001;39:207–14.

[8] Ejima H, Iwata T, Yoshie N. Macromolecules 2008;41:9846–8.

[9] Inagaki M, Takeichi T, Hishiyama Y, et al. Chem Phys Carbon 1999;26:246–333.

[10] Hatori H, Yamada Y, Shiraishi M, et al. Carbon 1992;30:305–6.

[11] Hatori H, Takagi H, Yamada Y. Carbon 2004;42:1169–73.

[12] Ohta N, Nishi Y, Morishita T, et al. TANSO 2008; No. 233: 174–180.

[13] Ohta N, Nishi Y, Morishita T, et al. Carbon 2008;46:1350–7.

[14] Ohta N, Nishi Y, Morishita T, et al. Ads Sci Technol 2008;26:373–82.

[15] Inagaki M, Ohmura M, Tanaike O. Carbon 2002;40:2502–5.

[16] Fuertes AB, Nevskaia DM, Centeno TA. Microp Mesop Mater 1999;33:115–25.

[17] Yamashita J, Shioya M, Kondo S, et al. Carbon 1999;37:71–8.

[18] Hatori H, Kobayashi T, Hanzawa Y, et al. J Appl Polym Sci 2001;79:836–41.

[19] Takeichi T, Yamazaki Y, Zuo M, et al. Carbon 2001;39:257–65.

[20] Inagaki M, Morishita T, Kuno A, et al. Carbon 2004;42:497–502.

[21] Park HB, Lee YM. Adv Mater 2005;17:477–83.

[22] Kaburagi Y, Aoki H, Yoshida A. TANSO 2012; No. 253: 95–99.

[23] Sun T, Feng L, Gao X, et al. Acc Chem Res 2005;38:644–52.

[24] Wang S, Zhu Y, Xia F, et al. Carbon 2006;44:1848–50.

[25] Fuertes AB. J Membr Sci 2000;177:9–16.

[26] Fuertes AB, Menendez I. Sep Purif Technol 2002;28:29–41.

[27] He S, Meng Y. Thin Solid Films 2009;517:5625–9.

[28] Kim C, Choi YO, Lee WJ, et al. Electrochim Acta 2004;50:883–7.

[29] Kim C. J Power Sources 2005;142:382–8.

[30] Ju YW, Choi GR, Jung HR, et al. Electrochim Acta 2008;53:5796–803.

[31] Guo Q, Zhou X, Li X, et al. J Mater Chem 2009;19:2810–6.

[32] Ji L, Jung KH, Medford AJ, et al. J Mater Chem 2009;19:4992–7.

[33] Rose M, Kockrick E, Senkovska I, et al. Carbon 2010;48:403–7.

[34] Wang M, Huang ZH, Wang L, et al. New J Chem 2010;34:1843–5.

[35] Kim BH, Yang KS, Kim YA, et al. J Power Sources 2011;196:10496–501.

[36] Gierszal KP, Jaroniec M. J Am Chem Soc 2006;128:10026–7.

[37] Gierszal KP, Jaroniec M, Liang C, et al. Carbon 2007;45:2171–7.

[38] Reculusa S, Agricole B, Derre A, et al. Adv Mater 2006;18:1705–8.

[39] Pang J, Li X, Wang D, et al. Adv Mater 2004;16:884–6.

[40] Liang C, Hong K, Guiochon GA, et al. Angew Chem Int Ed 2004;43:5785–9.

[41] Tanaka S, Nishiyama N, Egashira Y, et al. Chem Commun 2005:2125–7.

[42] Liang C, Dai S. J Am Chem Soc 2006;18:5316–7.

[43] Meng Y, Gu D, Zhang F, et al. Angew Chem Int Ed 2005;44:7053–9.

[44] Liang C, Li Z, Dai S. Angew Chem Int Ed 2008;47:3696–717.

[45] Wan Y, Shi Y, Zhao D. Chem Mater 2008;20:932–45.

[46] Gorka J, Zawislak A, Choma J, et al. Carbon 2008;46:1159–61.

[47] Tanaka S, Doi A, Nakatani N, et al. Carbon 2009;47:2688–98.

[48] Song L, Feng D, Fredin NJ, et al. ACS Nano 2010;4:189–98.

[49] Wang X, Zhu Q, Mahurin SM, et al. Carbon 2010;48:557–70.

[50] Shao Y, Wang X, Engelhard M, et al. J Power Sources 2010;195:4375–9.

[51] Jin J, Nishiyama N, Egashira Y, et al. Chem Commun 2009:1371–3.

[52] Hsieh CT, Chen JM, Kuo RR, et al. Appl Phys Lett 2004;84:1186–8.

[53] Xiao X, Cheng YT, Sheldon BW, et al. J Mater Res 2008;23:2174–8.

[54] Wang LJ, Hong FCN. Appl Surf Sci 2005;240:161–74.

[55] Wang LJ, Hong FCN. Microp Mesop Mater 2005;77:167–74.

[56] Noda M, Yukawa H, Matsushima M, et al. Diamond Relat Mater 2009;18:426–8.

[57] Vick D, Tsui YY, Brett MJ, et al. Thin Solid Films 1999;350:49–52.

[58] Kiema GK, Brett MJ. J Electrochem Soc 2003;150:E342–7.

[59] Bongiorno G, Podesta A, Ravagnan L, et al. J Mater Sci: Mater Electron 2006;17:427–41.

[60] Tzeng SS, Hung KH, Ko TH. Carbon 2006;44:859–65.

[61] Zhang JN, Huang ZH, Lv R, et al. Langmuir 2009;25:269–74.

[62] Endo M, Muramatsu H, Hayashi T, et al. Nature 2005;433:476.

[63] Muramatsu H, Hayashi T, Kim YA, et al. Chem Phys Lett 2005;414:444–8.

[64] Kim YA, Muramatsu H, Hayashi T, et al. Chem Vap Deposition 2006;12:327–30.

[65] Tao Y, Muramatsu H, Endo M, et al. J Am Chem Soc 2010;132:1214–5.

[66] Kim DY, Yang CM, Yamamoto M, et al. J Phys Chem C 2007;111:17448–50.

[67] Miyamoto J, Hattori Y, Noguchi D, et al. J Am Chem Soc 2006;128:12636–7.

[68] Wei J, Zhu H, Li Y, et al. Adv Mater 2006;18:1695–700.

[69] Kaempgen M, Chan CK, Ma J, et al. Nano Lett 2009;9:1872–6.

[70] Kang SJ, Kocabas C, Kim HS, et al. Nano Lett 2007;7:3343–8.

[71] Boccaccini AR, Cho J, Roether JA, et al. Carbon 2006;44:3149–60.

[72] Cho J, Konopka K, Rozniatowski K, et al. Carbon 2009;47:58–67.

[73] Zhang M, Fang SL, Zakhidov AA, et al. Science 2005;309:1215–7.

[74] Wei Y, Jiang KL, Feng XF, et al. Phys Rev B 2007:76; 045423.

[75] Liu K, Sun Y, Liu P, et al. Nanotechnology 2009;20:335705.

[76] Futaba DN, Hata K, Yamada T, et al. Nat Mater 2006;5:987–94.

[77] Matsuo T, Nishi T. Carbon 2000;38:709–14.

[78] Ohta N, Nishi Y, Morishita T, et al. New Carbon Mater 2008;23:216–20.

[79] Zhuang X, Wan Y, Feng C, et al. Chem Mater 2009;21:706–16.

[80] Tavanai H, Jalili R, Morshed M. Surf Interface Anal 2009;41:814–9.

[81] Jimenez V, Diaz JA, Sanchez P, et al. Chem Eng J 2009;155:931–40.

[82] Saufi SM, Ismail AF. Carbon 2004;42:241–59.

[83] Sakata Y, Muto A, MdA Uddin, et al. Sep Purif Technol 1999;40:97–100.

[84] Centeno TA, Fuertes AB. Carbon 2000;38:1067–73.

[85] Merritt A, Rajagopalan R, Foley HC. Carbon 2007;45:1267–78.

[86] Jones CW, Koros WJ. Carbon 1994;32:1419–25.

[87] Jones CW, Koros WJ. Ind Eng Chem Res 1995;34:158–63.

[88] Hatori H, Yamada Y, Shiraishi M, et al. TANSO 1995; No. 167: 94–100.

[89] Geiszler VC, Koros WJ. Ind Eng Chem Res 1996;35:2999–3003.

[90] Suda H, Haraya K. J Phys Chem B 1997;101:3988–94.

[91] Fuertes AB, Centeno TA. Microp Mesop Mater 1998;26:23–6.

[92] Fuertes AB, Centeno TA. J Membr Sci 1998;144:105–11.

[93] Ogawa M, Nakano Y. J Membr Sci 1999;162:189–98.

[94] Hatori H, Takagi H, Yamada Y. Carbon 2004;42:1169–73.

[95] Itoh N, Haraya K. Catal Today 2000;56:103–11.

[96] Singh-Ghosal A, Koros WJ. J Membrane Sci 2000;174:177–88.

[97] Liu Q, Wang T, Liang C, et al. Chem Mater 2006;18:6283–8.

[98] Hayashi J, Yamamoto M, Kusakabe K, et al. Ind Eng Chem Res 1995;34:4364–70.

[99]   Hayashi J, Mizuta H, Yamamoto M, et al. Ind Eng Chem Res 1996;35:4176–81.
[100] Yamamoto M, Kusakabe K, Hayashi J, et al. J Membr Sci 1997;133:195–205.
[101] Kusakabe K, Yamamoto M, Morooka S. J Membr Sci 1998;149:59–67.
[102] Dye AL, Rhead MM, Trier CJ. J Microsc 1997;187:134–8.
[103] Barara-Rodrigues PM, Mays TJ, Moggridge GD. Carbon 2003;41:2231–46.
[104] Menendez I, Fuertes AB. Carbon 2001;39:733–40.
[105] Wang J, Chen Q, Renschler CL, et al. Anal Chem 1994;66:1988–92.
[106] Wang J, Pamidi PVA, Renschler CL, et al. J Electroanal Chem 1996;404:137–42.
[107] Zheng D, Hu C, Peng Y, et al. Electrochim Acta 2009;54:4910–5.
[108] Kong J, Franklin NR, Zhou C, et al. Science 2000;287:622–5.
[109] Someya T, Small J, Kim P, et al. Nano Lett 2003;3:877–81.
[110] Jang YT, Moon SI, Ahn JH, et al. Sens Actuators B 2004;99:118–22.
[111] Bekyarova E, Kalinina I, Itkis ME, et al. J Am Chem Soc 2007;129:10700–6.
[112] Yue Y, Hu G, Zheng M, et al. Carbon 2012;50:107–14.
[113] Chmiola J, Gogotsi Y, Largeot C, et al. Prepr Symp Am Chem Soc, Div Fuel Chem 2008;53:867.
[114] Chmiola J, Largeot C, Taberna PL, et al. Science 2010;328:480–3.
[115] Jagannathan S, Liu T, Kumar S. Compos Sci Technol 2010;70:593–8.
[116] Qie L, Chen WM, Wang ZH, et al. Adv Mater 2012;24:2047–50.
[117] Shanahan PV, Xu L, Liang C, et al. J Power Sources 2008;185:423–7.
[118] Li M, Han G, Yang B. Electrochem Commun 2008;10:880–3.
[119] Shao Y, Zhang S, Kou R, et al. J Power Sources 2010;195:1805–11.
[120] Pushparaj VL, Shaijumon MM, Kumar A, et al. Proc Natl Acad Sci USA 2007;104: 13574–7.
[121] Zhou C, Kumar S, Doyle CD, et al. Chem Mater 2005;17:1997–2002.
[122] Du C, Pan N. J Power Sources 2006;160:1487–94.
[123] Paul S, Kim DW. Carbon 2009;47:2436–41.
[124] Rowell MW, Topinka MA, McGehee MD, et al. Appl Phys Lett 2006;88:233506.
[125] Li J, Hu L, Wang L, et al. Nano Lett 2006;6:2472–7.
[126] Li Z, Handel HR, Dervishi E, et al. Langmuir 2008;24:2655–62.
[127] Kang SJ, Song Y, Yi Y, et al. Carbon 2010;48:520–4.

Carbon Materials for Electrochemical Capacitors

11

Development of renewable energy resources, such as wind power, solar energy, and geothermal energy, is demanded urgently. In the devices to convert these natural energies to electrical energy and also to store the generated electrical energy, carbon materials are playing an important role. Electrochemical capacitors are expected to be one of the devices that will be used for the storage of electrical energy, and can be represented by the electric double-layer capacitor (EDLC). The elemental process in EDLCs is the physical adsorption/desorption of electrolyte ions, cations and anions, onto the surface of electrodes, in contrast to lithium-ion rechargeable batteries, where redox reactions between the electrode material and electrolyte are the elemental process for energy storage. For the electrode material of EDLCs, nanoporous carbons with a high surface area are usually employed. Since carbon materials usually have functional groups containing foreign atoms, such as oxygen and nitrogen, redox reactions between these functional groups and the electrolyte may give an additional capacitance, pseudo-capacitance. Also, capacitors with asymmetrical electrode constructions have been developed, where different materials are used for the cathode and anode. In Table 11.1, electrochemical capacitors are classified on the basis of electrode construction.

The advantages of electrochemical capacitors using carbon materials for the electrodes can be summarized as follows:

1. The ability to charge/discharge quickly.
2. A long cycle life because of the physical adsorption/desorption of electrolyte ions and no chemical reactions included.
3. A high efficiency in the charge/discharge cycle.
4. The lack of heavy metals, such as Cd or Pb, which makes them environmentally friendly.

On the basis of these advantages, in addition to their light weight, low cost, and maintenance-free operation, they are now used in various electronic devices, as described in various reviews [1–6]. However, high energy density comparable to rechargeable batteries is strongly desired for electrochemical capacitors.

Prerequisite for readers: Chapter 3.8, Section 2 (Electrochemical capacitors), in *Carbon Materials Science and Engineering: From Fundamentals to Applications*, Tsinghua University Press.

Advanced Materials Science and Engineering of Carbon.

Table 11.1 Electrochemical Capacitors

Symmetry of Electrodes	Mechanism of Energy Storage	Capacitors	
Symmetrical	Physical adsorption/desorption of electrolyte ions onto surface of carbon electrodes (Electric double-layer formation)	Supercapacitors	Electric double-layer capacitor (EDLC)
	Redox reaction of electrolyte ions with functional groups of carbon surface (Addition of pseudo-capacitance)		
Asymmetrical	Carbon materials with different pore structures in cathode and anode (Electric double-layer formation)		Asymmetrical EDLC
	Carbon materials in cathode (electric double-layer formation) and carbon loaded by metals in anode (Addition of pseudo-capacitance due to metallic particles)		
	Carbon materials in cathode (electric double-layer formation) and carbon materials intercalated by Li in anode (intercalation of Li ions)	Lithium-ion capacitors	

In this chapter, carbon materials used for electrochemical capacitors are reviewed, dividing electrode construction into symmetrical and asymmetrical. Theoretically, capacitance of supercapacitors depends strongly on the surface area of the carbon material used for the electrodes because energy storage is based on the physical adsorption of electrolyte ions. However, capacitance values reported in the literature cannot be compared directly, because the capacitance observed is determined by different factors: the electrolyte, the solvent, the concentration of electrolyte, the charge/discharge rate, and the temperature of measurement, in addition to factors related to carbon materials [6]. This chapter, therefore, concentrates mainly on describing the carbon materials that have been tested, and the possibilities for a general understanding of capacitive performance are presented. Here, the electrode on which the surface electric double-layer is formed by cation adsorption is called the anode, and that on which it is formed by anion adsorption is called the cathode.

11.1 **Symmetrical supercapacitors**

11.1.1 **Activated carbons**

In order to achieve better EDLC performance, the preparation conditions of carbon have been explored using different precursors and activation processes [7–19]. The effect of carbonization temperature on EDLC performance was studied on poly(vinylidene chloride) (PVDC) [7]. Capacitances in non-aqueous $LiClO_4/PC$ (propylene carbonate) and 30 mass% H_2SO_4 aqueous electrolytes were measured as a function of Brunauer-Emmett-Teller (BET) surface area, S_{BET}. Only carbons having S_{BET} higher than 1500 m^2/g gave capacitance in $LiClO_4/PC$ solution. Carbon materials prepared from different precursors (peach stone, furfuryl resin, saran, etc.) were subjected to capacitance measurement in 7 mol/L H_2SO_4 and KOH by a slow cyclic voltammetry [8]. The carbons with high S_{BET} of 2100 and 2700 m^2/g were able to retain capacitance at a temperature as low as −40 °C with a 10 and 25% reduction respectively, a much smaller decrease than with carbons having a lower S_{BET}. Mesophase pitch mixed with KOH was carbonized at 700 °C by changing the mixing ratio of KOH/pitch from 1.5 to 4.5 [11]. With the increase in ratio, S_{BET} increased from 1310 to 2860 m^2/g and capacitance from 80 to 130 F/g in 1 mol/L H_2SO_4. Heat treatment of mesophase pitch before carbonization at 560–750 °C was shown to have an influence on pore structure of the resultant activated carbon and, as a consequence, on EDLC performance [18,19]. For phenol-formaldehyde-derived carbons, the capacitance observed in 1 mol/L H_2SO_4 at a rate of 10 mV/s increased from about 90 to 105 F/g with increasing amount of activation reagent KOH, but it led to increases in the leakage current from 80 to 167 mA and in the resistance of the electrode from 22 to 75 Ω [13]. Charcoals prepared from cypress chips by carbonization in superheated steam gave a capacitance of c. 280 F/g in 1 mol/L H_2SO_4 at 20 mA/g, and very high retention in capacitance of more than 90% at 1000 mA/g [16]. Carbons prepared from seaweed (sodium alginate) by carbonization at 600 °C showed a high cycle performance in 1 mol/L H_2SO_4; the capacitance decreased from 198 to 190 F/g after 6000 cycles of charge/discharge at a sweep rate of 2 mV/s [17].

For various activated carbon fibers (ACFs) prepared from pitches, phenol resins, and polyacrylonitrile (PAN), capacitances in 1 mol/L H_2SO_4 and 1 mol/L $LiClO_4/PC$ were measured [20–22]. EDLC performance of mesophase-pitch-based and isotropic-pitch-based carbon fibers activated using KOH has been studied in 1 mol/L $TEABF_4/PC$ [23]. Mesophase-pitch-based ACFs gave higher capacitance, 40–45 F/g at 1–10 mA/cm^2, than isotropic-pitch-based ACFs, despite the latter giving a high S_{BET}, higher than 2000 m^2/g, where S_{BET} of the former was smaller than 70 m^2/g. Mesoporosity was introduced into microporous ACFs, which were prepared from a mixture of novolac-type phenol resin with nickel acetylacetonate dihydrate (0.1 mass% Ni) by spinning, carbonization, and activation in steam [24,25]. ACFs prepared with Ni salt had S_{BET} of 710 to 1660 m^2/g, depending on activation period, which is comparable to ACFs prepared without Ni addition, but the former had a larger mesopore volume, up to 0.86 cm^3/g, than the latter (0.16 cm³/g) [24]. In Figure 11.1A, the capacitance calculated from overall potential-time curves measured in 1 mol/L LiClO₄/PC in the potential range of 2.0–4.0 V vs Li/Li⁺ is plotted against S_{BET}. In Figures 11.1B and 11.1C, the capacitances calculated by dividing the curves into two ranges, corresponding to cation and anion adsorption, are plotted, in the range of 3.0 to 2.0 V (anode capacitance) and in the range of 3.0 to 4.0 V (cathode capacitance) [25]. Cation adsorption gives smaller capacitance than anion adsorption for both ACFs, with and without Ni salt, but the difference between these two ACFs is more marked in cation adsorption. These results suggest that solvated Li⁺, of larger size (0.82 nm), may be less easy to be adsorbed into micropores in the anode than ClO₄⁻, of smaller size (0.52 nm), into the cathode (ion sieving of micropores), and that mesopores may assist the diffusion of Li⁺ to micropores. ACFs with a relatively large amount of mesopores show good rate performance, much smaller reduction in capacitance with increasing current density.

When ACFs are used as electrodes in EDLCs, the low bulk density of less than 0.2 g/cm³ is disadvantageous. Electrode sheets with bulk densities of 0.36–0.81 g/cm³ have been prepared by hot-briquetting isotropic pitch fibers at 400 °C under 20 MPa, and the sheet capacitance in 1 mol/L KCl increased to 7.20 F, from 2.69 F for ACF without briquetting, although the capacitance on the basis of ACF weight was nearly the same, about 28 F/g [26].

The results published on activated carbons, including activated carbon fibers, suggest that pore-size distribution and surface chemistry of electrode carbons have to be taken into account, in addition to surface area, for achieving high performance of EDLCs. On the contribution of surface areas due to micropores and mesopores to EDLC capacitance, the following equation was proposed, observed capacitance, C_{obs}, being divided into two components, capacities due to the micropore surface,

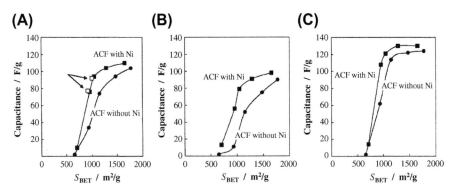

FIGURE 11.1 Dependence of Capacitance in 1 mol/L LiClO₄/PC on S_{BET} for ACFs Prepared From Phenol Resin With or Without Ni Complex

(A) Overall capacitance, (B) capacitance due to cation adsorption, and (C) that due to anion adsorption

Courtesy of Prof. S. Shiraishi of Gunma University, Japan

C_{micro}, and those due to the surface of larger pores (i.e. mesopores and macro-pores), C_{ext}:

$$C_{obs} = C_{micro} \times S_{micro} + C_{ext} \times S_{ext} \tag{11.1}$$

By using microporous surface area, S_{micro}, and external surface area, S_{ext}, determined from N_2-adsorption isotherms, a linear relationship between C_{obs}/S_{ext} and S_{micro}/S_{ext} has been observed for a number of activated carbons and activated carbon fibers [27]. In Figure 11.2, the corresponding plots are shown for various activated carbons, covering a wide range of S_{BET}, S_{micro}, and S_{ext} in 1 mol/L H_2SO_4 and 1 mol/L TEMABF$_4$/PC at two current densities [28,29]. The relationship is well approximated by applying least squares calculation to be linear both in the non-aqueous and aqueous electrolytes at different current densities of 100–1000 mA/g. The results show that the contributions from micropores (C_{micro}) and larger pores (C_{ext}) to C_{obs} are different.

In Table 11.2, data of C_{micro} and C_{ext} published in the literature [27,28,30–34] are summarized, together with the carbon material, electrolyte, test cell, current density, and method of surface-area determination. The conditions for the measurement of capacitance were very different, and the evaluation methods of surface areas were also different. So, direct comparison among the data of C_{micro} and C_{ext} might lead to an incorrect understanding. In Table 11.2, therefore, the ratio C_{ext}/C_{micro} is listed. As a whole, C_{ext}/C_{micro} is about 2 in the aqueous electrolyte, and about 4–7 in the non-aqueous electrolyte, and the charge/discharge rate has more marked effect on C_{micro} than on C_{ext}.

An EDLC composed of 1050 °C-treated activated carbon electrodes and an ionic liquid electrolyte showed capacitance of about 107 F/g at 20 mV/s and 60 °C [36]. A capacitor using an ionic liquid electrolyte was demonstrated to have high stability in capacitance and cycling at a high operating potential, up to 4.5 V [37].

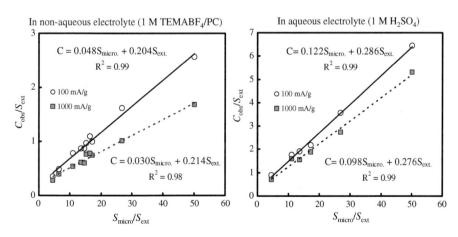

FIGURE 11.2 Capacitance at Different Current Densities as Functions of S_{ext} and S_{micro}

Table 11.2 Data Reported on the Contribution of Microporous and External Surfaces to EDLC Capacitance

Electrode Carbon	EDLC Measurement				Capacitance Contribution (F/cm²)			Reference
	Electrolyte	Test Cell	Current	Surface Area	C_{micro}	C_{ext}	C_{ext}/C_{micro}	
Microbeads ACFs	30 mass% KOH	2-electrode cell	5 mA	DFT	0.195	0.74	3.8	25
					0.145	0.075	0.5	
Coal-derived carbons	6 mol/L KOH / 1 mol/L H_2SO_4	2-electrode cell	160 mA/g / 2 mV/s	C_6H_6-adsorption	0.101	0.091	0.9	28
					0.098	0.231	2.4	
Cypress charcoals	1 mol/L H_2SO_4	3-electrode cell	50 mA/g / 1000 mA/g	α_s-plot	0.17	0.28	1.4	29
					0.13	0.18	1.2	
Activated carbons	1 mol/L H_2SO_4	3-electrode cell	100 mA/g / 1000 mA/g	α_s-plot	0.12	0.29	2.4	26
					0.10	0.28	2.8	
Glassy carbons	6 mol/L KOH	2-electrode cell	40 mA/g / 3000 mA/g	BET and t-plot	0.18	0.0039	0.16	30
					0.062	0.23	3.7	
Activated carbons	1 mol/L TEMABF$_4$/AN	3-electrode cell	20 mV/s	BET and DFT	0.076	−0.086	—	31
Activated carbons	1.2 mol/L TEMABF$_4$/AN	2-electrode cell	10 mV/s	α_s-plot	0.121	−0.112	—	32
Activated carbons	1 mol/L TEMABF$_4$/PC	2-electrode cell	100 mA/g / 1000 mA/g	α_s-plot	0.05	0.20	4.3	26
					0.03	0.21	7.2	
Air-oxidized carbon spheres	1 mol/L TEMABF$_4$/PC	2-electrode cell	100 mA/g / 1000 mA/g	α_s-plot	0.017	0.187	11.0	27
					0.008	0.116	14.5	

To improve EDLC performance, double-walled carbon nanotubes (DWCNTs) were mixed into activated carbon, and capacitance in 1.5 mol/L TEABF$_4$/AN (aceto-nitrile) was measured [38]. The DWCNT content of 15 mass% appeared to be a good compromise between stored energy and delivered power, and gave a capacitance of 93 F/g, equivalent series resistance of 0.4 Ωm^2, and stable cyclability over 10,000 cycles at 100 mA/cm^2.

Carbons with boron content of 1–3.4 mass% and S_{BET} of 850–1360 m^2/g were prepared from glucose-borate complexes under hydrothermal conditions, and gave a capacitance of about 230 F/g in 1 mol/L H$_2$SO$_4$ and about 100 F/g in 1 mol/L Na$_2$SO$_4$ at 2 mV/s [39]. Boron was found to be mainly in the form of C-B-O bonding, and to give a broad peak corresponding to pseudo-capacitance in a cyclic voltammogram (CV).

11.1.2 Templated carbons

For microporous carbons prepared by using zeolite templates, EDLC performance has been studied in 1 mol/L TEABF$_4$/PC [40]. CVs showed non-rectangular patterns, owing to pseudo-capacitance mainly resulting from oxygen functional groups on the carbon surface, but this did not correlate to the oxygen content of the carbons. Capacitance measured for these carbons is plotted as a function of current density in Figure 11.3, compared with an ACF. For most zeolite-templated microporous carbons, very good rate performance is observed, high and almost constant capacitance

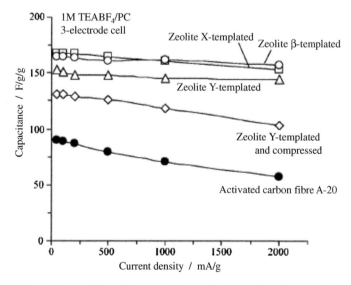

FIGURE 11.3 Dependence of Capacitance on Current Density for Zeolite-templated Carbons in 1 mol/L TEABF$_4$/PC

Courtesy of Prof. T. Kyotani of Tohoku University, Japan

of 160 F/g even at a current density as high as 2 A/g. This excellent rate performance was thought to be due to the three-dimensionally ordered and highly connected pores, giving a low ion-transfer resistance in micropores. On microporous carbons prepared using acetonitrile vapor as the carbon precursor and zeolite Y as the template, high capacitance of 146 F/g in TEABF$_4$/AN with low current of 10 mA was reported [41]. Microporous carbons were prepared via two-step process, impregnation of furfuryl alcohol (FA) and then chemical vapor deposition (CVD) of acetonitrile at 700 °C, with a zeolite 13X template, followed by KOH activation [42]. Capacitance in 1.5 mol/L TEABF$_4$/AN was 160 F/g (energy density of 30 Wh/kg) at 0.25 A/g, with a high retention of about 94% up to 2 A/g.

With the template method using mesoporous silica, the first impregnation of FA resulted in a bimodal mesoporous carbon with population maxima at 2.9 and 16 nm in width, but one with a unimodal distribution for mesopores of 2.8 nm was obtained after an additional impregnation [43]. Capacitance in 2 mol/L H$_2$SO$_4$ changed from about 200 to 110 F/g with current density from 1 to 150 mA/cm^2, and the results were divided into three regimes: capacitances due to micropores, those due to structural mesopores, and those due to complementary mesopores. The total surface areas of the same kinds of silica-templated carbons have been determined by different techniques, in the range of 1330–2250 m^2/g, and the capacitance in 2 mol/L H$_2$SO$_4$ and 6 mol/L KOH was 159–187 and 138–162 F/g, respectively, at 1 mA/cm^2 [44]. On similar silica-templated carbons, good cyclability was reported in 1 mol/L H$_2$SO$_4$, 6 mol/L KOH, and 1 mol/L TEABF$_4$/AN, about 77–97% of the initial capacitance being retained after 3000 cycles at 500 mA/g [45]. Templated carbons prepared using the mesoporous silicas MCM-48 and SBA-15 from carbon precursors of propylene and sucrose were tested in 1 mol/L H$_2$SO$_4$, 6 mol/L KOH, and 1 mol/L TEABF$_4$/AN at slow rates [46]. On silica-templated carbons prepared from sucrose, pitch, and propylene, capacitances measured in 1 mol/L H$_2$SO$_4$ and 1 mol/L TEABF$_4$/AN were proportional to micropore volume (0.06 to 0.36 cm^3/g), and were measured using CO$_2$ at room temperature [47,48]. Mesoporous carbons were prepared from FA at 900 °C using three silica templates, MCM-48, SBA-15, and modified SBA-15, which resulted in different mesopore arrangements: three-dimensional cubic, two-dimensional rod-like, and two-dimensional channel-like mesopore symmetries, respectively [49]. These ordered mesoporous carbons showed good rate performance in 30 mass% KOH, particularly the carbons with two-dimensional mesopore symmetry, giving about 90% retention by the change in sweep rate from 5 to 50 mV/s. Boron-doped mesoporous carbons prepared from sucrose with boric acid using SBA-15 gave high capacitance in both 6 mol/L KOH and 1 mol/L H$_2$SO$_4$, higher than pristine carbon [50], as shown in Figure 11.4. Nanoporous carbons prepared from FA using a metal-organic framework (MOF-5) as template had a high S_{BET} of 3040 m^2/g, and they were tested in 1 mol/L H$_2$SO$_4$ electrolyte [51]. By coupling with a conventional activation process, micropores were able to be increased in these nanoporous carbons [52,53].

EDLC performance of MgO-templated mesoporous carbons has been studied in aqueous and non-aqueous electrolytes [54–58]. The carbons are characterized by a

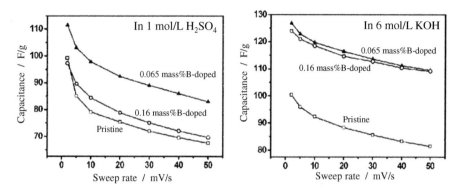

FIGURE 11.4 Rate Performances of Boron-doped Activated Carbons Prepared by Template Carbonization

From [50]

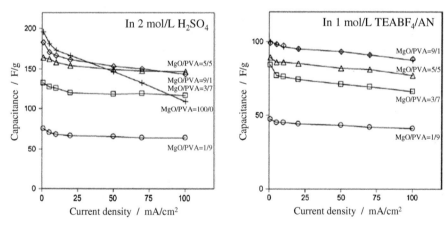

FIGURE 11.5 Dependence of Capacitance on Current Density in Aqueous and Non-aqueous Electrolyte Solutions for MgO-templated Carbons

PVA, poly(vinyl alcohol)

From [55]

high percentage retention at high charge/discharge rate, as shown in Figure 11.5. With increasing current density from 1 to 100 mA/cm^2, the capacitance reduces by approximately 15% in both electrolytes [55]. High retention of capacitance was related to the mesoporous surface area, S_{meso}, of the carbon used in the anode [54]. By using needle-like nanoparticles of Mg(OH)$_2$ as template, tunnel-like mesopores were introduced in the resultant carbon, which led to an improvement in rate performance in KOH electrolyte [56,57].

Mesoporous carbons have been prepared using Ni(OH)$_2$ as a template and phenol resin as the carbon precursor, and the pore structure was characterized by micropores of 1–2 nm in width, mesopores of 5–50 nm, and macropores of 60–100 nm [59]. Their CVs were rectangular even at a high sweep rate of 200 mV/s, and high capacitance retention of 90% was obtained by the increase in sweep rate from 20 to 100 mV/s. Boron and nitrogen co-doped mesoporous carbons have been prepared from gels formed from citric acid and boric acid together with $NiCl_2$ [60].

11.1.3 **Other carbons**

Using microporous carbons prepared from various metal carbides by chlorination (carbide-derived carbons), EDLC characteristics have been studied in various electrolytes [61–64], and discussed in relation to the effects of electrolyte ion size and the organic solvents used [65,66]. In an aqueous H_2SO_4 electrolyte, the carbon derived from ZrC gave 190 F/g at 5 mV/s [63]. For TiC-derived microporous carbons, the capacitance measured in 1.5 mol/L $TEABF_4$/AN and normalized by S_{BET} increased with decreasing average pore size [67]. B_4C-derived carbons gave a good rate performance in 6 mol/L KOH, about 86% retention by the change from 2 to 50 mV/s [68]. Micro-supercapacitors have been fabricated using thin films of carbide-derived microporous carbons, and the volumetric capacitance decreased rapidly with increasing thickness of the film [69].

Defluorination of polytetrafluoroethylene (PTFE) using alkali metals was found to give porous carbons (PTFE-derived carbons) with comparable S_{micro} and S_{meso}, which have been used as electrodes in EDLCs of 1 mol/L H_2SO_4 and 1 mol/L $LiClO_4$/PC electrolyte at 20 mA/g [70–72]. The capacitance of PTFE-derived carbons is much higher than that of ACFs prepared from phenol resin in steam [71], as shown in Figure 11.6A. The preparation conditions, defluorination reagent, and γ-ray irradiation all influenced EDLC capacitance. A carbon prepared using lithium naphthalenide ($LiC_{10}H_8$) in dimethoxyethane (DME) gave higher capacitance than ACFs in 1 mol/L $LiClO_4$/PC at 40 mA/g, as shown in Figure 11.6B [72]. The PTFE-derived carbon showed much better rate performance than ACFs in 1 mol/L $LiClO_4$/PC [72].

Capacitance of a non-activated carbon xerogel was 112 F/g in 30 mass% H_2SO_4, and this increased to 170 F/g after activation by CO_2 at 1050 °C [73]. This change was due to marked increases in both S_{micro}, from 530 to 1290 m^2/g, and S_{meso}, from 170 to 530 m^2/g. In order to apply carbon aerogels to EDLC electrodes, activation was shown to be necessary to develop micropores in the primary particles [74]. CO_2 activation at 1000 °C resulted in the development of micropores in comparable amount with mesopores, and about two-times larger capacitance in 0.8 mol/L $TEABF_4$/PC electrolyte. Modification of the carbon aerogel surface by a surfactant (sodium oleate) gave higher capacitance at a high current density of 48 mA/cm^2, probably owing to wettability improvement. Micropores in carbon aerogels increased by increasing the ratio of resorcinol to catalyst (R/C ratio) up to 1500, but this led to a slight decrease in capacitance in 30% H_2SO_4 [75]. Pore structure in a resorcinol-acetaldehyde cryogel has been controlled by changing the pH of the solution, i.e. by changing R/C [76]. A mesopore-rich carbon prepared in the pH range 7.3–7.7 gave

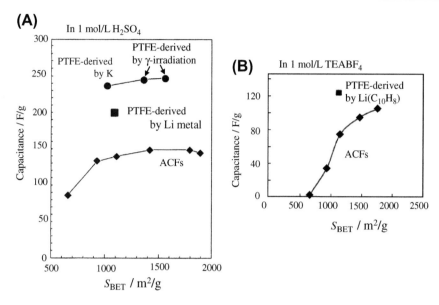

FIGURE 11.6 Relationship Between Capacitance and S_{BET} for PTFE-derived Carbons Defluorinated in Different Conditions, in Comparison with ACFs

In (A) aqueous and (B) non-aqueous electrolyte solutions

Courtesy of Prof. S. Shiraishi of Gunma University, Japan

high EDLC capacitance due to anion adsorption. By doping ammonium borate into organic cryogels, a mesoporous carbon co-doped with boron and nitrogen was prepared by carbonization at 1050 °C, giving a slightly higher capacitance in TEABF$_4$/ PC/DMC (dimethylcarbonate) [77].

EDLC performance of exfoliated carbon fibers, prepared from 3000 °C-treated mesophase-pitch-based carbon fibers, was studied in 1 mol/L H$_2$SO$_4$ [78]. In Figure 11.7A, their capacitance is plotted against S_{BET}, compared with ACFs. Exfoliated carbon fibers do not have so high an S_{BET} as commercial ACFs, but show relatively high capacitance. In contrast to ACFs, in which micropores are predominant, pore structure in exfoliated carbon fibers is characterized by a large amount of mesopores, which can be inferred from a marked hysteresis in N$_2$-adsorption/desorption isotherms, as shown in Figure 11.7B. By air-oxidation (activation) of exfoliated carbon fibers, capacitance increases markedly, in spite of a slight increase in S_{BET}. For exfoliated carbon fibers, a huge EDLC capacitance of 450–555 F/g was observed in concentrated 18 mol/L H$_2$SO$_4$ [79,80]. This high capacitance was thought to come from pseudo-capacitance due to intercalation of sulfuric acid molecules into the graphite gallery of exfoliated carbon fibers, in addition to electric double-layer capacitance, but no peak corresponding to redox reaction (intercalation) was observed on the CV. Stable galvanostatic charge/discharge was achieved for these exfoliated carbon fibers over 7000 cycles at 500 mA/g.

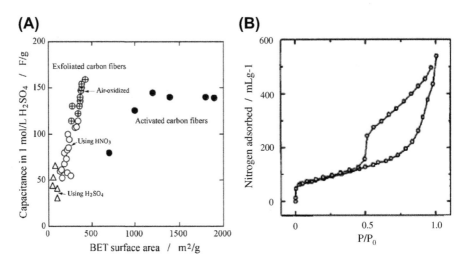

FIGURE 11.7 Characteristics of Exfoliated Carbon Fibers

(A) Relationship between capacitance in 1 mol/L H_2SO_4 and S_{BET}, in comparison with ACFs, and (B) adsorption/desorption isotherm of N_2 at 77 K

From [78]; courtesy of Prof. M. Toyoda of Oita University, Japan

From graphene synthesized by CVD on Cu foil and by the reduction of graphite oxide, ultrathin supercapacitors were constructed by arranging the layers of graphene particles perpendicularly to the current collector in a polymer-gel (PVA-H_3PO_4) electrolyte (all solid-state two-dimensional in-plane supercapacitors) [81].

11.1.4 Carbons containing foreign atoms

Carbon materials have various functional groups on the surface, most of them containing oxygen and/or nitrogen. Some of these functional groups are electrochemically active and contribute to the observed capacitance as pseudo-capacitance [82]. Also, nanoparticles of some metals and metal oxides loaded onto the carbon surface give additional pseudo-capacitance due to faradaic reactions.

The content of oxygen-containing functional groups was determined by a titration method (Boehm method) for ACFs prepared from phenol resin and activated carbons from a petroleum coke [83]. The ACF rich in carboxyl functional groups (total amount of oxygen-containing groups is 1.76 mmol/g) shows a clear hump in CV, which suggests the occurrence of faradic reaction, but the same ACF rich in phenolic hydroxyl groups (0.79 mmol/g) does not show this hump, as illustrated in Figure 11.8A. Capacitance, C_0, in 1 mol/L H_2SO_4 was calculated by extrapolation of the relationship between observed capacitance and current density to zero current density, which was assumed to be the capacitance excluding the effect of pore structures. A linear relationship between C_0 and the total amount of oxygen-containing groups was obtained for two carbon materials, as shown in

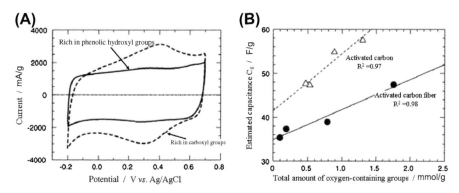

FIGURE 11.8 Characteristics of ACFs Rich in Either Carboxyl or Phenolic Hydroxyl Groups

(A) Cyclic voltammograms, and (B) relationship between the extrapolated capacitance, C_0, and the total amount of oxygen-containing groups, in comparison with activated carbon

Courtesy of Prof. H. Oda of Kansai University, Japan

Figure 11.8B. Oxygen-plasma treatment in an atmosphere containing 10 vol% O_2 increased S_{BET} of an ACF from 1570 to 2100 m²/g, the capacitance from 110 to 142 F/g in 0.5 mol/L H_2SO_4 at 5 mV/s [84]. On CV, faradaic redox peaks were markedly observed. Among the oxygen-containing functional groups measured by titration, a quinone group was thought to responsible for the capacitance increase.

Introduction of oxygen-containing functional groups onto PAN-based carbon fibers was effective for increasing capacitance in 2 mol/L HNO_3 at 90 °C [85]. Concerning the fibers, CO-desorbing complexes measured by temperature-programmed desorption (TPD) were thought to cause pseudo-capacitance in aqueous electrolytes, but CO_2-desorbing complexes to give a negative effect [85,86]. On various carbon materials, including anthracite and different carbon fibers, the capacitance in 1 mol/L H_2SO_4 was measured after activation by either KOH, NaOH, CO_2, or steam at 650–750 °C [87], and the capacitance was closely related to the concentration of CO-desorbing complexes. A part of pseudo-capacitance is expected to disappear during cycling. Hot HNO_3 treatment of multi-walled carbon nanotubes (MWCNTs) was shown to give pseudo-capacitance in 6 mol/L KOH, but only on the first charging [88]. On the other hand, capacitors composed of carbon-fiber fabrics treated either in HNO_3 at 450 °C or in oxygen at 250 °C exhibited excellent cyclability up to 100 cycles with 99.5% coulombic efficiency [85,89].

Capacitance of mesophase-pitch-derived carbons in both acidic and basic aqueous electrolytes (1 mol/L H_2SO_4 and 6 mol/L KOH) was found to decrease with increasing heat-treatment temperature of the carbons, on a parallel with the reduction of oxygen-containing surface groups [90]. The contribution of pseudo-capacitance due to CO-desorbing complexes was more significant in the acidic electrolyte than in the basic one. Electrochemical pretreatment of PAN-based ACFs in $NaNO_3$ increased the content of oxygen-containing functional groups and consequently increased the capacitance from 87 to 152 F/g in 0.5 mol/L H_2SO_4 [91]. Surfactant treatment of activated carbons was also effective to increase pseudo-capacitance [92].

Nitrogen-containing carbon materials have also attracted attention because high EDLC capacitance has been obtained, even though they did not have a large surface area. Nitrogen-containing carbons were prepared via a template carbonization method using expandable fluorine mica ($Na_{0.6-0.7}Mg_{2.80-2.85}Si_4O_{10}F_2$) as the template and either quinoline or pyridine as intercalates (carbon precursors) [93]. The resultant carbons were thin flakes and contained 10–18 mass% N. Though S_{BET} was very low at 30–150 m^2/g, the capacitance measured in 1 mol/L H_2SO_4 at 20 mA/g was relatively high at 100–180 F/g, which was thought to be the result of pseudo-capacitance due to nitrogen. Mesoporous carbon, which was prepared from quinoline-polymerized pitch via a silica template and contained 5.1 mass% N, gave a high capacitance of 289 F/g in 1 mol/L H_2SO_4 at 20 mA/g [94]. Nitrogen-containing carbons derived from melamine ($C_3H_6N_6$) showed nonrectangular-shaped CVs, giving enhanced capacitance on the negative-potential side due to desorption of cations in aqueous electrolytes [95,96]. Nitrogen content decreased gradually with increasing carbonization temperature, but capacitance showed a broad maximum of about 200 F/g at around 750–850 °C. However, no definite relationship was obtained between the capacitance in aqueous electrolytes and the nitrogen content in the electrode carbon, probably because the observed capacitance was a sum of EDLC capacitance due to the pore structure of the electrode carbon and pseudo-capacitance due to nitrogen-containing functional groups. Nitrogen-containing carbons have also been prepared by a simple carbonization of commercially available melamine foams at 600–1200 °C [97]. A scanning electron microscopy (SEM) image of carbon foam and its non-rectangular CV in 1 mol/L H_2SO_4 are shown in Figure 11.9. These carbon foams exhibited relatively high capacitance and good rate performance in 1 mol/L H_2SO_4, 241 F/g at 200 mA/g and 222 F/g at 1000 mA/g, although their surface areas were negligibly small. Carbon spheres containing about 10 mass% N,

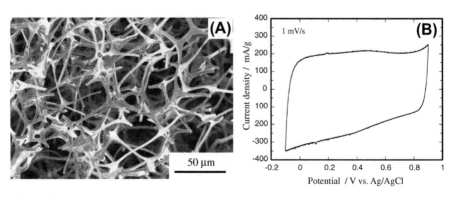

FIGURE 11.9 Melamine-derived Carbon Foam

(A) SEM image, and (B) cyclic voltammogram in 1 mol/L H_2SO_4

Courtesy of Dr M. Kodama of AIST, Japan

which were prepared via a silica colloid template method from a melamine-formaldehyde mixture at 800 °C, showed a good rate performance in 5 mol/L H_2SO_4, 211 F/g at 1 A/g and 200 F/g even at 20 A/g [98]. Mesoporous carbon spheres (c. 1.2 $\mu m\phi$) with 6 mass% N were prepared from a melamine-formaldehyde resin by using a fumed silica template, and the capacitance in 1 mol/L $LiPF_6/(EC + DMC)$ (EC: ethylenecarbonate) was about 159 F/g at 0.5 A/g and about 130 F/g at 2.0 A/g [99]. The addition of boric acid to melamine-formaldehyde resin was effective to keep the nitrogen content high even after carbonization, and consequently to improve capacitance (500 F/g at 100 mA/g) [100].

Nitrogen-containing activated carbons have been prepared by carbonization of polyacrylonitrile, oxidized polyvinylpyridine cross-linked with divinylbenzene, and coal tar pitch blended with them in steam at 800 °C [101]. Nitrogen content and S_{BET} of the resultant carbons were 7.2–1.9 mass% and 750–1420 m^2/g, respectively. Nitrogen in the carbon was assigned as pyridinic, pyrrolic and/or pyridonic, quaternary, and pyridine-oxide. EDLC capacitance in 1 mol/L H_2SO_4 was relatively high at 113–176 F/g at 100 mA/g, and that in a non-aqueous 1 mol/L $TEABF_4/$ AN was 67–94 F/g. For nitrogen-containing activated carbons prepared by the same process, the capacitance measured in 1 mol/L H_2SO_4 at 100 mA/g was found to be proportional to the nitrogen content, from 113 to 180 F/g with change in nitrogen content from 1.9 to 7.2 mass% [48]. The preparation and capacitive performance was reported for nitrogen-containing carbons prepared from electrospun poly(amic acid) [102], 2,3,6,7-tetracyano-1,4,5,8-tetraazanaphthalene [103], silk fibroin [104], and polyaniline [105]. Both urea and melamine treatment of a wood-derived activated carbon reduced S_{BET} from 2200 to 720–1400 m^2/g, but effectively improved rate performance of capacitance in 1 mol/L H_2SO_4, with about 80% retention with increasing current density from 0.05 to 1 A/g, although the original activated carbon gave only 15% retention [106]. Ammoxidation and urea treatment of carbons prepared from brown coals was effective in enhancing capacitance, despite the nitrogen content being small [107,108]. Surface modification of activated carbons by heat treatment in He/NO at 800 °C suppressed capacitance decline under high voltage operation (0–3 V at 70 °C) [109].

Co-doping of boron and nitrogen into carbon was performed by carbonization of a mixture of melamine-formaldehyde with boric acid at 800–1200 °C, and capacitive performance in 40 mass% H_2SO_4 was studied [110,111]. Doping of boron and nitrogen was confirmed by X-ray photoelectron spectroscopy (XPS). Capacitance of the carbon prepared at 1000 °C gave 500 F/g at 100 mA/g.

Amorphous ruthenium oxide, $RuO_2\cdot H_2O$, gives a high electrochemical capacitance (about 800 F/g) in aqueous electrolytes, but is extremely expensive. Loading of 35 mass% Ru into a carbon aerogel by CVD was reported to give 260 F/g [112]. By 7.1 mass% Ru-loading into an activated carbon with S_{BET} of 3000 m^2/g by the hydrolysis of $RuCl_3$, capacitance in 0.5 mol/L H_2SO_4 increased from 260 to 308 F/g [113]. Loading was also carried out by electroless deposition of Ru metal onto an activated carbon (S_{BET} of 1200 m^2/g) in a $RuCl_3$ aqueous solution and then heating at 100 °C in an oxygen atmosphere [114]. Capacitance in 1 mol/L H_2SO_4 increased from 98 F/g for the pristine carbon to 260 F/g for the 20 mass% Ru-loaded one, although the rate performance became a little worse. CVs of the RuO_2-loaded activated carbons are rectangular, with a redox peak for RuO_2 being difficult to detect. Ru-loading onto a carbon black was done through colloidal aqueous solution prepared from $RuCl_3 \cdot \times H_2O$ by adding an appropriate amount of $NaHCO_3$ [115]. Ru-loaded carbon black gave maximum capacitance in 1 mol/L H_2SO_4 from 221 to 599 F/g with increase in the amount of Ru from 20 to 80 mass%. Loading of RuO_2 on graphene layers has been carried out to improve capacitance in 1 mol/L H_2SO_4 [116]. Replacement of expensive Ru by other, inexpensive, metals, such as MoO_3 [117–119], Fe_3O_4 [120], metallic Fe and Co [121], and MnO_2 [122–126], has also been reported to give some benefit. Electrospun PAN-derived carbon nanofibers loaded by MnO_2 via dip-coating gave a high capacitance in Na_2SO_4 electrolyte: 365 F/g at 0 °C and 546 F/g at 75 °C with 1 A/g [125]. Loading of MnO_2 with conductive polypyrrole (PPy) onto carbon nanofibers was effective to improve capacitive performance in a KCl electrolyte [124]. A film prepared by layer-by-layer deposition of graphene and MnO_2 via electrostatic adsorption gave high capacitance in 0.1 mol/L Na_2SO_4 [126].

11.1.5 Carbon nanotubes and nanofibers

The theoretical surface area of single-walled carbon nanotubes (SWCNTs) can be as high as 2630 m^2/g, and SWCNTs are expected to have a homogeneous pore size of 1–3 nm if they are opened at the tips [127]. Therefore, carbon nanotubes might be an ideal electrode material, having homogeneous size and morphology of micropores. There has been much work in applying carbon nanotubes (CNTs) and nanofibers (CNFs) to EDLC electrodes [128–144]. However, the capacitance values reported for CNTs and CNFs scattered in a wide range, and most of them were not as high as expected.

For EDLC measurements, the use of CNT sheets (buckypaper) seems to be useful for understanding the proper electrochemical characteristics of CNTs themselves [128,130], though the volumetric capacitance is very low because of the low bulk density of the sheet [135]. The necessity of opening tips by oxidation was shown by using SWCNTs, and the central channels in the tubes, with the size of 2–5 nm, were considered to be mainly responsible for the EDLC capacitance [88]. In Figure 11.10, capacitance measured for a SWCNT sheet is plotted against current density during charge/discharge, in comparison with an ACF [127]. SWCNTs show an excellent rate performance, almost constant up to a high current density, which is explained by the presence of mesopores among the SWCNTs. A MWCNT forest grown on a Si substrate was shown to give a constant capacitance of 12 F/g over the wide range of current density of 0.1–200 A/g in 1.96 mol/L TEABF4/PC [133].

By KOH activation of the CNTs prepared by catalytic CVD, the capacitance in 1 mol/L $LiCl_4$/(EC + EDC) electrolyte doubled [134]. Strong oxidation of MWCNTs in 69% HNO_3 increased the capacitance in 6 mol/L KOH, owing to pseudo-capacitance

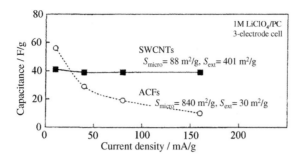

FIGURE 11.10 Rate Performance of SWCNTs in 1 mol/L LiClO$_4$/PC, in Comparison with ACFs

From [127]

of surface functional groups [131]. Capacitance of MWCNTs was increased in both 4 mol/L H$_2$SO$_4$ and 7 mol/L KOH by KOH activation and subsequent ammoxidation, and also self-discharge properties were improved [138].

SWCNTs that were rapidly grown by a water-assisted CVD method and vertically aligned ("super-grown" SWCNTs) [141] were successfully formed into densely packed sheets, in which tubes were highly aligned in one direction [142]. With these sheets, a high EDLC capacitance of 20 F/g was obtained in TEABF$_4$/PC, and the energy density per single electrode mass was estimated to be 69.4 Wh/kg, much higher than values reported before. These SWCNT sheets showed a unique butterfly-shaped CV in 1 mol/L TEABF$_4$/PC, as described in Chapter 2. The result suggested that two different mechanisms for electric energy storage coexist in the sheets, owing to the coexistence of metallic and semiconducting nanotubes [143]. For commercial SWCNTs and DWCNTs, similar butterfly-shaped CVs have been observed [144].

A composite of MWCNTs with PPy gave a high capacitance in 1 mol/L H$_2$SO$_4$, much higher than that of either MWCNT or PPy itself, maintaining a box-type CV [88]. Composites of SWCNTs with thin graphite flakes improved capacitance in 1 mol/L KCl [145]. Simple oxidation of CNFs in air was efficient to increase both surface area and capacitance [146]. By KOH activation of CNFs, S_{BET} increased from 140 m^2/g for the original to 1200 m^2/g, and the capacitance increased from about 6 to 29 F/g in 1 mol/L TEABF$_4$/PC [147].

For CNFs prepared from a polymer blend of phenolic resin with polyethylene after activation in a steam, a high S_{BET} of 2090 m^2/g, high capacitance of 120 F/g, and good cycle performance were obtained in an ionic liquid 1-ethyl-3-methylimidazolium tetrafluoroborate (EMImBF$_4$) and in 0.5 mol/L EMImBF$_4$/PC solution [148].

A capacitor made with electrodes of CNTs (10–30 nm in diameter prepared by catalytic CVD) and the electrolyte of a room-temperature molten salt (lithium bis(trifluoromethane sulfone)imide in acetamide), gave excellent EDLC performance: rectangular CV, high capacitance (22, 26, and 30 F/g at 20, 40, and 60 °C, respectively), high retention of capacitance at a high scan rate (22 F/g at 1 mV/s to 18 F/g at 50 mV/s), and high cyclability (about 90% retention after 1000 cycles) [149].

11.2 Asymmetrical supercapacitors

Asymmetrical supercapacitors have been constructed using different combinations of microporous and mesoporous carbons as cathode and anode [150,151]. Two series of asymmetrical capacitors, expressed by AC-5/XX and XX/AC-5, are employed, the former consisting of a microporous activated carbon, AC-5, as the anode and different carbons, XX, as the cathode, and the latter consisting of XX as the anode and AC-5 as the cathode. Symmetrical capacitors of each carbon XX/XX have also been constructed for comparison.

In Figure 11.11, the capacitance, C, measured in 1 mol/L TEMABF$_4$/PC at 100 and 1000 mA/g, is plotted against the S_{BET} ratio of anode to cathode carbons, S_{BET}(anode)/S_{BET}(cathode), for the asymmetrical capacitors AC-5/XX and XX/AC-5. In the figure, the capacitance measured for symmetrical capacitors is also plotted at the position of S_{BET} ratio of 1.0. The left-hand side from the position for the symmetrical capacitor (S_{BET} ratio = 1) shows how the capacitance changes with respect to anode carbon, and the right-hand side shows the effect of the cathode on capacitance.

On the right-hand side (i.e. AC-5/XX) at 100 mA/g (Figure 11.11A), capacitance decreases gradually as the S_{BET} ratio increases from 1.0, i.e. as S_{BET} of the cathode carbon decreases. In the range of S_{BET} ratio 1.0–1.7, capacitance decreases slightly with increasing S_{BET} ratio, though capacitance of the symmetrical capacitor using the same carbon decreases markedly. By increasing the charge/discharge rate to 1000 mA/g, capacitance becomes almost constant, independent of cathode carbons, as shown in Figure 11.11B. In the range of S_{BET} ratios greater than 1.7, capacitance appears to decrease rapidly, though only one data-point is reported. On the left-hand side (i.e. XX/AC-5), capacitance decreases rapidly with decreasing S_{BET} ratio both at 100 and 1000 mA/g, and is very similar to that for symmetrical XX/XX. In Figure 11.12A, performance rating expressed by the ratio of capacitance at 1000 mA/g to that at 100 mA/g, C_{1000}/C_{100}, is plotted against S_{BET} ratio. On the right-hand side (AC-5/XX), C_{1000}/C_{100} is very low and almost constant at about 55%. On the left-hand side (XX/AC-5), however, C_{1000}/C_{100} depends strongly on the S_{BET} ratio, i.e. on anode carbon.

The results of asymmetrical EDLCs shown in Figure 11.11 and Figure 11.12A suggest that capacitance and rate performance are predominantly governed by the pore structure of the anode carbon. In Figure 11.12B, C_{1000}/C_{100} is plotted against S_{meso} of anode carbon instead of against the S_{BET} ratio. The good linear relationship suggests that the rate performance is governed particularly by S_{meso} of the anode carbon. Accordingly, high-S_{meso} anode carbon is desired in order to have high rate performance, and so using mesoporous carbon of relatively high S_{meso} of 1700 m^2/g provides high C_{1000}/C_{100} close to 90%.

Asymmetrical supercapacitors have been constructed using nitrogen-containing carbons for the anode or cathode. By using cellulose-derived carbon fabrics as the anode, which were pyrolyzed at 400 °C followed by ammoxidation at 250 °C, capacitance enhancement was more pronounced in 7 mol/L KOH than in 4 mol/L H$_2$SO$_4$ [152].

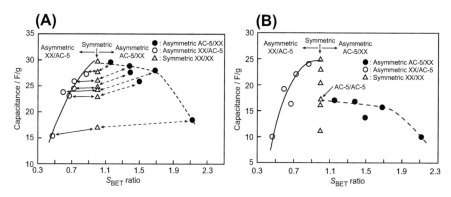

FIGURE 11.11 Changes in Capacitance With the Ratio in BET Surface Area of Anode to Cathode, S_{BET}(anode)/S_{BET}(cathode) (S_{BET} ratio)

(A) At 100 mA/g, and (B) at 1000 mA/g

From [151]

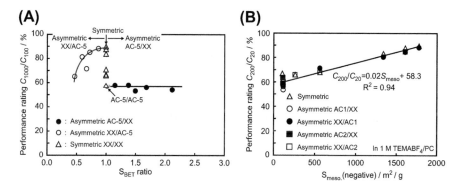

FIGURE 11.12 Asymmetrical and Symmetrical Supercapacitors

Performance rating, C_{1000}/C_{100}, (A) against S_{BET} ratio, and (B) against mesoporous surface area, S_{meso}, of anode carbon

From [151]

Ammoxidation of the cellulose precursor before carbonization was not effective, although a larger amount of nitrogen (2.5 mass%) was introduced. Ammoxidation of coal before carbonization resulted in high capacitance of 175 F/g in 4 mol/L H_2SO_4 for the cathode, but ammoxidation after activation resulted in a high capacitance in 7 mol/L KOH for the anode [153]. Improvement in self-discharge properties was also found for MWCNTs by ammoxidation [154]. Asymmetrical capacitors constructed using an anode of a mesoporous pitch-derived activated carbon and a cathode of the same material after 1000 °C-treatment had a good cyclability in 1 mol/L H_2SO_4 at 500 mA/g: 220 F/g for the first cycle, 210 F/g after 10,000 cycles, and 200 F/g after 20,000 cycles [155].

11.3 Asymmetrical capacitors

Asymmetrical capacitors consisting of different charge-storage mechanisms have been proposed: electric double-layer formation at the cathode and faradaic charge-transfer reaction with Li^+ at the anode (lithium-ion capacitor, LIC) [156–158]. A cell composed of an activated carbon with S_{BET} of 2200 m^2/g at the cathode and a carbon with a low S_{BET} of 250 m^2/g at the anode, which was prepared by heat treatment of the activated carbon with pitch at 700 °C, was reported to have high cycle performance in 1 mol/L $LiPF_6$/(EC/DEC) (DEC: diethylcarbonate) [156]. For an asymmetrical capacitor composed of a cathode of commercial activated carbon (S_{BET} of 1500 m^2/g) and an anode of non-graphitizing carbon (S_{BET} of 4.5 m^2/g), high energy and power densities (81.4 Wh/L and 8567 W/L) were obtained in a wide potential range of 1.5–4.3 V in 1 mol/L $LiPF_6$/(EC+DEC) [157]. In order to extend the potential range, a preliminary charging process was needed by using an auxiliary Li electrode in the cell, which was called doping of Li into the anode, consisting of non-graphitizing carbon. Energy and power densities became 1.14- and 2.3-times higher by extending the potential range from 2.5–4.3 V to 1.5–4.3 V. Asymmetrical capacitors using an activated carbon (S_{BET} of 1500 m^2/g) as cathode and a non-graphitizing carbon (S_{BET} of c. 3 m^2/g) or KOH-activated pitch-derived carbon (S_{BET} of 46 m^2/g) as the anode were tested in various non-aqueous electrolyte solutions of lithium salt [158]. A combination of natural graphite for the anode and commercial activated carbon for the cathode gave high energy density in a potential range of 1.5–4.5 V in 1 mol/L $LiPF_6$/EC/DMC [159].

An asymmetrical capacitor was constructed using an activated carbon for the anode and amorphous hydrous manganese oxide, α-$MnO_2 \cdot nH_2O$, for the cathode in neutral 1 mol/L KCl [160]. By operating at the potential range of 0–2 V, energy density of 28.8 Wh/kg was obtained. An asymmetrical capacitor of an activated carbon/ MnO_2 couple showed energy density of 10 Wh/kg between 0 and 2.2 V in 0.65 mol/L K_2SO_4 [161]. For improvement in cyclability, however, the cell voltage was recommended to be lower than 1.5 V to avoid evolution of hydrogen and oxygen. A composite of α-$MnO_2 \cdot nH_2O$ with carbon nanotubes (15 mass%) coupled with an activated-carbon anode was reported to extend the operating cell voltage up to 2 V, where hydrogen formed by water decomposition was thought to be adsorbed into micropores of the carbon and electrochemically oxidized during the discharge cycle [162]. Amorphous MnO_2 prepared by a simple co-precipitation method has been used as a cathode in 0.1 mol/L K_2SO_4, by coupling with an activated carbon with S_{BET} of 2300 m^2/g [163]. Long life of the cell over 195,000 cycles was achieved, with stable performance, by removing dissolved oxygen from the electrolyte. A stack of 40 electrodes was tested with sufficient stability. Asymmetrical capacitors constructed with a cathode of this MnO_2 and an anode of an activated carbon gave high capacitance in 0.1 mol/L aqueous solutions of Mg(II), Ca(II), and Ba(II) nitrates: 37, 32, and 29 F/g, respectively [164]. These capacitors showed rectangular CVs in a potential window from 0 to 2 V, as shown in Figure 11.13A, and high cyclability over 5000 cycles at 300 mA/g, as shown in Figure 11.13B.

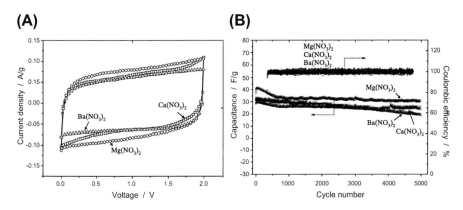

FIGURE 11.13 Asymmetrical Capacitors Composed of an MnO$_2$ Cathode and Activated Carbon Anode in 0.1 mol/L Mg(NO$_3$)$_2$, Ca(NO$_3$)$_2$, and Ba(NO$_3$)$_2$ Aqueous Solutions

(A) CVs, and (B) cycle performances

From [164]

Asymmetrical capacitors have been constructed by using various couples of carbon and electrochemically active materials in different electrolytes: activated carbon and LiMn$_2$O$_4$ in Li$_2$SO$_4$ [165], mesoporous NiO as the cathode and Ni(OH)$_2$-templated mesoporous carbon as the anode in 6 mol/L KOH [166], activated carbon and NaMnO$_2$ in Na$_2$SO$_4$ [167], and activated carbon and K$_{0.27}$MnO$_2$·0.6H$_2$O in K$_2$SO$_4$ [168]. In Figure 11.14A, CVs of the asymmetrical capacitor (mesoporous NiO/mesoporous carbon) are shown with different potential windows [166]. Up to 0.8 V, the main mechanism for charge storage seems to be the formation of an electric double-layer with a small hump due to redox reactions (pseudo-capacitance). With increasing potential to 1.5 V, the main storage mechanism becomes redox reaction on the surface of the NiO electrode. Since the pseudo-capacitance due to redox reactions on NiO is the main process for storage for a potential window of 1.0 to 1.5 V, capacitive performance is governed by the NiO electrode and the capacitance value depends strongly on the sweep rate, even though the capacitance is relatively high, as shown in Figure 11.14B. For this asymmetrical capacitor, high energy density and power density were expected, because it was possible to apply a wide potential window and a stable CV was confirmed at least up to 1000 cycles [164].

Asymmetrical capacitors have been constructed with a carbon anode and a conductive polymer cathode [169–171]. Capacitive performance of the couple of an active carbon and poly(4-fluorophenyl-3-thiophene) was studied in 1 mol/L TEACF$_3$SO$_3$/AN in the configuration of a prototype cell with an electrode area of 60 cm^2 [169]. A combination of an active carbon and poly(3-methylthiophene) has been studied in ionic liquid electrolytes [170,171]. Electropolymerization of poly(3-methylthiophene) at room temperature for the positive electrode was effective to increase capacitance in ionic liquids [171]. Asymmetrical electrochemical capacitors with polyfluorene as the cathode and an activated carbon (S_{BET} of 1600 m^2/g) as the

(A) **(B)**

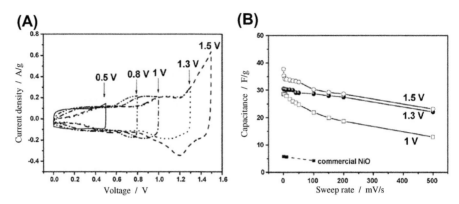

FIGURE 11.14 Asymmetrical Capacitor Composed of Mesoporous NiO and Carbon

(A) CVs, and (B) rate performances in different potential ranges

From [166]

anode gave 34 F/g in 1 mol/L $TEABF_4$/PC, higher capacitance and better cyclability than the symmetrical supercapacitor using the activated carbon [172].

An asymmetrical capacitor has been proposed using activated carbon as the cathode and $Li_4Ti_5O_{12}$ as the anode with $LiPF_6$/(EC + DMC) electrolyte solution [173]. It showed a sloping voltage profile in a range of 1.5 and 3 V, which was because of faradaic reaction between the anode material and Li^+ ions in the electrolyte, and good cyclic performance with about 10% loss after 5000 cycles. Prototypes of this asymmetrical capacitor with 500 F were built, which had an energy density of 10.4 Wh/kg, a power density of 793 W/kg at 95% discharge efficiency, long cycle life up to 10^5 cycles, and a weight of 43 g [174]. The performance of this prototype cell was compared with that of a conventional symmetrical supercapacitor, a lithium-ion rechargeable battery, and two other types of asymmetrical capacitors composed of $LiCoO_2$ or $LiMn_2O_4$ anodes and an activated-carbon cathode [175].

11.4 Carbon-coating of electrode materials

Carbon-coated tungsten and molybdenum carbides, WC and Mo_2C, were found to give a high capacitance in 1 mol/L H_2SO_4 [176]. WC and Mo_2C were prepared from mixtures of hydroxyl propyl cellulose (the carbon precursor) and K_2WO_4 or K_2MoO_4 (the carbide precursor) via a xerogel, followed by heat treatment at 800–1050 °C in Ar. In Figure 11.15A, CV of the first cycle is shown for the carbon-coated WC prepared at 1000 °C. In the same figure, CVs of the first and hundredth cycles for a WC powder of about 1 μm in size are also shown for comparison. In Figure 11.15B, X-ray diffraction (XRD) patterns are shown for the carbon-coated WC and the reference WC powder after the first cycle. When WC powder without carbon coating was used as the electrode, a redox peak is clearly observed at the

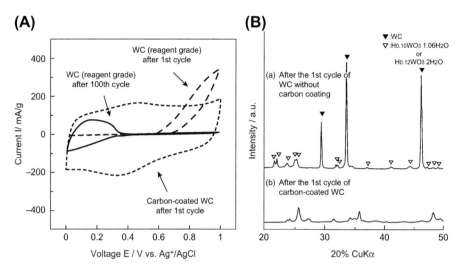

FIGURE 11.15 Characteristics of WC Electrodes With or Without Carbon Coating

(A) CVs, and (B) XRD patterns after first charge-discharge of carbon-coated WC

From [176]

first cycle. After the hundredth cycle, this peak becomes broad and shifts to the low-voltage side. WC changed only partly to oxy-hydroxides, such as $WO_{0.88}(OH)_{4.12}$ and $WO_{1.84}(OH)_{2.22}$, after the first cycle (Figure 11.15B); it needed more than 100 cycles to complete the transformation to oxy-hydroxides. For the carbon-coated WC, of small particle size, on the other hand, the CV is rectangular in shape with a faint redox peak, no appreciable change in CV being detected during the cycles. Diffraction peaks corresponding to WC disappeared completely after the first cycle and many small diffraction lines were observed, which were thought to be due to tungsten oxy-hydroxides, (Figure 11.15B). For carbon-coated Mo_2C, a similar structural change from carbide to oxy-hydroxide was reasonably supposed to occur in the first charge/discharge cycle.

By selecting the optimal temperature for the preparation of carbon-coated WC and Mo_2C, high gravimetric and volumetric capacitance was achieved at 50 mA/g [176]: 200 F/g and 370 F/cm^3 for carbon-coated WC, and 230 F/g and 750 F/cm^3 for carbon-coated Mo_2C. For carbon-coated Mo_2C, in particular, a high capacitance of about 730 F/cm^3 was maintained even after 500 cycles.

Coating of porous carbon inhibited the crystal growth of carbides, which made the complete transformation of carbide to hydroxide possible at the first cycle of charge-discharge in H_2SO_4 electrolyte, and also disturbed the agglomeration of metal hydroxides during charge-discharge cycles. Small particles coated by porous carbon are advantageous in attaining high efficiency for charge/discharge cycles, as is the case with electrode materials in lithium-ion rechargeable batteries (Chapter 12).

11.5 Concluding remarks

A huge number of papers have been published on electrochemical capacitors using carbon materials, including measurements of capacitance as a part of the characterization of porous carbons synthesized. Regrettably, direct comparison among these reported results is difficult because of the wide variety of measurement conditions, as mentioned at the beginning of this chapter. In some papers, the calculation of capacitance values from the data measured was not correctly done. Specifications for the procedure to determine the characteristics of capacitors are strongly required, which will have to be accepted by the international community. As recommended in a 2010 review [6], the following items should be specified:

1. For the electrolyte solution, specifications and concentration of electrolytes, solvent of non-aqueous solutions.
2. For the electrode sheet, particle size of sample carbons; specifications of conductive additives and binders; mixing ratio with electrode materials; thickness, size, and density of the sheet; and current collectors.
3. For the electrochemical measurements, measuring methods (cyclic voltammetry or galvanostatic charge/discharge cycle); test cell (2-electrode or 3-electrode); potential range; current density by area, mass, or volume; formula for capacitance calculations; and measuring temperature.

The use of both aqueous and non-aqueous solutions is strongly desired. In addition, it might be beneficial to select a reference porous carbon, of which the capacitance value and pore structure are well established, and recommend the researchers to present its capacitance value in each paper to demonstrate the reliability and/or reproducibility of the capacitance measurements.

Various carbon materials have been tested as electrode materials for electrochemical capacitors, as described above. The pore structure of a carbon electrode is well understood to be important and needs to be controlled. However, it is not yet established what pore structure is the most appropriate. In order to obtain high capacitance, microporous carbon is recommended as the electrode material, but high rate performance cannot be expected; in other words, microporous carbons are suitable for use at a low current density. To have high rate performance, mesoporous carbon is desired, mainly because of the low diffusion resistance of electrolyte ions in the electrode, but micropores are needed to attain usable capacitance. There has been no systematic study to optimize the combination of microporous and mesoporous surface areas for obtaining high capacitive performance. The analysis of data by the procedure explained in relation to Figure 11.2 is one possibility for understanding the contributions of micropores and mesopores. However, the analysis procedure has to be consistent both for the determination of capacitance and for the determination of surface areas.

Template carbonization processes are effective for controlling pore structure in carbon materials, giving high concentration and uniform size of pores, as discussed in Chapter 7. In order to satisfy the requirements of electrode materials in capacitors, it might be worthwhile to use preparation procedures combined with a template process; for example, conventional activation for mesoporous carbons synthesized via a silica template. Although CNTs and CNFs are interesting for use in capacitors, the detailed structural characterization of these carbons is strongly required to discuss further development of the application of these materials. When CNTs and CNFs are used as webs or mats, the spaces and/or pores between the fibrous particles have to be characterized correctly, some of them being able to work as mesopores while some reduce the density of the electrode.

Capacitive performance of exfoliated carbon fibers [73] suggests the possibility to extend exfoliation to other carbon materials. Since exfoliation is known to change the morphology of graphitized vapor-grown carbon fibers completely [177], it might be interesting to study their capacitive performance to be able to reproduce the reported high values [79,80]. Exfoliation of natural graphite has been widely used to prepare flexible graphite sheets and graphene (Chapter 3). Exfoliated graphite may give an interesting performance as an electrode, but it is important to take into account whether graphene, i.e. a single layer of carbon hexagons, is really needed.

Asymmetrical supercapacitors should be investigated further for understanding the fundamental factors to control capacitive performance and also for the development of practical capacitors. As shown in section 11.2, capacitive performance was predominantly governed by the pore structure of the anode carbon in 1 mol/L TEMABF$_4$/PC. However, it depends on the electrolyte used which electrode governs the capacitive performance. In the case of TEMABF$_4$ electrolyte, the anode governs the performance probably because the cation, TEMA$^+$, is much larger than the anion, BF4$^-$. In the case of H$_2$SO$_4$, however, the cathode might govern the performance because the anion, SO$_4{}^{2-}$, is much larger than the cation, H$^+$. More detailed studies have to be done for various electrolytes. The results on asymmetrical supercapacitors suggest that the two electrodes do not need to be composed of the same porous carbon. Therefore, more detailed studies might give hints for cost saving for practical capacitor devices.

Asymmetrical capacitors based on the redox reaction of un-doping/re-doping of Li$^+$ ions at the carbon anode combined with electric double-layer formation at the carbon cathode have been proposed and studied, as described above. In 2011, the redox reaction of un-doping/re-doping of Br$^-$ and I$^-$ ions in Et$_4$NBF$_4$/PC at the carbon anode was shown to give pseudo-capacitance [178], suggesting the possibility of constructing an asymmetrical capacitor combined with a carbon anode. In addition, multi-valence metal salts, like Mg(II) and Al(III) ions, are worthwhile testing for electrolytes to improve the efficiency of charge/discharge, as nitrates of alkaline earth metals have been used [164].

References

[1] Parsons R. Chem Rev 1990;90:813–26.
[2] Conway BE. Electrochemical Supercapacitors. Kluwer Academic/Plenum 1999.
[3] Frackowiak E, Beguin F. Carbon 2001;39:937–50.
[4] Pandolfo AG, Hollenkamp AH. J Power Sources 2006;157:11–27.

[5] Beguin F, Frackowiak E, editors. Carbon Materials for Electrochemical Energy Storage Systems. CRC Press: 2009.

[6] Inagaki M, Konno H, Tanaike O. J Power Sources 2010;195:7880–903.

[7] Endo M, Kim YJ, Takeda T, et al. J Electrochem Soc 2001;148:A1135–40.

[8] Toupin M, Belanger D, Hill I, et al. J Power Sources 2005;140:203–10.

[9] Braun A, Baertsch M, Schnyder B, et al. J Non-Cryst Solids 1999;260:1–4.

[10] Lozano-Castello DL, Cazorla-Amoros D, Linares-Solano A, et al. Carbon 2003;41: 1765–75.

[11] Weng TC, Teng H. J Electrochem Soc 2001;148:A368–73.

[12] Kierzek K, Frackwiak E, Lota G, et al. Electrochim Acta 2004;49:515–23; Erratum, 1169–1170.

[13] Teng H, Chang Y, Hsich C. Carbon 2001;39:1981–7.

[14] Alonso A, Ruiz V, Blanco C, et al. Carbon 2006;44:441–6.

[15] Asakura R, Kondo T, Morita M, et al. TANSO 2004; No. 215: 231–235.

[16] Ito E, Mozia S, Okuda M, et al. New Carbon Mater 2007;22:321–6.

[17] Raymundo-Piñero E, Leroux F, Béguin F. Adv Mater 2006;18:1877–82.

[18] Zhai D, Li B, Du H, et al. J Solid State Electrochem 2011;15:787–94.

[19] Zhai D, Li B, Kang F, et al. Microp Mesop Mater 2010;130:224–8.

[20] Endo M, Maeda T, Takeda T, et al. J Electrochem Soc 2001;148:A910–4.

[21] Endo M, Kim YJ, Maeda T, et al. J Mater Res 2001;16:3402–10.

[22] Kim Y, Horie Y, Matsuzawa Y, et al. Carbon 2004;42:2423–32.

[23] Kim YJ, Horie Y, Ozaki S, et al. Carbon 2004;42:1491–500.

[24] Shiraishi S, Kurihara H, Oya A. Carbon Sci 2001;1:133–7.

[25] Shiraishi S, Kurihara H, Shi L, et al. J Electrochem Soc 2002;149:A855–61.

[26] Nakagawa H, Shudo A, Miura K. J Electrochem Soc 2000;147:38–42.

[27] Shi H. Electrochim Acta 1996;41:1633–9.

[28] Wang L, Toyoda M, Inagaki M. New Carbon Mater 2008;23:111–5.

[29] Wang L, Inagaki M, Toyoda M. TANSO 2009; No. 240: 230–238.

[30] Gryglewicz G, Machnikowski J, Lorenc-Grabowska E, et al. Electrochim Acta 2005;50: 1197–206.

[31] Ito E, Mozia S, Okuda M, et al. New Carbon Mater 2007;22:321–6.

[32] Wen Y, Cao G, Cheng J, et al. J Electrochem Soc 2005;152:A1770–5.

[33] Barbieri O, Hahn M, Koetz R. Carbon 2005;43:1303–10.

[34] Jaenes A, Kurig H, Lust E. Carbon 2007;45:1226–33.

[35] Wang L, Fujita M, Inagaki M. Electrochim Acta 2006;51:4096–102.

[36] Lazzari M, Mastragostino M, Soavi F. Electrochem Commun 2007;9:1567–72.

[37] Balducci A, Dugas R, Taberna PL, et al. J Power Sources 2007;165:922–7.

[38] Portet C, Taberna PL, Simon P, et al. Electrochim Acta 2005;50:4174–81.

[39] Ito T, Ushiro M, Fushimi K, et al. TANSO 2009; No. 239: 156–161 [in Japanese].

[40] Nishihara H, Itoi H, Kogure T, et al. Chem Eur J 2009;15:5355–63.

[41] Portet C, Yang Z, Korenblit Y, et al. J Electrochem Soc 2009;156:A1–6.

[42] Wang H, Gao Q, Hu J, et al. Carbon 2009;47:2259–68.

[43] Fuertes AB, Pico F, Rojo JM. J Power Sources 2004;133:329–36.

[44] Centeno TA, Sevilla M, Fuertes AB, et al. Carbon 2005;43:3012–5.

[45] Fuertes AB, Lota G, Centeno TA, et al. Electrochim Acta 2005;50:2799–805.

[46] Jurewicz K, Vix-Guterl C, Frackowiak E, et al. J Phys Chem Solids 2004;65:287–93.

[47] Vix-Guterl C, Frackowiak E, Jurewicz K, et al. Carbon 2005;43:1293–302.

[48] Frackowiak E, Lota G, Machnikowski J, et al. Electrochim Acta 2006;51:2209–14.

[49] Xing W, Qiao SZ, Ding RG, et al. Carbon 2006;44:216–24.

[50] Wang DW, Li F, Chen ZG, et al. Chem Mater 2008;20:7195–200.

[51] Liu B, Shioyama H, Jiang H, et al. Carbon 2010;48:456–63.

[52] Gorka J, Zawislak A, Choma J, et al. Carbon 2008;46:1159–74.

[53] Xia K, Gao Q, Jiang J, et al. Carbon 2008;46:1718–26.

[54] Morishita T, Soneda Y, Tsumura T, et al. Carbon 2006;44:2360–7.

[55] Fernandez JA, Morishita T, Toyoda M, et al. J Power Sources 2008;175:675–9.

[56] Zhang WF, Huang ZH, Cao GP, et al. J Power Sources 2012;204:230–5.

[57] Zhang WF, Huang ZH, Zhou CJ, et al. J Mater Chem 2012;22:7158–63.

[58] Zhang WF, Huang ZH, Cao GP, et al. J Phys Chem Solids 2012;73:1428–31.

[59] Wang DW, Li F, Liu M, et al. Angew Chem Int Ed 2008;47:373–6.

[60] Guo H, Gao Q. J Power Sources 2009;186:553–6.

[61] Jaenes A, Permann L, Arulepp M, et al. Electrochem Commun 2004;5:313–8.

[62] Chmiola J, Yushin G, Dash RK, et al. Electrochem Solid State Lett 2005;8:A357–60.

[63] Chmiola J, Yushin G, Dash R, et al. J Power Sources 2006;158:765–72.

[64] Lin R, Taberna PI, Chmiola J, et al. J Electrochem Soc 2009;156:A7–12.

[65] Lust E, Nurk G, Jaenes A, et al. J Solid State Electrochem 2003;7:91–105.

[66] Arulepp M, Permann L, Leis J, et al. J Power Sources 2004;133:320–6.

[67] Chmiola J, Yushin G, Gogotsi Y, et al. Science 2006;313:1760–3.

[68] Wang H, Gao Q. Carbon 2009;47:820–8.

[69] Chmiola J, Largeot C, Taberna PL, et al. Science 2010;328:480–3.

[70] Shiraishi S, Kurihara H, Tsubota H, et al. Electrochem Solid State Lett 2001;4:A5–8.

[71] Yamada Y, Tanaike O, Liang TT, et al. Electrochem Solid State Lett 2002;5:A283–5.

[72] Shiraishi S, Aoyama Y, Kurihara H, et al. Mol Cryst Liq Cryst 2002;388:543–9.

[73] Lin C, Ritter JA, Popov BN. J Electrochem Soc 1999;146:3639–43.

[74] Fang B, Wei YZ, Maruyama K, et al. J Appl Electrochem 2005;35:229–33.

[75] Salinger R, Fischer U, Herta C, et al. J Non-Cryst Solids 1998;225:81–5.

[76] Kuwahara Y, Yano H, Kanematsu Y, et al. TANSO 2008; No. 235: 263–267 [in Japanese].

[77] Sepehri S, Garcia BB, Zhang Q, et al. Carbon 2009;47:1436–43.

[78] Toyoda M, Tani Y, Soneda Y. Carbon 2004;42:2833–7.

[79] Soneda Y, Toyoda M, Hashiya K, et al. Carbon 2003;41:2680–3.

[80] Soneda Y, Yamashita J, Kodama M, et al. Appl Phys A 2006;82:575–8.

[81] Yoo JJ, Balakrishnan K, Huang J, et al. Nano Lett 2011;11:1423–7.

[82] Conway BE, Birss V, Wojtowicz J. J Power Sources 1997;66:1–4.

[83] Oda H, Yamashita A, Minoura S, et al. J Power Sources 2006;158:1510–6.

[84] Okajima K, Ohta K, Sudoh M. Electrochim Acta 2005;50:2227–31.

[85] Nian YR, Teng H. J Electrochem Soc 2002;149:A1008–14.

[86] Nian Y, Teng H. J Electroanal Chem 2003;540:119–27.

[87] Bleda-Martinez MJ, Macia-Agullo JA, Lozano-Castello D, et al. Carbon 2005;43: 2677–84.

[88] Frackowiak E, Jurewicz K, Delpeux S, et al. J Power Sources 2001;97–98:822–5.

[89] Hsieh C, Teng H. Carbon 2002;40:667–74.

[90] Ruiz V, Blanco C, Raymundo-Pinero E, et al. Electrochim Acta 2007;52:4969–73.

[91] Hu C, Wang C. J Power Sources 2004;125:299–308.

[92] Qu D. J Power Sources 2002;109:403–11.

[93] Kodama M, Yamashita J, Soneda Y, et al. Mater Sci Eng B 2004;108:156–61.

[94] Kodama M, Yamashita J, Soneda Y, et al. Chem Lett 2006;35:680–1.

[95] Hulicova D, Yamashita J, Soneda Y, et al. Chem Mater 2005;17:1241–7.
[96] Hulicova D, Kodama M, Hatori H. Chem Mater 2006;18:2318–26.
[97] Kodama M, Yamashita J, Soneda Y, et al. Carbon 2007;45:1105–7.
[98] Li W, Chen D, Li Z, et al. Electrochim Commun 2007;9:569–73.
[99] Li W, Chen D, Li Z, et al. Carbon 2007;45:1757–63.
[100] Arai Y, Kinumoto T, Tsumura T, et al. TANSO 2012; No. 251: 11–14 [in Japanese].
[101] Lota G, Grzyb B, Machnikowaka H, et al. Chem Phys Lett 2005;404:53–8.
[102] Kim C, Choi YO, Lee WJ, et al. Electrochim Acta 2004;50:883–7.
[103] Kawaguchi M, Itoh A, Yagi S, et al. J Power Sources 2007;172:481–6.
[104] Kim YJ, Abe Y, Yanagiura T, et al. Carbon 2007;45:2116–25.
[105] Shiraishi S, Mamyouda H. TANSO 2008; No. 232: 61–66 [in Japanese].
[106] Seredych M, Hulicova-Jurcakova D, Lu GQ, et al. Carbon 2008;46:1475–88.
[107] Jurewicz K, Babel K, Ziolkowski A, et al. Fuel Process Technol 2002;77–78:191–8.
[108] Jurewicz K, Pietrzak R, Nowicki P, et al. Electrochim Acta 2008;53:5469–75.
[109] Shiraishi S. Key Eng Mater 2012;497:80–6.
[110] Arai K, Kinumoto T, Tsumura T, et al. TANSO 2012; No. 251: 11–14 [in Japanese].
[111] Arai K, Kinumoto T, Tsumura T, et al. TANSO 2013; No. 256: 2–7 [in Japanese].
[112] Miller JM, Dunn B, Tran TD, et al. J Electrochem Soc 1997;144:L309–11.
[113] Sato Y, Yomogida K, Nanaumi T, et al. Electrochem Solid-State Lett 2000;3:113–6.
[114] Ramani M, Haran BS, White RE, et al. J Power Sources 2001;93:209–14.
[115] Kim H, Popov BN. J Power Sources 2002;104:52–61.
[116] Wu Z-S, Wang D-W, Ren W, et al. Adv Funct Mater 2010;20:3595–602.
[117] Takasu Y, Matsuo C, Ohnuma T, et al. Chem Lett 1998 1998:1235–6.
[118] Takasu Y, Nakamura T, Murakami Y. Chem Lett 1998 1998:1215–6.
[119] Takasu Y, Ohnuma T, Sugimoto W, et al. Electrochem 1999;67:1187–8.
[120] Wu NL, Wang SY, Han CY, et al. J Power Sources 2003;113:173–8.
[121] Ushiro M, Yoneda A, Iwasa N, et al. TANSO 2009; No. 238: 121–125 [in Japanese].
[122] Wang JG, Yang Y, Huang ZH, et al. Electrochim Acta 2011;56:9240–7.
[123] Wang JG, Yang Y, Huang ZH, et al. Electrochim Acta 2012;75:213–9.
[124] Wang JG, Yang Y, Huang ZH, et al. J Mater Chem 2012;22:16943–9.
[125] Wang JG, Yang Y, Huang ZH, et al. J Power Sources 2013;224:86–92.
[126] Zhai D, Li B, Du H, et al. Carbon 2012;50:5034–43.
[127] Shiraishi S, Kurihara H, Okabe K, et al. Electrochem Commun 2002;4:593–8.
[128] Liu C, Bard J, Wudl F, et al. Electrochem Solid State Lett 1999;2:577–8.
[129] Ma RZ, Liang J, Wei BQ, et al. J Power Sources 1999;84:126–9.
[130] Barisci JN, Wallace GG, Baughman RH. Electrochim Acta 2000;46:509–17.
[131] Frackowiak E, Metenier K, Bertagna V, et al. Appl Phys Lett 2000;77:2421–3.
[132] An KA, Kim WS, Park YS, et al. Adv Mater 2001;13:497–500.
[133] Honda Y, Haramoto T, Takeshige M, et al. Electrochem Solid State Lett 2007;10: A106–10.
[134] Jiang Q, Qu MZ, Zhou GM, et al. Mater Lett 2002;57:988–91.
[135] Emmenegger CH, Mauron PH, Sudan P, et al. J Power Sources 2003;124:321–9.
[136] Barisci JN, Wallace GG, Chattopadhyyay D, et al. J Electrochem Soc 2003;150: E409–15.
[137] Du C, Yeh J, Pan N. Nanotechnology 2006;16:350–3.
[138] Jurewicz K, Babel K, Pietrzak R, et al. Carbon 2006;44:2368–75.
[139] Shiraishi S, Kibe M, Yokoyama T, et al. Appl Phys A 2006;82:585–91.
[140] Heller I, Kong J, Williams KA, et al. J Am Chem Soc 2006;128:7353–9.

[141] Hata K, Futaba DN, Mizuno K, et al. Science 2004;306:1362–4.
[142] Futaba DN, Hata K, Yamada T, et al. Nat Mater 2006;5:987–94.
[143] Kimizuka O, Tanaike O, Yamashita J, et al. Carbon 2008;46:1999–2001.
[144] Yamada Y, Kimizuka O, Tanaike O, et al. Electrochem Solid State Lett 2009;12: K14–6.
[145] Cheng Q, Tang J, Ma J, et al. Phys Chem Chem Phys 2011;13:17615–24.
[146] Snyder JF, Wong EL, Hubbard CW. J Electrochem Soc 2009;156:A215–24.
[147] Yoon S, Lim S, Song Y, et al. Carbon 2004;42:1723–9.
[148] Shiraishi S, Miyauchi T, Sasaki R, et al. Electrochem 2007;75:619–21.
[149] Xu B, Wu F, Chen R, et al. J Power Sources 2006;158:773–8.
[150] Wang L, Morishita T, Toyoda M, et al. Electrochim Acta 2007;53:882–6.
[151] Wang L, Toyoda M, Inagaki M. Adsorpt Sci Technol 2008;46:491–500.
[152] Jurewicz K, Babel K, Ziolkowski A, et al. Electrochim Acta 2003;48:1491–8.
[153] Jurewicz K, Babel K, Zioekowski A, et al. J Phys Chem Solids 2004;65:269–73.
[154] Jurewicz K, Babel K, Pietrzak R, et al. Carbon 2006;44:2368–75.
[155] Ruiz V, Blanco C, Granda M, et al. Electrochim Acta 2008;54:305–10.
[156] Yoshino A, Tsubata T, Shimoyamada M, et al. J Electrochem Soc 2004;151:A2180–2.
[157] Aida T, Yamada K, Morita M. Electrochem Solid State Lett 2006;9:A534–6.
[158] Aida T, Murayama I, Yamada K, et al. J Power Sources 2007;166:462–70.
[159] Khomenko V, Raymundo-Pinero E, Beguin F. J Power Sources 2008;177:643–51.
[160] Hong MS, Lee SH, Kim SW. Electrochem Solid State Lett 2002;5:A227–30.
[161] Brousse T, Toupin M, Belanger D. J Electrochem Soc 2004;151:A614–22.
[162] Khomenko V, Raymundo-Pinero E, Beguin F. J Power Sources 2006;153:183–90.
[163] Brousse T, Taberna PL, Corsnier O, et al. J Power Sources 2007;173:633–41.
[164] Xu C, Du H, Li B, et al. J Electrochem Soc 2009;156:A73–8.
[165] Wang YG, Xia YY. Electrochem Commun 2005;7:1138–42.
[166] Wang DW, Li F, Cheng HM. J Power Sources 2008;185:1563–8.
[167] Qu QT, Shi Y, Tian S, et al. J Power Sources 2009;194:1222–5.
[168] Qu QT, Li L, Tian S, et al. J Power Sources 2010;195:2789–94.
[169] Laforgue A, Simon P, Fauvarque JF, et al. J Electrochem Soc 2001;148:A1130–4.
[170] Lafogue A, Simon P, Fauvarque JF, et al. J Electrochem Soc 2003;150:A645–51.
[171] Arbizzani C, Beninati S, Lazzari M, et al. J Power Sources 2007;174:648–52.
[172] Machida K, Suematsu S, Ishimoto S, et al. J Electrochem Soc 2008;155:A970–4.
[173] Amatucci GG, Badway F, Pasquier AD, et al. J Electrochem Soc 2001;148:A930–9.
[174] Pasquier AD, Plitz I, Gural J, et al. J Power Sources 2003;113:62–71.
[175] Pasquier AD, Plitz I, Menocal S, et al. J Power Sources 2003;115:171–8.
[176] Morishita T, Soneda Y, Hatori H, et al. Electrochim Acta 2007;52:2478–84.
[177] Yoshida A, Hishiyama Y, Inagaki M. Carbon 1990;28:539–43.
[178] Tanaike O, Yamada Y, Yamada K, et al. ECS Trans 2011;33:71–6.

Carbon Materials in Lithium-ion Rechargeable Batteries

12

Energy storage and conversion are important areas of research for realization of a low-carbon society, and there have been many challenges. The lithium-ion rechargeable (secondary) battery (LIB) is one of the devices for electrical energy storage, and its performance depends strongly on carbon materials. In LIBs, highly crystalline natural graphite is used for intercalation/deintercalation of Li^+ ions in the anode, as illustrated in Figure 12.1, and amorphous porous carbon coats both anode and cathode materials to improve the charge/discharge performance, and carbon black (mostly acetylene black) is mixed into cathode and anode sheets to gain sufficient electrical conductivity. Various carbon materials have been used as anode material, such as natural graphite, graphitized materials prepared from cokes, carbon fibers, and mesocarbon microbeads, and also non-graphitic carbons derived from different resins. At present, natural graphite is used in most commercially available batteries, mainly because of the low cost, high capacity, and stable charge/discharge performance. On the natural graphite flakes, carbon coating through various processes has been found to improve the performance of LIBs. Carbon coating of the cathode materials $LiFePO_4$ and $Li_4Ti_5O_{12}$ has also been shown to be effective in improving charge/discharge performance.

Research and development of advanced electrode materials and electrolyte solutions for the next generation of lithium-ion batteries have been reviewed; these materials have to meet the demands necessary to make an electric vehicle fully commercially viable [1–3]. Electrode materials for LIBs have been reviewed with a focus on nanocarbon materials, carbon nanotubes (CNTs) and graphene [4]. High energy and power densities with no compromise in safety are required. LIBs have been reviewed as one of the electrochemical energy-storage and conversion systems, including electric double-layer capacitors and fuel cells [5].

In this chapter, carbon materials used in LIBs are reviewed, focusing on the electrodes, the carbon materials for anodes, and the carbon coating for both anode and cathode materials.

Prerequisite for readers: Chapter 3.8 Section 1 (Rechargeable batteries) in *Carbon Materials Science and Engineering: From Fundamentals to Applications*, Tsinghua University Press.

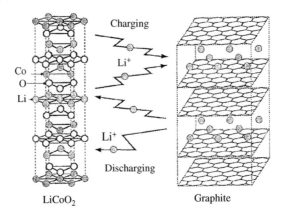

FIGURE 12.1 Movement of Li⁺ Ions Between Graphite Anode and LiCoO₂ Cathode During Charging and Discharging in an LIB

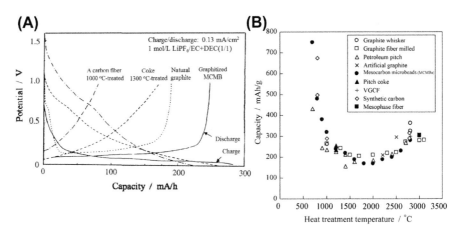

FIGURE 12.2 LIB Performances of Different Carbon Materials

(A) Charge/discharge curves, and (B) relationship between discharge capacity and heat-treatment temperature (HTT)

12.1 Anode materials

12.1.1 Materials

Various carbon materials have been used for the anode of LIBs. In Figure 12.2A, charge/discharge curves of different carbon materials are compared, although the curves depend strongly on the charge/discharge conditions and also on the preparation conditions of the electrode carbon. Cokes and carbon fibers have high irreversible capacity, in addition to low discharge capacity. The mesophase spheres formed in pitches and separated (mesocarbon microbeads, MCMBs) have a high discharge

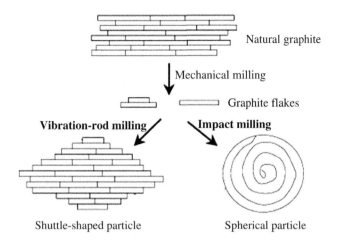

Natural graphite

Mechanical milling

Graphite flakes

Vibration-rod milling   **Impact milling**

Shuttle-shaped particle   Spherical particle

FIGURE 12.3 Morphology Change of Natural Graphite Aggregates by Milling

From [17]

capacity and relatively low irreversible capacity after high-temperature treatment (graphitization). Therefore, graphitized MCMBs have frequently been used as an anode material in commercial LIBs. As shown in Figure 12.2B, discharge capacity of various carbon materials depends strongly on heat-treatment temperature (HTT). Carbon materials heat-treated at low temperatures below 1000 °C show very high discharge capacity, but their irreversible capacities are also very high. Even for MCMBs, therefore, heat treatment at high temperatures above 2800 °C is needed to have high discharge capacity and low irreversible capacity. The charge/discharge performance of MCMBs and their mechanism of action in LIBs have been studied in detail [6–11]. For the commercial LIBs of the moment, however, natural graphite is preferred, which has high discharge capacity comparable to or better than that of graphitized MCMBs under optimum conditions. Lithium-ion transfer at the interface between the electrolyte and the various anode carbons, including graphite, has been discussed [12–14].

For natural graphite, the effect of milling (jet and turbo milling) on the electro-chemical performance was studied; the change in particle morphology, mainly due to the reduction of particle size and thickness, improved the performance [15]. Electro-chemical intercalation of Li ions has been discussed in relation to particle morphol-ogy of natural graphite, showing an improved performance of spherical aggregates with a size of 12 μm [16]. After mechanical milling to make the size of graphite flakes small, an additional vibration-rod or impact milling was found to be effective to change morphology of the aggregates, as shown schematically in Figure 12.3 [17].

Graphitized carbon nanospheres with an onion-type texture, which were prepared by chemical vapor deposition (CVD) and graphitized at 2800 °C, had better per-formance in the anode than MCMBs at a high charge/discharge rate [18]. Detailed structure of the spheres was studied by using high-resolution transmission electron microscopy (TEM) analysis: the appearance changed to polyhedrons after heat treatment above 2000 °C and the development of graphitic structure was clearly observed, though graphitization depended strongly on the particle size of MCMB [19,20].

Coke heat-treated at around 2200 °C (named ICOKE) was found to have excellent pulse-charge/discharge characteristics and long cycle life, though discharge capacity was relatively low [21]. In Figure 12.4, the discharge curve of ICOKE heat-treated at 2200 °C is compared with that of natural graphite. ICOKE shows a higher discharge capacity than natural graphite when high rates, more than 3C, are applied, although the capacity of ICOKE is lower at low rates.

Mesoporous carbon "nanosheets" prepared from thermoplastic phenolic-formaldehyde resin (as carbon source) and copper nitrate (as template precursor) at 600 °C gave a reversible capacity of 748 mAh/g at a current density of 20 mA/g, and 460 mAh/g even at 1 A/g [22]. So-called graphene nanosheets, formed from an artificial graphite by exfoliation through graphite oxide (GO) and sonication in ethanol, were tested as an anode material without any reduction and purification of GO [23]. Anode performance has been reported on mixtures of metal oxide nanoparticles, CuO, Fe_3O_4, and graphene nanosheets [24,25].

12.1.2 Carbon coating of graphite

Carbon coating onto anode materials has been reported to be effective to improve charge/discharge performance in LIBs. It reduces the irreversible capacity of the first charge/discharge cycle and improves the cyclability, but the effect depends on the

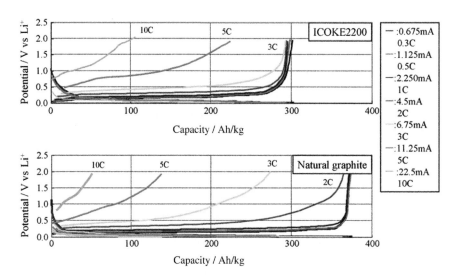

FIGURE 12.4 Discharge Performance of ICOKE Heat-treated at 2200 °C (ICOKE2200) and Natural Graphite with Different Charge Currents

From [21]

amount and the nature of coated carbon. By using a carbon-coated graphite anode, it is possible to use electrolyte solutions containing propylene carbonate (PC).

Carbon coating of anode materials has been carried out by CVD [17,26–36], chemical vapor infiltration (CVI) [37,38], and heat treatment of a mixture of organic precursors at high temperature under an inert atmosphere [39–50]. The anode materials subjected to carbon coating were natural graphite [26–33,40–47], synthetic graphite [39,49,50], carbon cloth [33], MCMBs [39,49,50], and non-graphitizing carbons, such as sucrose- and phenol-derived carbons [37,46]. Particle morphology of natural graphite was modified by selecting the milling method, as shown in Figure 12.3 [17].

Carbon deposition has been successfully performed at 950–1000 °C on natural graphite under fluidizing in a flow of carrier gas containing toluene vapor [17,26,28,29,31] and ethylene [27]. The amount of coated carbon was controlled in the range of 8.6–17.6 mass% by changing deposition time. The carbon deposited was composed of well-oriented small crystallites and had a density of about 1.86 g/cm^3. After carbon deposition at 1000 °C, the intensity ratio of D-band to G-band in the Raman spectrum, I_D/I_G, increased to 0.71 from 0.07 for the pristine graphite, suggesting the deposition of disordered carbon [27]. Propane gas has also been used for CVD at 1000–1200 °C for natural graphite under tumbling by rotating the reaction tube [30]. The coated carbon appeared to be disordered by TEM observation. Carbon coating of synthetic graphite particles has been done in a flow of methane at 1000 °C [35]. Carbon derived from sucrose at 1100 °C has been coated by CVD of ethylene at 300–700 °C [36]. Carbon coating was successfully performed by pressure-pulsed chemical vapor infiltration (PCVI) using CH$_4$/H$_2$ gas on a non-graphitizing carbon substrate, carbon beads of c. 3 μm diameter [37], and on a carbon-fiber paper [38].

Heating of a powder mixture of graphite with poly(vinyl alcohol) (PVA) at 900 °C resulted in carbon-coated natural graphite powder [45–47]. In Figure 12.5, scanning

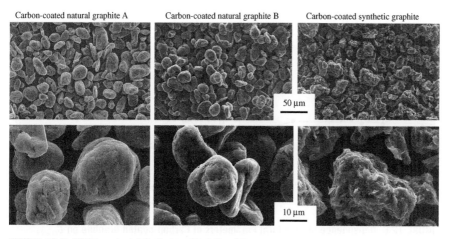

FIGURE 12.5 SEM Images of Carbon-coated Graphites

From [46]

electron microscopy (SEM) images are shown for carbon-coated natural and synthetic graphite particles. Morphology of the particles did not change appreciably before and after the carbon coating, except that edges of the particles became round and size increased slightly, and no marked coagulation of particles was observed [45]. The thickness of the carbon layer, which was calculated from the difference in particle size distribution before and after carbon coating, is proportional to the amount of coated carbon, C_{coated} (Figure 12.6A), and so controllable by changing the mixing ratio of PVA. The coated layer was proved to be disordered and porous from measurements of immersion density (Figure 12.6B), BET surface area, S_{BET} (Figure 12.6C), and Raman intensity ratio, I_D/I_G (Figure 12.6D), of the particles.

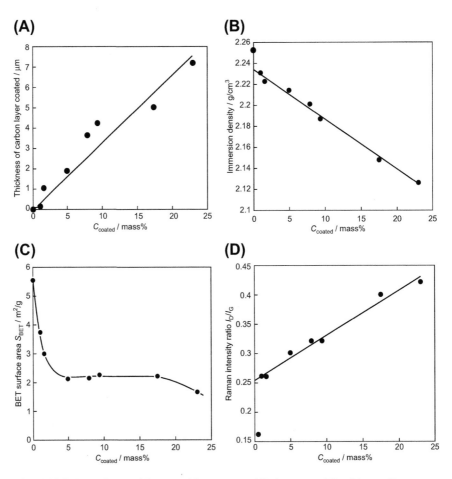

FIGURE 12.6 Dependence of Structural Parameters of Carbon-coated Graphite on C_{coated}

(A) Thickness of carbon layer coated, (B) immersion density, (C) S_{BET}, and (D) I_D/I_G

From [46]

Carbon coating of natural graphite spheres [40,44] and of MCMB [48,50] has been performed by mixing with poly(vinyl chloride) (PVC). Synthetic graphite particles have been coated in a tetrahydrofurane/acetone solution of coal tar pitch [41]. Mixtures of pitch and phenol have also been used for carbon coating of various graphite samples at 1200 °C [40]. Natural graphite particles have been coated by polyurea through the reaction between 2,4-toluene-diisocyanate and water at 60 °C, and then heated at 850 °C to obtain carbon-coated graphite [48].

In Figure 12.7, discharge and irreversible capacities for the first cycle are plotted against C_{coated} with respect to the carbon precursors PVA and PVC, and carbonization temperatures [45]. Although PVC gives a higher carbon yield than PVA, the same changes in capacities with C_{coated} and with carbonization temperature are obtained for the two carbon precursors. A small amount (<5 mass%) of carbon coating at 900 °C tends to increase irreversible capacity, though discharge capacity decreases or does not change appreciably. Carbonization at a temperature higher than 1100 °C gives a very high irreversible capacity and low discharge capacity. Therefore, more than 5 mass% of carbon has to be coated on natural graphite at a carbonization temperature of 700–1000 °C in order to achieve a small irreversible capacity while keeping discharge capacity more than 350 mAh/g.

Carbon coating of a carbon prepared from sucrose at 1100 °C by ethylene CVD at 700 °C reduced irreversible capacity from 150 to 60 mAh/g, accompanied by an increase in discharge capacity from 532 to 571 mAh/g [36]. An increase in the first-cycle coulombic efficiency from 78 to more than 91% was reported by carbon coating of synthetic graphite at 800 °C [27]. The reduction in irreversible capacity due to carbon coating by propylene CVD has been discussed in relation to the active surface area [33].

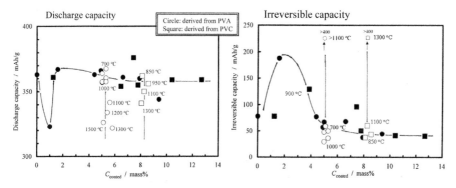

FIGURE 12.7 Dependence of Discharge and Irreversible Capacities on C_{coated}

Comparing PVA and PVC precursors in different mixing ratios and at different carbonization temperatures, in 1 mol/L LiClO$_4$/(EC + PC) solution at a charge/discharge current density of 1.56 mA/cm^2 (0.5C/0.5C). EC, ethylene carbonate

From [46]

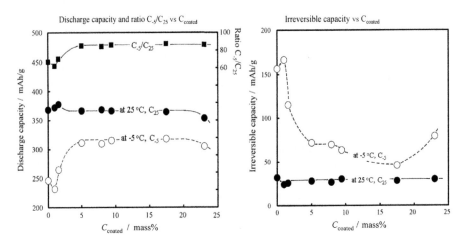

FIGURE 12.8 Dependence of Discharge Capacity at 25 and −5 °C (C_{25} and C_{-5}) and the First-cycle Irreversible Capacity on C_{coated}

In 1 mol/L LiClO$_4$/(EC + DMC) solution with charge/discharge current density of 0.5C/0.5C. DMC, dimethyl carbonate; EC, ethylene carbonate

From [47]

The reduction of the first-cycle irreversible capacity was more marked at low temperatures. In Figure 12.8, discharge and irreversible capacities at 25 and −5 °C are plotted against C_{coated} for carbon-coated natural graphite prepared at 900 °C, together with the discharge capacity ratio, C_{-5}/C_{25} [46]. The increase in discharge capacity and the decrease in irreversible capacity by carbon coating more than 5 mass% are observed more markedly at −5 °C than at 25 °C. To give this advantage at −5 °C, C_{coated} in a range of 5–20 mass% at a carbonization temperature of 700–1000 °C was shown to be optimum [46].

The procedure of carbon coating affects the improvement in cyclic performance of graphite, as shown in Figure 12.9 [44]. Cyclic performance is compared between pristine and carbon-coated graphite, the latter prepared via three different processes, as shown in Figure 12.9A. Cyclability is improved, particularly by step-wise heating of the mixture of natural graphite with PVC (process 2), as shown in Figure 12.9B. Carbon-coated graphite showed excellent rate capability at a range of 0.1C–1.2C. For MCMBs coated with carbon by process 2, excellent cyclability and rate capability were also obtained.

A marked decrease in irreversible capacity in the first cycle was obtained for non-graphitizing carbon papers by PCVI at 800 °C with a pulse of C$_3$H$_8$ (30%)/N$_2$ gas of 0.1 MPa for 1 s [48]. By carbon coating of 5 mass% after 3000 pulses of PCVI, irreversible capacity reduced from 500 to 150 mAh/g and discharge capacity increased from 400 to 600 mAh/g.

Ethylene carbonate (EC)-based electrolytes are currently used in commercially available LIBs. However, it is strongly desired to use electrolytes based on PC in

(A)   **(B)**

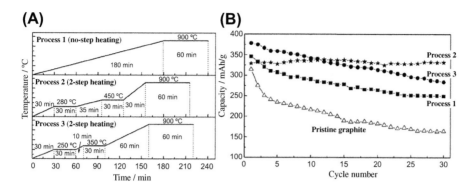

FIGURE 12.9 Step-wise Carbonization of the Mixture of Natural Graphite with PVC

(A) The carbon-coating process, and (B) cycle performance of the carbon-coated graphite in comparison with pristine graphite

Courtesy of Prof. H.M. Cheng of the Institute of Metal Research, China

order to improve low-temperature performance. Anode performance of carbon-coated graphite has been studied in different PC-containing solutions: PC/DMC (dimethyl carbonate) [17,26,28,29,31], PC/EC [40,42,45–47], PC/EC/BL (γ-butyrolactone) [39], and PC/EC/DEC (diethyl carbonate) [48,49], and also in polymer electrolytes [39,50]. In EC/PC = 3/1, irreversible capacity decreases to 29 mAh/g by carbon coating (5.0 mass%), which is comparable with that in electrolyte solution without PC. In EC/PC = 1/1, however, an extremely large irreversible capacity was observed. By a large amount of carbon coating of more than 18 mass%, low irreversible capacity was obtained even in EC/PC = 1/1 [46].

Natural graphite spheres coated with carbon nanofibers (CNFs) by acetylene CVD at 850 °C using Fe nanoparticles as a catalyst (natural graphite (NG)/CNF composite; Figure 12.10A) showed improvement in cyclability and rate capability of LIBs [51]. By depositing CNFs on the natural graphite spheres, cyclability is markedly improved, compared with both the spheres without coating and the mechanical mixture of the spheres with CNFs, as shown in Figure 12.10B.

12.1.3 Carbon coating of $Li_4Ti_5O_{12}$

The spinel-type lithium titanate ($Li_4Ti_5O_{12}$, LTO) has attracted great interest as an anode material because of:

1. Zero strain during charging and discharging.
2. Excellent cycle reversibility.
3. Fast Li^+ ion insertion and extraction ability.
4. High lithiation voltage plateau at 1.55 V vs Li/Li^+.

Its theoretical reversible capacity is 175 mAh/g. These merits work for the improvement of the safety of LIBs, and make $Li_4Ti_5O_{12}$ more competitive as a safe anode

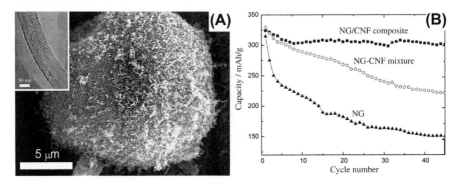

FIGURE 12.10 Natural Graphite/Carbon Nanofiber (NG/CNF) Composite

(A) SEM image of the composite and TEM image of CNF deposited (inserted), and (B) cycle performance of the composite, in comparison with NG and the mixture of NG and CNFs

Courtesy of Prof. H.M. Cheng of the Institute of Metal Research, China

material for long-life LIBs [52]. However, its inherent poor electrical conductivity (c. 10^{-13} S/cm) and moderate Li+-ion diffusion coefficient (10^{-8} cm^2/s) have to be improved for high-performance LIBs. Much research has focused on overcoming these problems through carbon coating [53–61], as well as reducing particle size, doping with other metals or metal oxides, and mixing with a conductive second phase [62,63].

Carbon coating of LTO has been carried out by CVD of toluene vapor at 650–900 °C in a fluidized-bed reactor [53]. Amorphous carbon was successfully coated on the surface of LTO particles at a thickness of 3–5 nm. The electrical conductivity of the LTO/C composite increased to 2.05 and 13.84 S/cm at coating temperatures of 800 and 900 °C, but the improvement in rate performance was not pronounced even by carbon coating at 900 °C. Fine particles of LTO synthesized under hydrothermal conditions were coated with carbon by heat treatment at 800 °C as a mixture of LTO with amphiphilic carbonaceous material, which was prepared from a green coke [61]. Particle morphology of the LTO was maintained after carbon coating, but its color changed from white to black. No appreciable coagulation of the particles was observed, as shown in Figure 12.11A, and the highly crystalline particles were coated by carbon homogeneously, as shown in Figure 12.11B. In Figure 12.12A, charge/discharge curves for carbon-coated LTOs in 1.0 mol/L LiPF$_6$/(EC + DEC) solution are compared with the pristine LTO at a slow rate of 0.1C, showing no pronounced difference in the voltage plateau or discharge capacity, which is close to the theoretical capacity. Even with increase of the charge/discharge rate up to 20C (current density at 3.5 A/g), the plateau stays flat for the LTO coated with 5.7 mass% carbon, though discharge capacity decreases, as shown in Figure 12.12B. LTO coated with 5.7 mass% carbon showed a high rate performance, delivering a discharge capacity as high as 160 mAh/g at 10C and 143 mAh/g at 20C, with 88 and 78% retention at 1C, respectively, but the capacity of the pristine LTO dropped

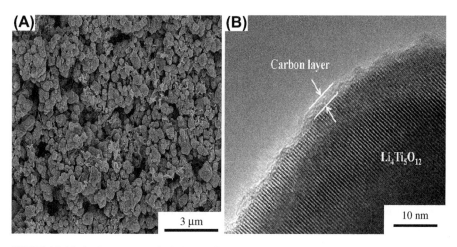

FIGURE 12.11 Carbon-coated Li$_4$Ti$_5$O$_{12}$ Particles (5.7 mass% C-coated)

(A) SEM image and (B) TEM image

Courtesy of Prof. C. Y. Wang of Tianjin University, China

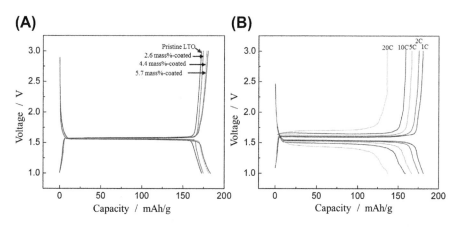

FIGURE 12.12 Charge/Discharge Curves of Carbon-coated Li$_4$Ti$_5$O$_{12}$

(A) Effect of the amount of coated carbon at 0.1C rate, and (B) effect of charge/discharge rate for 5.7 mass% C-coated LTO

Courtesy of Prof. C. Y. Wang of Tianjin University, China

significantly with increasing discharge rate, down to 102 mAh/g at 20C (58% retention). Cycle performance was also improved by carbon coating; 96% retention at 1C, 95% at 5C, and 91% at 20C after 100 cycles.

For spherical particles of LTO prepared by a spray-drying process of slurry and following calcination at 900 °C, carbon coating was carried out by mixing with pitch at 750 °C [55]. The 3.25 mass% carbon-coated LTO showed a high rate performance, delivering a discharge capacity as high as 170.2 mAh/g at a rate of 1C, and 81.7 mAh/g at 100C, and a capacity retention of 94% after 100 cycles at 5C. By using an ionic liquid, 1-ethyl-3-methylimidazolium dicyanamide, LTO particles were coated with carbon containing nitrogen, resulting in some improvement in LIB performance [56]. Carbon-coated LTO has been prepared from a mixed gel of CH_3COOLi and $Ti(OC_4H_9)_4$ with citric acid [57], and with graphene sheets [58], by calcination at 800 °C.

LTO/C composites were prepared by mixing carbon black with the LTO precursor (anatase-type TiO_2 and $LiCO_3$) in H_2/Ar flow, in which carbon black particles were embedded between LTO particles [62]. An improvement in cycle performance was observed for these composites. Similar composites were prepared by heating a mixed slurry of LiOH, TiO_2, and poly(acrylic acid) to 800 °C, and high capacity retention of 87% was obtained after the fiftieth cycle at a rate of 20C [54]. An LTO/C composite was prepared by the solid state reaction of carbon-coated anatase-type TiO_2 with Li_2CO_3 [59,60]. An LTO/C composite has also been prepared by dispersing LTO nanoparticles in graphene [63].

Carbon coating for a tin phosphate ($Sn_2P_2O_7$) anode was effective in reducing the irreversible capacity from 570 to about 200 mAh/g without loss of the discharge capacity of about 540 mAh/g [64]. Nanoparticles of Sn were formed in mesopores of carbon prepared via carbonization of a mixture of MgO, SnO_2, and PVA (MgO-template carbonization) [65]. The space neighboring an Sn nanoparticle in a mesopore is thought to absorb the marked expansion due to alloying of Sn with Li, and the carbon shell surrounding the Sn nanoparticle to disturb its movement during alloying/de-alloying, i.e. charge/discharge cycles. Hollow Sn nanoparticles coated with carbon were prepared from allyltriphenyl tin at 700 °C, which gave a highly stable and reversible capacity of about 550 mAh/g [66]. Thin MoS_2 flakes dispersed in a carbon matrix were prepared under hydrothermal conditions and gave excellent battery performance, a high discharge capacity of 962 mAh/g, and excellent cycle stability [67].

Carbon-coated nanoparticles of α-Fe_2O_3, NiO, and CuO were prepared by oxidation of metal carbides coated with carbon at around 250 °C in air, and these were studied as an anode material of LIBs [68–70].

12.2 Cathode materials

12.2.1 Materials

$LiCoO_2$ and $LiMn_2O_4$ have been used in cathodes of LIBs. Lithium iron phosphate, $LiFePO_4$, with an olivine-type crystal structure (triphylite), has drawn attention as a candidate cathode material of LIBs for electric vehicles (EVs). The main advantages of $LiFePO_4$ are a flat voltage profile, low material cost, abundant material supply, and better environmental compatibility [71–75]. The drawbacks of $LiFePO_4$ include a relatively low theoretical capacity of 170 mAh/g and low density of 3.60 g/cm^3, in comparison with 274 mAh/g and 5.05 g/cm^3 of the $LiCoO_2$ currently used, in addition to poor electrical conductivity and low ionic diffusivity. Cathode materials have been reviewed with a focus on battery performance [76].

LiFePO$_4$ has been synthesized by different processes: solid-state reaction, co-precipitation, hydrothermal reaction, sol-gel process, etc. [73,75]. A microwave-assisted process [77], an electric-discharge-assisted process [78], and chemical lithiation of FePO$_4$ [79] have also been proposed. Its synthesis, however, has to be precisely controlled to avoid electrochemically inactive by-products, such as ferromagnetic Fe$_2$P and γ-Fe$_2$O$_3$, and also Li$_3$Fe$_2$(PO$_4$)$_3$ consisting of Fe(III). In order to obtain pure LiFePO$_4$ and a discharge capacity close to the theoretical one, the choice of a moderate calcination temperature between 500 and 600 °C and a homogeneous precursor composed of various compounds of lithium, iron, and phosphorus have been reported to be desirable [80]. Further, the particles of LiFePO$_4$ have to be small enough to allow high utilization of Li during charge/discharge processes, particularly at high rates. At high temperatures, undesirable grain growth of LiFePO$_4$ occurred, and at low temperatures amorphous residual phases containing Fe(III) were formed. In addition, the low electrical conductivity of 10^{-9} to 10^{-10} S/cm has to be greatly improved for the cathodes of LIBs. Carbon coating was shown to be effective to improve electrical conductivity and cathode performance of an LiFePO$_4$ electrode, better than conventional mixing with conductive materials.

12.2.2 Carbon coating of LiFePO$_4$

Carbon coating of LiFePO$_4$ particles has been reported to be effective to: (1) reduce Fe(III) to Fe(II), (2) keep the particles small, (3) increase electrical conductivity of the sheet formed as electrode, and (4) enhance ionic diffusivity [81–122]. Carbon-coated LiFePO$_4$ was prepared by mixing different organic compounds (carbon precursors) with a stoichiometric mixture of lithium, iron, and phosphorus compounds (LiFePO$_4$ precursor) and then heating at high temperatures to form crystalline LiFePO$_4$. During this high-temperature treatment, the carbon precursor is carbonized and functions as a reducing agent for Fe(III) to Fe(II), leading to carbon-coated LiFePO$_4$ particles without grain growth. LiFePO$_4$ particles were reported to be coated with carbon in most reports in the literature, but the products in some reports were not confirmed to be coated with carbon and so they are called LiFePO$_4$/C composites.

Carbon-coated LiFePO$_4$ and LiFePO$_4$/C composites have been prepared from the mixtures of various LiFePO$_4$ precursors with various carbon precursors, such as resorcinol-formaldehyde [81], sugar [83], sucrose [88,99,116], glucose [110,118], poly(ethylene glycol) [117], and stearic acid [120], by heating at 500–800 °C in an inert atmosphere. The particle size of LiFePO$_4$ crystals was kept in the range of 100–200 nm and carbon content was around 3–15 mass%. High electrochemical performance was confirmed for these materials, discharge capacity of 160 mAh/g and excellent cycle and rate performances. However, it was pointed out that the presence of carbon, even less than 1 mass%, caused a significant decrease in the density of the electrode sheet prepared [83]. Carbon coating of LiFePO$_4$ using sucrose was shown to be effective to increase the electrical conductivity of the sheet, by almost seven orders of magnitude with increasing carbon content to 31 mass% [88]. Thickness and uniformity of the carbon coating was also an important factor for better performance, which was demonstrated by the deposition of additional carbon onto carbon-coated $LiFePO_4$ particles [114]. The thickness of coated carbon increased from 2–6 to 4–8 nm with increasing carbon content from 1.25 to 2.28 mass%, so that discharge capacity at the seventh cycle increased from 137 to 151 mAh/g. Nanofibers of $LiFePO_4$/C composite prepared through electrospinning of a polyacrylonitrile/N,N'-dimethylformamide (PAN/DMF) solution containing the $LiFePO_4$ precursor, followed by stabilization and carbonization, gave a reversible capacity of 160 mAh/g in 1 mol/L $LiPF_6$/(EC+EMC) solution [122]. In Figure 12.13, carbon-coated $LiFePO_4$ prepared under hydrothermal conditions using glucose as the carbon precursor is shown [110]. Rod-like particles of $LiFePO_4$ with uniform diameter of 220 nm (Figures 12.13A and 12.13B) are coated by a carbon layer of 5–12 nm thickness (Figure 12.13C). This gave excellent cycle and rate performance. Porous $LiFePO_4$ particles containing mesopores and macropores have been synthesized, via a sol-gel method using Fe(III) citrate and LiH_2PO_4, in which the pore walls were coated with carbon [102]. A $LiFePO_4$/C composite has been prepared from the mixture of $FePO_4$/polyaniline composite particles with CH_3COOLi and 25 mass% sucrose by heat treatment at 700 °C in Ar flow containing 5% H_2 [106]. Peanut shell has also been used as a carbon precursor that was able to improve the charge/discharge performance [112].

Spray pyrolysis of an aerosol precursor resulted in spherical $LiFePO_4$ particles of 50–100 nm diameter covered by carbon layers 2–4 nm thick [111]. It gave a relatively high tap density of 1.2–1.6 g/cm^3, probably owing to the spherical morphology of the particles, and had an excellent rate performance over a wide range from 0.045 mA/cm^2 (C/25) to 5.5 mA/cm^2 (5C).

Carbon coating has been performed by mixing $LiFePO_4$ powder with pyromellitic acid and ferrocene, followed by heat treatment at 600 °C [95]. Better rate performance was obtained for the carbon-coated $LiFePO_4$ prepared by 6 mass% addition of pyromellitic acid, and the use of ferrocene with pyromellitic acid improved the rate performance more at rates below 1C. These experimental results are explained by the catalytic graphitization of coated carbon due to ferrocene, but the Raman

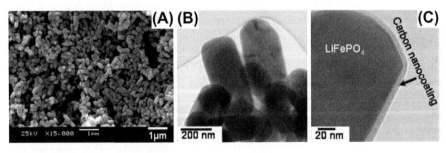

FIGURE 12.13 Carbon-coated LiFePO$_4$

(A) SEM image, (B and C) TEM images

From [111]

spectrum of carbon is not markedly different from that of the composite prepared without ferrocene.

A LiFePO$_4$/C composite was prepared at 550 °C from the mixture of a LiFePO$_4$ precursor and sucrose after mechanochemical activation in a planetary mill [84]. The composite exhibited better rate and cycle performance than that prepared without activation. Carbon coating using graphite, carbon black, and acetylene black at 700 °C in a 5 vol% H$_2$/Ar atmosphere after mechanochemical activation was shown to be effective to increase electrical conductivity to 10^{-2}–10^{-4} S/cm, and consequently to improve cycle performance, as shown in Figure 12.14 [97]. The addition of sucrose during mechanochemical activation was also effective to prepare carbon-coated LiFePO$_4$ at micrometer-size [104]. Carbon coating became more uniform and discharge capacity was slightly higher than by addition of acetylene black. The optimal process parameters for the synthesis of LiFePO$_4$/C composites through mechanochemical activation were reported to be high-energy ball milling for 2–4 h, followed by heat treatment at 700 °C for 20 h; the resultant composite gave a capacity of 174 mAh/g at 0.1C rate and 117 mAh/g at 20C [109]. Too low a temperature and insufficient heat-treatment time resulted in the formation of carbon residues containing hydrogen, and also in poor battery performance of the resultant LiFePO$_4$/C composite.

A sol-gel method using ethylene glycol, followed by heating at 700 °C in 5 vol% H$_2$/N$_2$, resulted in LiFePO$_4$/C composites containing various amounts of carbon and conductive FeP [105]. The composite containing 1.5 mass% carbon and 1.2 mass% FeP had discharge capacities of 155 and 111 mAh/g at rates of 0.1C and 1C, respectively.

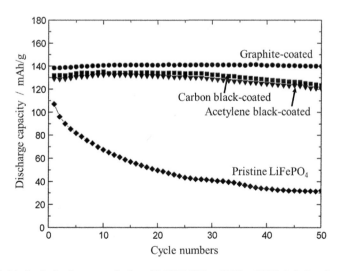

FIGURE 12.14 Cycle Performance in 1 mol/L LiPF$_6$/(EC + DMC + EMC) Solution for Carbon-coated LiFePO$_4$ Prepared From Different Carbon Precursors via Mechanochemical Activation

From [98]

The addition of the carbon precursor (glucose) to the $LiFePO_4$ precursor before hydrothermal treatment (in situ carbon coating) was shown to give better performance than the addition after synthesis (mixed coating) [107]. The effect of the carbon precursor—acetylene black, sucrose, and glucose—was examined at 650 °C using a $LiFePO_4$ precursor prepared by the sol-gel method [108]. As shown in the first charge/discharge curves in Figure 12.15, discharge voltage and discharge capacity depend on the carbon precursor, with glucose being the most efficient. This result shows clearly that complete coating of $LiFePO_4$ particles with carbon is important.

Several aromatic anhydrides have been used as the carbon precursor for carbon coating of amorphous $LiFePO_4$ at 750 °C [101]. Benzene-1,2,4,5-tetracarboxylic acid gave the best electrochemical performance to nano-sized $LiFePO_4$. Poly(ethylene oxide), polybutadiene, polystylene, and block copolymer (styrene-butadiene-styrene) were also tested as carbon precursors after carbonization at 600 °C, polystyrene being the most effective for improving electrochemical performance [113].

By a sol-gel process using an ethanol solution of Li_2CO_3, $FeCl_2$, H_3PO_4, and NH_4VO_3 and citric acid, a carbon-coated single phase of $LiFe_{1-x}V_xPO_4$ solid solution was obtained in the range of $0 \leq x \leq 0.07$ [123]. With increasing V content up to 7 mol%, electrical conductivity and apparent lithium-ion diffusion coefficient increased and, as a consequence, discharge capacity in 1 mol/L $LiPF_6$/(EC + DEC) solution increased at a high discharge rate of 10C. At $x = 0.09$, crystalline particles of $LiFe_{0.93}V_{0.07}PO_4$ coexisted with small VO_2 particles, which were thought to be fixed to the phosphate particles by the carbon layer coating, and gave a slightly higher discharge capacity.

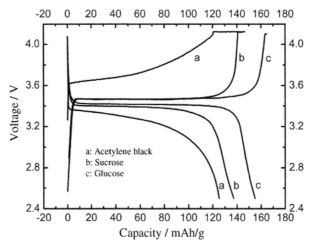

FIGURE 12.15 First Charge/Discharge Cycle in 1 mol/L LiPF₆/(EC + DMC) Solution for LiFePO₄/C Composites Prepared by Heat Treatment with Different Carbon Precursors

LiMnPO$_4$ has the same olivine-type structure and electrochemical activity as LiFePO$_4$ [71]. However, it has attracted little attention because of its poor cycle performance, despite a theoretical capacity of 170 mAh/g and high redox potential of 4.0 V vs Li$^+$/Li. Attempts to improve the electrochemical performance have been reported by preparing Li(Fe$_x$Mn$_{1-x}$)PO$_4$ [124,125] and the composite with CNTs [126]. Hollow microspheres of LiCoPO$_4$/C composite have been prepared by spray pyrolysis, of which the inner and outer surfaces were coated with carbon [127].

Electrochemical performance at low temperature has been compared in the electrolyte 1.0 mol/L LiPF$_6$/(EC + DMC) for Li$_3$V$_2$(PO$_4$)$_3$/C composites with LiFePO$_4$/C [121]. At −20 °C the former exhibited a stable discharge capacity of 104 mAh/g while the latter was 45 mAh/g, although at 23 °C the former had 127 mAh/g and the latter 142 mAh/g.

An improvement in cyclability of a LiCoO$_2$ cathode by mixing with multi-walled carbon nanotubes (MWCNTs) has been reported [128–130], as described in Chapter 2. The addition of graphene to a LiFePO$_4$ cathode gave better charge/discharge performance than that of conventional conductive additives, even better than MWCNTs [131]. a LiFePO$_4$/graphene composite was prepared by spray-drying of an aqueous suspension of LiFePO$_4$ nanoparticles and graphite oxide, followed by annealing at 600 °C, of which the cycle performance under 10C charging and 20C discharging was better than a LiFePO$_4$/C composite prepared using glucose as the carbon precursor [132].

12.3 **Concluding remarks**

Since the beginning of the development of LIBs, various non-graphitizing carbons, mesocarbon microbeads (MCMBs), natural graphite, CNTs, CNFs, and graphene have been studied as anode materials. However, fundamental research to improve LIB performance and to develop novel carbon materials for the anode is still active.

Carbon coating of anode materials, mostly natural graphite, leads to an increase in discharge capacity and a marked decrease in irreversible capacity, probably owing to the improvement in wettability with electrolyte solution. Carbon coating of natural graphite particles makes them possible to use in a PC-containing electrolyte solution. These improvements in carbon-coated natural graphite as the anode of LIBs are probably due to the change in the solid-electrolyte interface (SEI), which has to be studied in more detail. For graphite anodes, ethylene carbonate (EC) has to be used as the solvent for various electrolytes, LiPF$_6$, LiClO$_4$, etc. However, EC has a drawback in LIBs for EVs because of its relatively high melting point, 36 °C, and so linear carbonates, such as dimethyl carbonate (DMC), diethyl carbonate (DEC) and ethyl methyl carbonate (EMC), are mixed with EC in commercial electrolyte solutions. Propylene carbonate (PC) may be a compromise because its melting point is much lower (−49 °C), but continuous PC decomposition is known to occur at the graphite electrode and lead to exfoliation of the electrode. There have been many studies on suppression of the exfoliation of graphite in PC-based electrolytes by using various film-forming additives [133], a new lithium salt [134], high concentration of the electrolyte [135,136], and strong Lewis acid Ca(II) compounds [137]. Carbon coating of graphite particles is known to be effective, as described above.

For the cathode material $LiFePO_4$ in LIBs, carbon coating gives the following benefits:

1. The reduction of Fe(III) in the precursor to Fe(II) to form $LiFePO_4$ during its preparation, which makes the process simpler.
2. The inhibition of grain growth of $LiFePO_4$, which gives high utilization of Li ions.
3. An improvement in electrical conductivity of the cathode sheet.
4. An improvement in wettability of the cathode material with electrolyte solutions, such as EC + DEC, because of the hydrophobic surface nature of the carbon layer.

Consequently, carbon coating markedly improves the performance of the $LiFePO_4$ cathode, giving excellent cycle and rate performances with a capacity close to the theoretical one. In order to realize these merits, the process of mixing a carbon precursor with a $LiFePO_4$ precursor and heating at a high temperature under hydrothermal conditions (in situ carbon coating) was reported to be the most efficient. Grinding the $LiFePO_4$ precursor with the carbon precursor (mechanochemical activation) can also give carbon-coated $LiFePO_4$ after heating at a high temperature. For the anode material $Li_4Ti_5O_{12}$, carbon coating gives the same benefits as for cathode $LiFePO_4$, and again, its in situ synthesis under hydrothermal conditions is recommended.

Carbon is a very important material in LIBs, as an active material for the anode and as a coating material for both anode and cathode materials, as described in this chapter. In addition, carbon materials have been used as conductive additives for both electrodes, either acetylene black or ketjenblack, in most of the LIBs on the market, and also CNTs and CNFs are added into electrode sheets for mechanical reinforcing in addition to their use as conductive additives, which is explained in Chapter 2. Graphene has attracted attention not only as a conductive additive but also as a supporting material for electrochemically active materials, although much more detailed study is required, including quantitative analyses of graphene.

References

[1] Whittingham MS. MRS Bull 2008;33:411–9.
[2] Scrosati B, Garche J. J Power Sources 2010;195:2419–30.
[3] Marom R, Amalraj SF, Leifer N, et al. J Mater Chem 2011;21:9938–54.
[4] Liang M, Zhi L. J Mater Chem 2009;19:5871–8.
[5] Beguin F, Frackowiak E, editors. Carbons for Electrochemical Energy Storage and Conversion Systems. CRC Press; 2010.
[6] Dahn JR, Slaigh AK, Shi H, et al. Lithium Batteries. Elsevier; 1994, 1–47.
[7] Winter M, Besenhard JO, Spahr ME, et al. Adv Mater 1998;10:725–63.
[8] Yamamura J, Ozaki Y, Morita A, et al. J Power Sources 1993;43–44:233–9.

[9] Mabuchi A, Tokumitsu K, Fujimoto H, et al. J Electrochem Soc 1995;142:1041–6.
[10] Mabuchi A, Tokumitsu K, Fujimoto H, et al. J Electrochem Soc 1995;142:3049–51.
[11] Tatsumi K, Iwashita N, Sakaebe H, et al. J Electrochem Soc 1995;142:716–20.
[12] Ogumi Z, Abe T, Fukutsuk T, et al. J Power Sources 2004;127:72–5.
[13] Abe T, Fukuda H, Iriyama Y, et al. J Electrochem Soc 2004;151:A1120–3.
[14] Doi T, Iriyama Y, Abe T, et al. J Electrochem Soc 2005;152:A1521–5.
[15] Wang H, Ikeda T, Fukuda K, et al. J. Power Sources 1999;83:141–7.
[16] Zaghib K, Song X, Guerfi A, et al. J Power Sources 2003;124:505–12.
[17] Yoshio M, Wang H, Fukuda K. Angew Chem Int Ed 2003;42:4203–6.
[18] Wang H, Abe T, Maruyama S, et al. Adv Mater 2005;17:2857–60.
[19] Yoshizawa N, Hatori H, Yoshikawa K, et al. Mater Sci Eng B 2008;148:245–8.
[20] Yoshizawa N, Soneda Y, Hatori H, et al. Mater Chem Phys 2010;121:419–24.
[21] Fujimoto H. J Power Sources 2010;195:5019–24.
[22] Song R, Song H, Zhou J, et al. J Mater Chem 2012;22:12369–74.
[23] Guo P, Song H, Chen X. Electrochem Commun 2009;11:1320–4.
[24] Zhou J, Ma L, Song H, et al. Electrochem Commun 2011;13:1357–60.
[25] Zhou J, Song H, Ma L, et al. RSC Adv 2011;1:782–91.
[26] Yoshio M, Wang H, Fukuda K, et al. J Electrochem Soc 2000;147:1245–50.
[27] Natarajan C, Fujimoto H, Tokumitsu K, et al. Carbon 2001;39:1409–13.
[28] Wang H, Yoshio M. J Power Sources 2001;91:123–9.
[29] Wang H, Yoshio M, Abe T, et al. J Electrochem Soc 2002;149:A499–503.
[30] Han Y S, Lee JY. Electrochim Acta 2003;48:1073–9.
[31] Yoshio M, Wang H, Fukuda K, et al. J Mater Chem 2004;14:1754–8.
[32] Wang CW, Yi YB, Sastry AM, et al. J Electrochem Soc 2004;151:A1489–98.
[33] Beguin F, Chevallier F, Vix-Guterl C, et al. Carbon 2005;43:2160–7.
[34] Zhang HL, Liu SH, Li F, et al. Carbon 2006;44:2212–8.
[35] Ding YS, Li WN, Iaconetti S, et al. Surf Coat Technol 2006;200:3041–5.
[36] Buiel E, Dahn JR. J Electrochem Soc 1998;145:1977–81.
[37] Ohzawa Y, Yamanaka Y, Naga K, et al. J Power Sources 2005;146:125–8.
[38] Ohzawa Y, Okabe T, Kasugai T, et al. TANSO 2010; No.245: 192–195 [in Japanese].
[39] Kuribayashi I, Yokoyama M, Yamashita M. J Power Sources 1995;54:1–5.
[40] Tsumura T, Katanosaka A, Souma I, et al. Solid State Ionics 2000;135:209–12.
[41] Yoon S, Kim H, Oh SM. J Power Sources 2001;94:68–73.
[42] Ohta N, Nozaki H, Nagaoka K, et al. New Carbon Mater 2002;17:61–2.
[43] Wang G, Zhang B, Yue M, et al. Solid State Ionics 2005;178:905–9.
[44] Zhang HL, Li F, Liu C, et al. J Phys Chem C 2008;112:7767–72.
[45] Nozaki H, Nagaoka K, Hoshi K, et al. J Power Sources 2009;194:486–93.
[46] Ohta N, Nagaoka K, Hoshi K, et al. J Power Sources 2009;194:985–90.
[47] Hoshi K, Ohta N, Nagaoka K, et al. TANSO 2009; No. 240: 213–220.
[48] Zhou YF, Xie S, Chen CH. Electrochim Acta 2005;50:4728–35.
[49] Lee HY, Baek JK, Jang SW, et al. J Power Sources 2001;101:206–12.
[50] Imanishi N, Ono Y, Hanai K, et al. J Power Sources 2008;178:744–50.
[51] Zhang HL, Zhang Y, Zhang X-G, et al. Carbon 2006;44:2778–84.
[52] Yi TF, Jiang LJ, Shu J, et al. J Phys Chem Solids 2010;71:1236–42.
[53] Cheng L, Li XL, Liu HJ, et al. J Electrochem Soc 2007;154:A692–7.
[54] Hu X, Lin Z, Yang K, et al. J Alloys Compounds 2010;506:160–6.
[55] Jung HG, Kim J, Scrosati B, et al. J Power Sources 2011;196:7763–6.
[56] Zhao L, Hu YS, Li H, et al. Adv Mater 2011;23:1385–8.

[57] Wang J, Liu XM, Yang H, et al. J Alloys Compd 2011;509:712–8.

[58] Xiang H, Tian B, Lian P, et al. J Alloys Compd 2011;509:7205–9.

[59] Yuan T, Cai R, Shao ZP. J Phys Chem C 2011;115:4943–52.

[60] Chen JS, Liu H, Qiao SZ, et al. J Mater Chem 2011;21:5687–92.

[61] Guo X, Wang CY, Chen M. J Power Sources 2012;214:107–12.

[62] Yang L, Gao L. J Alloys Compd 2009;485:93–7.

[63] Shi Y, Wen L, Li F, et al. J. Power Sources 2011;196:8610–7.

[64] Kim E, Kim Y, Kim MG, et al. Electrochem Soild State Lett 2006;9:A156–9.

[65] Morishita T, Hirabayashi M, Nishioka Y, et al. J Power Sources 2006;160:638–44.

[66] Cui G, Hu YS, Zhi L, et al. Small 2007;3:2066–9.

[67] Chang K, Chen W, Ma L, et al. J Mater Chem 2011;21:6251–7.

[68] Zhou J, Song H, Chen X, et al. Chem Mater 2009;21:2935–40.

[69] Zhou J, Song H, Chen X, et al. Chem Mater 2009;21:3730–7.

[70] Zhou J, Song H, Fu B, et al. J Mater Chem 2010;20:2794–800.

[71] Padhi AK, Najundaswamy KS, Goodenough JB. J Electrochem Soc 1997;144:1188–94.

[72] Ohzuku T, Brodd RJ. J Power Sources 2007;174:449–56.

[73] Jugovic D, Uskokovic D. J. Power Sources 2009;190:538–44.

[74] Scrosati B, Garche J. J Power Sources 2010;195:2419–30.

[75] Zhang W J. J Power Sources 2011;196:2962–70.

[76] Fergus JW. J Power Sources 2010;195:939–54.

[77] Higuchi M, Katayama K, Azuma Y, et al. J Power Sources 2003;119–121:258–61.

[78] Needham SA, Calka A, Wang GX, et al. Electrochem Commun 2006;8:434–8.

[79] Prosini PP, Caeewska M, Suaccia S, et al. J Electrochem Soc 2002;149:A886–90.

[80] Yamada A, Chung SC, Hinokuma K. J Electrochem Soc 2001;148:A224–9.

[81] Huang H, Yin SC, Nazar LF. Electrochem Solid State Lett 2001;4:A170–2.

[82] Ravet N, Chouinard Y, Magnam JF, et al. J Power Sources 2001;97:503–7.

[83] Chen Z, Dahn JR. J Electrochem Soc 2002;149:A1184–9.

[84] Franger S, Le Cras F, Bourbon C, et al. Electrochem Solid State Lett 2002;5:A231–2.

[85] Ait Salah A, Mauger A, Zaghih K, et al. J Electrochem Soc 2003;153:A1692–701.

[86] Barker J, Saidi MY, Swoyer JL. Electrochem Solid State Lett 2003;6:A53–5.

[87] Doeff MM, Hu Y, McLarnon F, et al. Electrochem Solid State Lett 2003;6:A207–9.

[88] Bewlay SI, Konstantinov K, Wang GX, et al. Mater Lett 2004;58:1788–91.

[89] Yang J, Xu JJ. Electrochem Solid State Lett 2004;7:A515–8.

[90] Myung ST, Komaba S, Hirosaki N, et al. Electrochim Acta 2004;49:4213–22.

[91] Dominko R, Bele M, Gaberscek M, et al. J Electrochem Soc 2005;152:A607–10.

[92] Striebel K, Shim J, Srinivasan V, et al. J Electrochem Soc 2005;152:A664–70.

[93] Zhang SS, Allen JL, Xu K, et al. J Power Sources 2005;147:234–40.

[94] Yang J, Xu JJ. J Electrochem Soc 2006;153:A716–23.

[95] Doeff MM, Wilcox JD, Kostecki R. J Power Sources 2006;163:180–4.

[96] Gabrisch H, Wilcox JD, Doeff MM. Electrochem Solid State Lett 2006;9:A360–3.

[97] Shin HC, Cho WI, Jang H. Electrochim Acta 2006;52:1472–6.

[98] Nakamura T, Miwa Y, Tabuchi M, et al. J Electrochem Soc 2006;153:A1108–14.

[99] Roberts MR, Spong AD, Vitins G, et al. J Electrochem Soc 2007;154:A921–8.

[100] Kim K, Jeong JH, Kim IJ, et al. J Power Sources 2007;167:524–8.

[101] Ong CW, Lin YK, Chen J S. J Electrochem Soc 2007;154:A527–33.

[102] Dominko R, Bele M, Goupil JM, et al. Chem Mater 2007;19:2960–9.

[103] Kim JK, Cheruvally G, Ahn JH, et al. J Phys Chem Solids 2008;69:2371–7.

[104] Kim JK, Cheruvally G, Ahn JH, et al. J Phys Chem Solids 2008;69:1257–60.

[105] Lin Y, Gao MX, Zhu D, et al. J Power Sources 2008;184:444–8.

[106] Wang Y, Wang Y, Hosono E, et al. Angew Chem Int Ed 2008;47:7461–5.

[107] Liang G, Wang L, Ou X, et al. J Power Sources 2008;184:538–43.

[108] Chen ZY, Zhu HL, Ji S, et al. Solid State Ionics 2008;179:1810–5.

[109] Wang K, Cai R, Yuan T, et al. Electrochim Acta 2009;54:2861–8.

[110] Murugan AV, Muraliganth T, Manthiram A. J Electrochem Soc 2009;156:A79–83.

[111] Jaiswal A, Horne CR, Cheng O, et al. J Electrochem Soc 2009;156:A1041–6.

[112] Lu CZ, Fey GTK, Kao H M. J Power Sources 2009;189:155–62.

[113] Nien YH, Carey JR, Chen JS. J Power Sources 2009;193:822–7.

[114] Cho YD, Fey GTK, Kao HM. J Power Sources 2009;189:256–63.

[115] Palomares V, Goni A, de Muro IG, et al. J Power Sources 2010;195:7661–8.

[116] Kadoma Y, Kim JM, Abiko K, et al. Electrochim Acta 2010;55:1034–41.

[117] Fey GTK, Huang KP, Kao HM, et al. J Power Sources 2011;196:2810–8.

[118] Peng W, Jiao L, Gao H, et al. J Power Sources 2011;196:2841–7.

[119] Kang F, Ma J, Li B. New Carbon Mater 2011;26:161–70.

[120] Jugovic D, Mitric M, Kuzmanovic M, et al. J Power Sources 2011;196:4613–8.

[121] Rui XH, Jin Y, Feng XY, et al. J Power Sources 2011;196:2109–14.

[122] Toprakci O, Ji L, Lin Z, et al. J. Power Sources 2011;196:7692–9.

[123] Ma J, Li B, Du H, et al. J Electrochem Soc 2011;158:A26–32.

[124] Yamada A, Hosoya M, Chung SC, et al. J Power Sources 2003;119–121:232–8.

[125] Hong J, Wang F, Wang X, et al. J Power Sources 2011;196:3659–63.

[126] Kaymaksiz S, Kaskhedikar N, Sato N, et al. Electrochem Soc Trans 2010;25:187–97.

[127] Liu J, Conry TE, Song X, et al. J Mater Chem 2011;21:9984–7.

[128] Endo M, Kim YA, Hayashi T, et al. Carbon 2001;39:1287–97.

[129] Sheem K, Lee YH, Limd HS. J Power Sources 2006; 158–1425–1430.

[130] Sotowa C, Origi G, Takeuchi M, et al. Chem Sus Chem 2008;1:911–5.

[131] Su FY, You C, He YB, et al. J Mater Chem 2010;20:9644–50.

[132] Zhou X, Wang F, Zhu Y, et al. J Mater Chem 2011;21:3353–8.

[133] Jeong SK, Inaba M, Mogi R, et al. Langmuir 2001;17:8281–6.

[134] Xu W, Angell CA. Electrochem Solid State Lett 2000;3:366–8.

[135] Jeong SK, Inaba M, Iriyama Y, et al. Electrochem Solid State Lett 2003;6:A13–5.

[136] Jeong SK, Inaba M, Iriyama Y, et al. J Power Sources 2008;175:540–6.

[137] Takeuchi S, Miyazaki K, Sagane F, et al. Electrochim Acta 2011;56:10450–3.

Carbon Materials in Photocatalysis

13

Titanium dioxide, TiO_2, has three different crystalline phases: brookite, anatase, and rutile. One of these phases, anatase, was found to be able to decompose water molecules to hydrogen and oxygen under ultraviolet (UV) irradiation [1], opening the new field of photocatalysis. Much work on photocatalysts has been performed from various viewpoints, such as development of novel photocatalysts and improvement of their photocatalytic activity, and various applications have been proposed for environmental cleanup, such as air and water purification, hazardous waste remediation, and water disinfection. The strong and urgent demand for the conservation of the global environment and its protection from pollution has greatly promoted the research work on photocatalysis.

Coupling of photocatalysts with carbon materials has been recognized to be one of the possibilities to combine photocatalytic activity with adsorptivity, and carbon materials are found to have certain roles in reinforcing photocatalytic performance. The photocatalyst TiO_2 with anatase structure coupled with carbon can be classified into the following three types on the basis of the state of carbon: (1) TiO_2-loaded carbons, (2) carbon-doped TiO_2 (or carbon-modified TiO_2), and (3) carbon-coated TiO_2, as illustrated in Figure 13.1. TiO_2-loaded carbons have been prepared by loading of TiO_2 particles onto different activated carbons (ACs), and the processes have attained some success for the hybridization between adsorptivity and photoactivity. Mechanical mixing of TiO_2 particles with different ACs has also been found to give certain benefits. Carbon-doped anatase-type TiO_2 has been prepared, in which a part of either the oxygen or titanium atoms in the anatase crystal lattice is replaced by carbon atoms. This material was reported to have sensitivity even under visible light to some extent. Carbon-coated TiO_2 is completely different from other two composites, as illustrated in Figure 13.1. The carbon layer formed on the surface of TiO_2 particles is porous without any activation. Carbon also plays important roles in the fabrication of novel photocatalysts. Carbon materials used in the photocatalysis field have been reviewed in detail [2].

In this chapter, the carbon materials associated with photocatalysis are reviewed, classifying them into five kinds of carbon/photocatalyst composites:

1. TiO_2-loaded ACs.
2. Powder mixtures of TiO_2 with AC.
3. Carbon-doped TiO_2, including carbon-modified TiO_2.
4. Carbon-coated TiO_2.
5. Novel photocatalysts fabricated using carbon materials.

Prerequisite for readers: Chapter 3.9 (Carbon materials for environment remediation) in *Carbon Materials Science and Engineering: From Fundamentals to Applications*, Tsinghua University Press.

Advanced Materials Science and Engineering of Carbon.

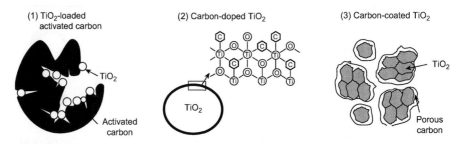

FIGURE 13.1 The Positional Relationship Between TiO$_2$ and Carbon in Various TiO$_2$/Carbon Composites

13.1 TiO$_2$-loaded activated carbons

In order to decompose pollutant molecules faster, it is better to concentrate the molecules on the surface of an adsorbent and then to transfer them to a photocatalyst TiO$_2$ particle. For this purpose, the loading of TiO$_2$ particles onto AC particles has been carried out in a number of studies [3–40]. ACs have been selected as supports because they are lightweight, nonpolar, nonreactive, nontoxic, and relatively cheap, in addition to their highly porous nature. The observed enhancement of apparent photocatalytic activity is understood to originate principally from the pollutant-rich environment at the TiO$_2$/AC interface: (1) pollutant concentration on the AC particles becomes higher than in solution mainly owing to adsorption, and (2) the AC helps to bring the hydrophobic organic pollutant molecules near the hydrophilic TiO$_2$ particles, particularly under UV irradiation. However, many studies have reported that there is an optimum in the TiO$_2$-loading amount and AC pore structure for achieving a higher photocatalytic activity than that of TiO$_2$ itself.

Spatial relationships among particles of AC, small particles of TiO$_2$ photocatalyst, and molecules of an organic pollutant are shown schematically in the dark and under UV irradiation in Figure 13.2. The pollutant molecules are assumed to be small enough to be adsorbed into the micropores. In most ACs, a large number of micropores exist on the surface of larger pores, mesopores and macropores, in which a large number of pollutant molecules are accommodated by physical adsorption. On the other hand, only a small number of pollutant molecules are adsorbed on the surface of TiO$_2$ nanoparticles. By depositing TiO$_2$ particles onto an AC particle, some micropores and mesopores become closed at their entrances (see Figure 13.2A) and this causes a marked decrease in the apparent surface area.

By UV irradiation onto such TiO$_2$-loaded AC particles (Figure 13.2B), oxidative •OH radicals are formed on the TiO$_2$ particles and can degrade pollutant molecules by oxidation. However, it has to be pointed out that these radicals are formed only on the surface of TiO$_2$ particles that is accessible to UV rays (active centers), and not on the surface in the "shadow"; radicals cannot migrate long distances and are confined to a region near the active centers in a TiO$_2$ particle [3]. In order for adsorbed pollutant molecules to be photocatalytically degraded, they have to diffuse along the

(A)

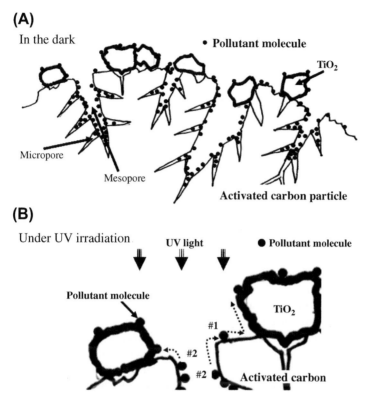

In the dark

• **Pollutant molecule**

TiO$_2$

Micropore

Mesopore

Activated carbon particle

(B)

Under UV irradiation

UV light

• **Pollutant molecule**

Pollutant molecule

TiO$_2$

#1

#2

#2

Activated carbon

FIGURE 13.2 Schematic Illustration of the Adsorption of Pollutant Molecules Onto a TiO$_2$-loaded AC Particle

(A) In the dark, and (B) under UV irradiation

surfaces of AC and TiO$_2$ particles, as shown by arrows in Figure 13.2B: the molecule #1 adsorbed on the AC surface diffuses to the carbon/TiO$_2$ interface, but this interface is not necessarily exposed to UV rays, and so it has to diffuse to an active center on the irradiated surface. The molecules accommodated on the pore surface of an AC particle have to migrate over a relatively long distance to reach TiO$_2$ particles (molecules #2 in Figure 13.2B). The driving force for such surface diffusion is mainly a concentration gradient between pollutant molecules on the irradiated TiO$_2$ surface and those on the surface of variable-sized pores in the AC particle. The molecules adsorbed deep inside micropores of AC diffuse with greater difficulty toward TiO$_2$ particles located on the external surface of the AC particle. Therefore, highly microporous ACs are not necessarily advantageous for the TiO$_2$-loaded AC to have superior photocatalytic performance [10]. The effect of substrate pore structure has been studied using four ACs (Brunauer-Emmett-Teller [BET] surface area, S_{BET}, of 770–1150 m^2/g) and a dip-hydrothermal method of photocatalyst preparation [4]. Better photocatalytic degradation of methyl orange (MO) has been achieved with

TiO$_2$-loaded ACs than with a simple mixture of TiO$_2$ and AC [40]. Mesoporosity of the substrate AC governed the amount of TiO$_2$ loaded and the MO adsorption rate, and thus also the photoactivity.

The decomposition rate observed on TiO$_2$-loaded ACs seems to be governed both by the surface diffusion of pollutant molecules and by the photocatalytic reaction rate. When adsorption occurs slowly, the change in relative pollutant concentration, c/c_0, with irradiation time depends on both adsorption and photodecomposition, particularly at the onset of UV irradiation. In Figure 13.3, changes in c/c_0 of phenol (Ph) remaining in solution with UV irradiation time are compared for TiO$_2$-loaded ACs, which were prepared by hydrolysis of tetraisopropyl orthotitanate and heat treatment at 650–900 °C [17]. Adsorption of Ph on the as-received AC, in the absence of TiO$_2$, occurs mainly during the first 1 h and achieves saturation after 3 h under UV irradiation. For the composites, however, adsorption and photodecomposition of Ph occur simultaneously, but the former is reasonably supposed to be the dominant process in the beginning and the latter in the later stage. The observed trend was approximated to consist of two linear processes, the change from one to the other occurring at around 1 h of irradiation. The photodecomposition rate constant was calculated from the slope of the linear region after 1 h irradiation, and adsorptivity was calculated from the intercept of the extrapolated linear region observed after 1 h to the ordinate (expressed as grams of Ph per gram composite). These parameters are shown for each composite in Figure 13.3, together with heat-treatment temperature (HTT). The highest decomposition rate was obtained for the composite heated at 700 °C, with a single anatase TiO$_2$ phase. The composites heated to 800 and 900 °C had a high Ph-adsorption capacity, comparable to the pristine AC, but low Ph decomposition rate constant, mainly because of partial transformation of anatase to rutile by heat treatment above 800 °C.

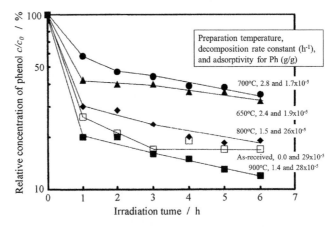

FIGURE 13.3 Changes in Relative Concentration, c/c_0, of Ph with Irradiation Time for TiO$_2$-loaded ACs

Loading of TiO_2 onto ACs has been carried out through two routes: (1) loading of TiO_2 particles prepared in advance, and (2) loading of a TiO_2 precursor, followed by conversion to TiO_2 particles. The latter is preferable for the preparation of fine and highly dispersed particles of TiO_2 on the AC surface; it has been done mostly by using either sol-gel or chemical vapor deposition (CVD) methods. Alkoxides (e.g. titanium tetraisopropoxide, tetrabutyl titanate) have often been used as TiO_2 precursors. Preparation of TiO_2-loaded ACs has been reviewed with a focus on the CVD method [36] and on the sol-gel methods [38].

Adhesion of TiO_2 particles to the AC surface appears to be important for the enhancement of photocatalytic activity and also for the practical application of these composites. By the CVD method using tetrabutyl titanate, nano-sized TiO_2 particles were shown to adhere to AC and to give high activity for photodecomposition (mineralization) of MO in water [26]. Addition of H_2O to titanium tetraisopropoxide vapor was also reported to facilitate the CVD process at a lower temperature and higher deposition rate [25]. Introduction of water vapor during the CVD process and adsorption onto AC in advance were reported to be crucial to obtain anatase-type TiO_2 on the surface of AC particles [16]. TiO_2-loaded ACs have also been prepared by dipping the support into a sol prepared by alkoxide hydrolysis, and then subjected to heat treatment at 300–500 °C [4–7]. In an alternative method, they were prepared by adding $TiCl_4$ drop-wise into an aqueous suspension of AC, followed by heat treatment at 500 °C in a N_2 atmosphere [32].

Loading of TiO_2 powder directly onto AC has also been carried out by mixing TiO_2 with an AC in aqueous suspension with stirring [22]. Loading onto an AC filter, prepared by gluing granular AC onto the surface of glass cloth, was performed by using a water suspension of 5 mass% TiO_2, and the resultant TiO_2-loaded AC filter was used for the removal of formaldehyde from air [41]. Photocatalyst loading by stirring TiO_2-loaded AC particles in a CCl_4 solution of pitch, followed by heat treatment at 750 °C, has also been reported [34]. In the resultant composite, TiO_2 particles on the AC were presumably coated by carbon, formed from the pitch during heat treatment, which might function to fix TiO_2 particles onto the AC surface.

TiO_2 particles were able to be loaded onto an AC by a spray-desiccation method, with only a small change in pore structure of the AC [42]. In another study, loading of TiO_2 onto the surface of AC was carried out by dipping AC particles in a peroxotitanate solution and then heating at 180 °C in a Teflon™-lined stainless-steel vessel, followed by calcination at 300–800 °C [43]. By using AC particles of 0.15–0.25 mm, separation of the particles from solution was much easier, and photocatalytic activity for the decomposition of MO remained almost the same even after five cycles. Loading of TiO_2 has also been carried out by plugging the pores of AC with paraffin [44,45]. By removing the paraffin at 250 °C in air after TiO_2 loading, the high surface area of the pristine AC could be recovered and high photocatalytic activity was obtained for the decomposition of methylene blue (MB).

TiO_2-loaded ACs have been produced also by mixing TiO_2 particles with a liquid- or solid-state carbon precursor [19,46,47]. By hydrolysis of tetraisopropyl orthotitanate, TiO_2 was precipitated on the surface of poly(vinyl butyral) (PVB) and the

TiO$_2$-loaded PVB was carbonized at high temperature in a flow of CO$_2$ [19]. TiO$_2$-loaded carbon microspheres with 25 μm diameter have been prepared from TiO$_2$-loaded cellulose microspheres, formed by one-step phase separation using cellulose xantate and sodium polyacrylate aqueous solutions with dispersed TiO$_2$ powder [46].

In order to discuss photocatalytic activity of TiO$_2$-loaded ACs in the context of practical application, the suitability for cyclic use must be tested in both adsorption and photodecomposition, with attention to fall-off of TiO$_2$ particles from the surface and an optimal TiO$_2$ loading. In Figure 13.4, temporal changes in c/c_0 of iminoctadine triacetate (IT), which is often used in rice and orchard fields as an insecticide and appears in water paths in the field, are shown for pristine AC and three TiO$_2$-loaded ACs [20]. The composite was first kept in the dark for 200 h to saturate IT adsorption, and then exposed to UV irradiation (first cycle). After about 800 h, the sample was removed from the test solution and dispersed again into a virgin 1.87×10^{-4} mol/L solution, and again kept in the dark for about 200 h, then irradiated by UV (second cycle). In the third cycle, the suspension was kept in the dark much longer, in order to ensure that the adsorption of IT reached saturation. For the pristine AC, c/c_0 was found to be almost constant in the dark and to scatter somewhat under UV irradiation. For the three TiO$_2$-loaded composites, on the other hand, an abrupt c/c_0 decrease was observed under UV irradiation, but c/c_0 changed only slightly in the dark. In each cycle, the rate constant (k) for IT decomposition was different for the three composites; the 5.8 mass%-loaded composite showing the largest decrease in each

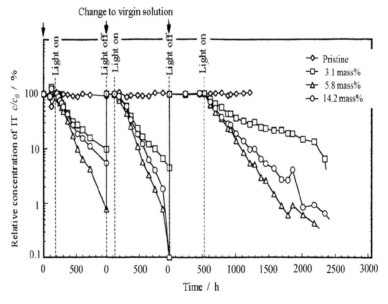

FIGURE 13.4 Change in c/c_0 of IT With Time Under On/Off Cycles of UV Rays for AC and TiO$_2$-loaded ACs

cycle and the highest photoactivity among the four samples. From scanning electron microscopy (SEM) observation, fine anatase particles covered the entire surface of the carbon substrate in two samples, those loaded by 5.8 and 14.2 mass% TiO$_2$, and in the latter the TiO$_2$ particles were packed more densely. Therefore, the latter is arguably overloaded by TiO$_2$ and the efficiency of photocatalysis seems to be low. For each TiO$_2$-loaded composite, the slope in Figure 13.4 decreases upon cycling, this being attributed to a reduction in number of TiO$_2$ particles during transfer from one solution to another.

Composites of different carbon nanotubes (CNTs) and nanofibers (CNFs) with TiO$_2$ powder, either by mixing or immobilizing TiO$_2$ particles, have been prepared and shown to give enhanced photocatalytic activity [48–54]. Single-walled CNTs exhibited enhanced and selective photocatalytic oxidation of Ph [49]. Heterostructured CNFs consisting of TiO$_{2-x}$N$_x$ and carbon have been prepared by carbonization of electrospun polyacrylonitrile nanofibers containing stabilized titanium oxoacetate [53].

The surface properties of ACs, including activated carbon fibers (ACFs), which are known to depend strongly on the preparation process, affect the loading of TiO$_2$ and consequently the adsorption of pollutant molecules. Experiments have shown, for example, that HNO$_3$ treatment results in more efficient TiO$_2$ loading by CVD, in comparison with other oxidation treatments [26,27]. Adsorptivity of ACs also depends strongly on the molecular morphology and size of pollutant. Since overly large molecules are difficult to be adsorbed in the micropores, the supply of the pollutant molecules to the TiO$_2$ surface occurs directly from solution, and not from the surface of the AC adsorbent [13].

Other photocatalysts have also been loaded onto carbon materials. Spherical AC particles containing ZnS and ZnO (photocatalysts) were prepared from a cation-exchange resin (polystyrene with sulfonate functional groups and cross-linked by divinylbenzene) and ZnCl$_2$ aqueous solution, followed by carbonization at 500–900 °C [56]. The photocatalytic performance of multi-walled CNT/WO$_3$ composites has also been studied [57]. A review focusing on CNT/TiO$_2$ composites has been published [55]

13.2 **Mixture of activated carbon and TiO$_2$**

A suspended mixture of AC with TiO$_2$ (TiO$_2$ + AC) was reported to give enhanced photocatalytic activity for the decomposition of pollutants in aqueous solution [58–62]. In Figure 13.5, concentration change of 4-chlorophenol solution is shown in the dark and then under UV irradiation for a commercial photocatalyst TiO$_2$, P25, without and with a commercially available AC (particle size around 60 μm, S_{BET} 775 m^2/g, and mean pore width ~0.8 nm) [58]. The concentration decreases faster for TiO$_2$ + AC than for TiO$_2$, revealing that mixing with AC accelerates photodecomposition of the pollutant. The concentration changes can be approximated by a pseudo-first-order equation, as shown in the linear relationship between $\ln(c_0/c)$ and

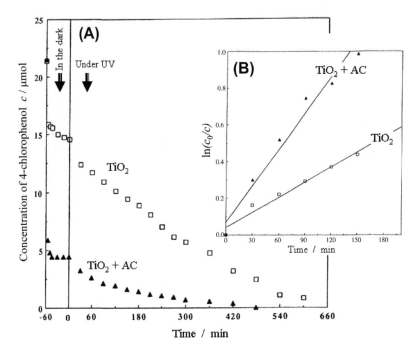

FIGURE 13.5 Change in Concentration, *c*, of 4-chlorophenol with UV Irradiation Time for TiO$_2$ and a Mixture of TiO$_2$ with AC

(A) *c* vs time, and (B) ln(c_0/c) vs time

Courtesy of Prof. J. Matos of Instituto Venezolano de Investigaciones Cientificas

irradiation time, *t*, in Figure 13.5B. A similar synergistic effect was observed in the decomposition of Ph and 2,4-dichlorophenoxyacetic acid using the same AC and TiO$_2$ mixture [59].

The transmission of UV rays to the surface of TiO$_2$ particles may be disturbed by the addition of AC in the suspension. However, experimental results show that such perturbation is largely compensated by a strong synergistic effect. This effect for Ph decomposition was found to depend on the mass ratio of TiO$_2$ to AC, in the range of 5/10 to 75/10 [61]. The synergistic effect was thought to be attributable to rapid transfer of the Ph molecules initially adsorbed on the AC to the surface of TiO$_2$; the driving force for this transfer is presumably the difference in the surface concentration of Ph between AC and TiO$_2$. A certain amount of Ph was found to remain in the AC, even after the concentration in solution became negligibly small. Since residual Ph molecules are mostly trapped in micropores of AC particles, a long irradiation time is required to eliminate the residual Ph [58]. Enhancement of photocatalytic activity of the mixture depended also on the AC used [60,62,63]. Different ACs have been prepared from a wood-derived lignocellulose under different conditions, heat treatment at different temperatures (450–1000 °C), and activation by either CO$_2$, ZnCl$_2$, H$_3$PO$_4$, or KOH [64]. The results indicate that the AC must be carefully selected in order to achieve high photodecomposition efficiency, because the pore structure of the AC is strongly influenced by the carbonization and activation conditions.

The suspended mixture of AC and TiO$_2$ can be reused for photodecomposition of 4-chlorophenol without reduction of the beneficial synergistic effect. The same effect was confirmed in a pilot plant with a total solution volume of 247 L, using sunlight [59,65].

For decomposition of dichloromethane, however, a suspension of TiO$_2$ with AC showed lower photoactivity than the TiO$_2$ immobilized on the AC surface (TiO$_2$-loaded AC) [13]. A mixture (TiO$_2$ + AC) gave higher photocatalytic activity for mineralization of 3,5-dichloro-N-(1,1-dimethyl-2-propynyl) benzamide (propyzamide) than TiO$_2$ without AC, but lower than TiO$_2$-loaded AC [4]. Mixtures of TiO$_2$ with either AC alone or AC with ferric oxide (Fe$_2$O$_3$) were shown to accelerate the oxidation of NO to NO$_2$ or HNO$_3$ in a flow-type reaction chamber [29].

13.3 **Carbon-doped TiO$_2$**

Carbon is known to be able to replace either oxygen or titanium in TiO$_2$ crystal lattice and also to precipitate at the grain boundary of TiO$_2$, the former giving visible-light sensitivity for photocatalyst TiO$_2$ particles and the latter acting as a photosensitizer and transferring electrons to the TiO$_2$ conduction band. Theoretical calculation using density functional theory predicted that the substitution of oxygen by carbon and the formation of oxygen vacancies are favored under oxygen-poor conditions, but the interstitial carbon and carbon substitution of Ti are preferred under oxygen-rich conditions [66]. Other theoretical calculations suggested that carbon doping is the most promising among the strategies of preparation of TiO$_2$ photocatalysts doped by anions (C, N, and S), because of a significant overlap between the O 2p state and the dopant states near the valence band edge [67]. However, the development of visible-light activity in the carbon-doped TiO$_2$ was also reported to be an aromatic compound deposited on the TiO$_2$ surface, i.e. neither substituted nor interstitial carbon in the lattice [68].

Carbon-doped TiO$_2$ replacing oxygen in the anatase lattice by carbon was first synthesized by mild oxidation of TiC at 350 °C, followed by annealing under O$_2$ flow at 600 °C, and the substituted carbon content was estimated to be 0.32 at% [69]. The resultant carbon-doped TiO$_2$ was visible-light sensitive and its photoactivity was examined by decomposition of gaseous 2-propanol (IPA), as shown in Figure 13.6. Another study reported that, to obtain carbon-doped TiO$_2$, oxidation at 350 °C for more than 50 h was recommended without any annealing in O$_2$ atmosphere [70]. It was also reported that carbon-doped TiO$_2$ with carbon content of about 0.7 mass% and high photocatalytic activity can be obtained from TiC by oxidation at 350 °C [71]. This material was gray-white in color and its diffuse reflectance spectrum showed a red shift by 36 nm compared with pure anatase-type TiO$_2$. Its photoactivity under visible-light irradiation was confirmed by the decomposition of trichloroacetone. Carbon-doped TiO$_2$ has also been prepared by hydrolysis of TiCl$_4$

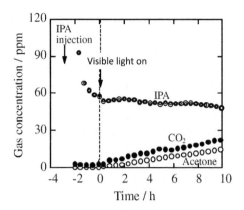

FIGURE 13.6 Changes in Concentrations of 2-propanol (IPA) and of its Decomposition Products, CO$_2$ and Acetone, with Time Under Visible-light Irradiation

From [69]

with tetrabutylammonium hydroxide in aqueous solution at room temperature, followed by calcination in air [72], by hydrothermal treatment of the mixture of glucose and amorphous TiO$_2$ [73] or Ti(SO$_4$)$_2$ [74], by surface oxidation of Ti metal film in a natural gas flame [75], and by sol-gel synthesis from titanium butoxide [76].

Carbon-doped TiO$_2$ thin films have been fabricated by radio-frequency magnetron sputtering of a Ti metal target under CO$_2$/Ar gas mixture at 2 Pa [77], and by ion-assisted electron-beam evaporation using rutile-type TiO$_2$ in either CO$_2$ or CO gas [78]. The film with a carbon content of 1.25 at% gave the highest visible-light photoactivity for MB decomposition and also superhydrophilicity. Films incorporating carbon have also been deposited on Si and quartz substrates by sputtering Ti and graphite targets in Ar/O$_2$ plasma [79]. By ultrasonic spray pyrolysis of an aqueous solution containing TiCl$_4$, ZrOCl$_2$, and CH$_3$(CH$_2$)$_{17}$NH$_2$, carbon-doped Ti$_{0.91}$Zr$_{0.09}$O$_2$ was obtained, and its visible-light activity was confirmed by the decomposition of rhodamine B [80]. Co-doping of C and N into TiO$_2$ has been attempted by calcination of the precipitates from titanium isopropoxide in NH$_3$ flow at 500 °C [81]. The product was visible-light active, nitrogen had replaced some oxygen atoms in the TiO$_2$ lattice, but carbon was precipitated at the grain boundary. Visible-light-sensitive TiO$_2$ co-doped with carbon and nitrogen has been synthesized under hydrothermal conditions at 180 °C from biomass derivatives (furfural) and titanium isopropoxide [82].

Replacement of Ti in an anatase-type TiO$_2$ lattice by carbon atoms was possible by heating a mixture of TiO$_2$, thiourea, and urea in a lidded alumina crucible at 400 and 500 °C under aerated flow [83]. The resultant powder was a single phase of anatase with dark orange color and its C1s peak in X-ray photoelectron spectroscopy (XPS) indicated the substitution of Ti^{4+} by C^{4+} (C^{4+}-doped TiO$_2$). Carbon content in the powder was reported to be 0.2–0.4 at% on the surface of particles. Its photocatalytic activity was confirmed by decomposition of MB in water under visible light: this was about four times higher than that of the pristine TiO$_2$. Co-doping of C^{4+} and S^{4+} into TiO$_2$ with rutile structure has been carried out by calcination of TiO$_2$ powder with thiourea similarly to the method described above [84]. The resultant photocatalyst was dark yellow and photocatalytic decomposition of MB and 2-methylpyridine under visible light was confirmed.

Carbon-containing TiO$_2$ photocatalysts in which a part of the carbon replaced O in the lattice and some was precipitated at the grain boundaries of TiO$_2$ crystals (carbon-modified TiO$_2$), have been fabricated through a sol-gel process [85,86] and by CVD of titanium tetraisopropoxide [87]. The photoactivity evaluated by p-chlorophenol decomposition under visible light was much higher than that of commercial photocatalysts, including P25 [85]. Successful modification of the surface of TiO$_2$ particles with carbon has been accomplished by the heat treatment of commercial photocatalyst powder in a flow of n-hexane vapor [88–90] and of ethanol vapor [91–95]. The resultant photocatalyst had enhanced adsorptivity for azo dyes, and thus a high photocatalytic activity. Marked reduction in turbidity was observed due to a change in surface properties of TiO$_2$ particles from hydrophilic to hydrophobic, by the reduction in hydroxyl-group concentration and the formation of –CH$_3$ groups. The resulting photocatalysts were evaluated by "practical efficiency," which was defined as $(A_{\text{C-modified}}/A_{\text{pristine}})(1/B)$, where $A_{\text{C-modified}}$ is the fraction of Ph decomposed by carbon-modified TiO$_2$, A_{pristine} is that by the pristine TiO$_2$, and B is turbidity of carbon-modified TiO$_2$ in nephelometric turbidity units. In Figure 13.7, practical efficiency determined under UV and visible-light irradiation is plotted against carbon content. Under UV irradiation (Figure 13.7A), the parameter increases with increasing carbon content, most markedly for the photocatalyst prepared at 450 °C, and tends to saturate. Under visible-light irradiation (Figure 13.7B), however, there was almost no change in the parameter up to about 0.6 mass% C, and then it increased

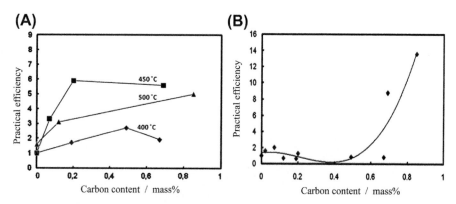

FIGURE 13.7 Changes in Practical Efficiency for Ph Decomposition with Carbon Content for Carbon-modified TiO$_2$ Prepared at Different Temperatures

(A) Under UV irradiation, and (B) under visible-light irradiation

Courtesy of Prof. A. W. Morawski of the West Pomeranian University of Technology, Szczecin, Poland

to about 14 times as carbon content reached 0.8 mass%. These results suggest that the carbon content should be controlled in the preparation process, in order to get appreciable visible-light photoactivity.

A carbon-modified TiO_2 has been prepared by heat treatment of an ethanol/water solution of *n*-tetrabutyl titanate and dodecylamine at 80 °C with subsequent calcination at 500 °C, in which carbon atoms had arguably substituted partly Ti and partly O [96]. Modification of TiO_2 nanotubes has been carried out by mixing tetraethylammonium hydroxide in aqueous solution followed by calcination at 350–450 °C [97].

13.4 Carbon-coated TiO_2

Carbon-coated TiO_2 has been prepared by mixing TiO_2 powder or a TiO_2 precursor with a carbon precursor either in powder or in aqueous solution, follwed by heat treatment at a temperature of 700–1100 °C in an inert atmosphere [98–110].

Transmission electron microscopy (TEM) observation proved carbon layer formation on the surface of TiO_2 particles, about 2 nm thick for a sample containing 2 mass% C (Figure 13.8A) and about 10 nm for 10 mass% C (Figure 13.8B), which were prepared from a powder mixture of commercial TiO_2 (ST-01) with poly(vinyl alcohol) (PVA) at 900 °C [99]. After heat treatment at 700 °C without mixing with PVA, particles of the pristine TiO_2 were sintered and showed marked grain growth, but the carbon-coated TiO_2 showed no marked sintering of the particles even after heating to 900 °C. The pristine TiO_2 showed broad diffraction lines of anatase-type crystal structure, but they became sharp with increasing HTT and a part of the anatase phase was transformed to rutile phase above 800 °C, the transformation being

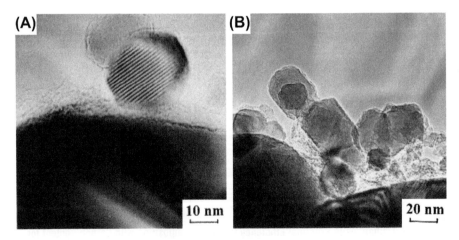

FIGURE 13.8 TEM Images of Carbon-coated TiO_2 Particles

(A) 2 mass% C-coated, and (B) 10 mass% C-coated

From [99]

complete at 900 °C. For mixtures of TiO$_2$ with PVA, however, diffraction lines of anatase phase were broad even after heat treatment at 800 °C. Above 900 °C, all diffraction lines became sharp, but phase transformation to rutile was markedly suppressed and only a small amount of rutile was detected even at 1100 °C. This suppressing effect seemed to become detectable with carbon content above 5 mass%. A similar change in X-ray diffraction (XRD) pattern was observed for TiO$_2$ precipitated from titanium tetraisopropoxide (TTIP) in water and in 2 mass% PVA aqueous solutions [100].

Carbon content and apparent S_{BET} of carbon-coated TiO$_2$ were found to depend strongly on the carbon precursors—PVA, poly(ethylene terephthalate) (PET), and hydroxypropyl cellulose (HPC)—and also heat-treatment conditions [104]. The apparent S_{BET} depends strongly on carbon content, suggesting that carbon is responsible for the high surface area observed. Adsorptivity for some model pollutants—MB, reactive black 5 (RB5), Ph, and IT—depended on S_{BET} of carbon-coated TiO$_2$, and also on the pollutants, MB being adsorbed in relatively large amounts but Ph only a small amount [104].

As shown schematically in Figure 13.9, pollutant molecules in solution have to be adsorbed into the carbon layer first and diffuse toward the surface of the photocatalyst particle. When the carbon layer on the TiO$_2$ surface can adsorb a large amount of the pollutant molecules, they accumulate near the surface of the TiO$_2$ particle, which surely accelerates the photodecomposition. This is thought to be the case for pollutants that are easily absorbed into the carbon layer, like MB. In the case of pollutant molecules that are poorly adsorbed into the carbon layer, like Ph, however, high photocatalytic activity cannot be expected for carbon-coated

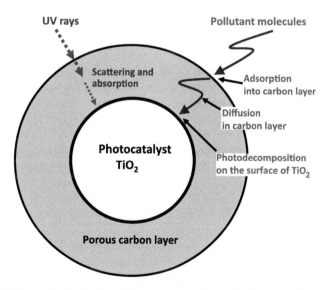

FIGURE 13.9 Schematic Illustration of Photodecomposition of Pollutant on Carbon-coated TiO$_2$ Under UV Irradiation

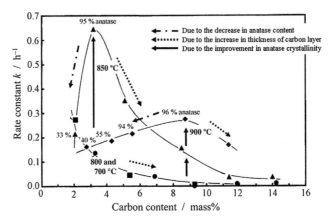

FIGURE 13.10 Relationship Between _k_ for MB Decomposition and Carbon Content of Carbon-coated TiO$_2$ at Different HTTs

Anatase content, f_A, in TiO$_2$ is indicated for each experimental point, and no indication means 100% anatase

From [2]

TiO$_2$. In the case of carbon-coated TiO$_2$, UV rays are weakened by scattering and absorption in the carbon layer before arriving at the surface of the TiO$_2$ particle, as illustrated in Figure 13.9.

In order to consider the role of the carbon layer in the photocatalytic activity of carbon-coated TiO$_2$, k for MB decomposition is plotted against carbon content as a function of HTT in Figure 13.10. The relationship between k and carbon content at different HTTs has to be discussed taking into account the crystallinity and content of anatase phase in TiO$_2$, in addition to the carbon content (corresponding to the average thickness of the carbon layer on the surface of TiO$_2$). The gradual decrease in k at 700 and 800 °C with increasing carbon content is caused by attenuation of the strength of UV arriving on the photocatalyst TiO$_2$ particle due to the increase in thickness of the carbon layer. Below 800 °C, the improvement in crystallinity of the anatase phase is not yet pronounced and no rutile phase appears. Therefore, no difference is observed at 700 and 800 °C. Above 850 °C, the crystallinity of the anatase phase improves markedly and so k becomes much larger than the values below 800 °C at the same carbon content, although anatase starts to transform to rutile at a small content of carbon. Above 4 mass% carbon, the transformation to rutile is suppressed and k decreases with increasing carbon content (i.e. with increasing carbon thickness). Below 4 mass% C, k decreases rapidly with decreasing carbon content, mainly because of rapid decrease in anatase content, f_A, as shown at each experimental point in Figure 13.11B. At 900 °C, the phase transformation is accelerated and a large carbon content up to about 9 mass% is needed for the suppression of phase transformation. The k for the carbon-coated TiO$_2$ with a carbon content of 10–12 mass% prepared at 900 °C is larger than that at 850 °C, and much larger

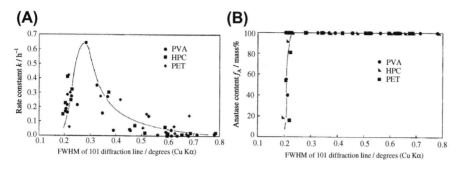

(A) Rate constant k / h^{-1} — FWHM of 101 diffraction line / degrees (Cu Kα)
- PVA
- HPC
- PET

(B) Anatase content f_A / mass% — FWHM of 101 diffraction line / degrees (Cu Kα)
- PVA
- HPC
- PET

FIGURE 13.11 Crystallinity of Anatase Phase in Carbon-coated TiO$_2$

(A) Dependence of k for MB decomposition, and (B) anatase content, f_A, on crystallinity measured by full width at half maximum intensity (FWHM) of the 101 diffraction line of TiO$_2$

From [104]

than those at 700 and 800 °C, because of the improvement in crystallinity of the anatase phase and no rutile formation. Since the phase transformation to rutile cannot be suppressed at 900 °C with a carbon content below 9 mass%, k decreases with decreasing carbon content and becomes smaller than that at 850 °C. Below 3 mass%, k after 900 °C-treatment is thought to be smaller than those after 800 and 700 °C-treatment because of a pronounced decrease in anatase content.

For anatase-type TiO$_2$ without carbon coating, a marked dependence of photoactivity on crystallinity of the anatase phase has been observed [111–113]. For carbon-coated TiO$_2$, a very similar relationship between k for MB decomposition and full width at half maximum intensity (FWHM) of the 101 diffraction line of the anatase phase has been obtained [104]; note that a smaller FWHM means higher crystallinity. As shown in Figure 13.11A, k shows a maximum at a FWHM of about 0.3°. By high-temperature treatment for the preparation of carbon-coated TiO$_2$, FWHM decreases gradually from that of the pristine TiO$_2$, c. 0.8°, i.e. a gradual improvement in crystallinity of the anatase phase. With decreasing FWHM, k increases, particularly below 0.4°, and then decreases abruptly, resulting in a maximum at around 0.3° (Figure 13.11A). The abrupt decrease in k below 0.3° is caused mainly by the decrease in anatase content due to the phase transformation to rutile phase (Figure 13.11B) [104,111]. As discussed above, the phase transformation from anatase to rutile is suppressed by carbon coating, but it occurs suddenly when crystallinity of anatase reaches a certain level, at a FWHM of about 0.3°.

In order to have a large k, i.e. high photocatalytic activity, the balance among different factors is important. In Figure 13.10, the photocatalyst TiO$_2$ with carbon content of about 3.5 mass% after heat treatment at 850 °C gave the largest k, owing to the effect that the transformation to rutile was suppressed, the anatase phase had high crystallinity, and the carbon layer was thin enough to transmit UV rays to the surface of the photocatalyst particle. Before this maximum of k (smaller carbon content), the phase transformation to rutile was not suppressed sufficiently because of the thinner layer of coated carbon, resulting in an abrupt reduction of k. After this maximum, the thicker layer of carbon coating adversely affected transmission of UV rays.

Adsorptivity of the carbon layer, which can be estimated from apparent S_{BET}, is also important to concentrate the pollutant molecules in the close vicinity of TiO_2 particles. It was shown using four pollutants—MB, RB5, Ph, and IT—that a pollutant with higher adsorptivity to carbon-coated TiO_2 particles gave a larger k, as shown in Figure 13.12 [109].

For practical application of carbon-coated TiO_2 photocatalysts, the endurance of photoactivity and the repeating usability of the photocatalysts are very important. Carbon coating on TiO_2 particles has been demonstrated to work effectively to fix photocatalyst particles by using an organic binder for cyclic use, because the carbon layer on the particles prevents direct contact between the organic binder and the photocatalyst [99,103,110]. An acryl resin film containing photocatalyst TiO_2 (ST-01) lost weight rapidly under UV irradiation and reached 39% loss after 870 h, which was mainly owing to the photocatalytic decomposition of the resin by TiO_2. However, a film containing a carbon-coated TiO_2 showed only 5% loss even after 870 h of irradiation, which was almost the same as the film without any photocatalyst [99]. Carbon-coated TiO_2 photocatalysts fixed in two ways, by fixing photocatalyst particles on an adhesive tape and by forming into a film with a binder of polytetrafluoroethylene, have been shown to be able to be used repeatedly [110].

When contaminated water is purified using dispersed photocatalysts, the particles must be separated from the purified water, usually by sedimentation. A small amount of carbon coating to TiO_2, even 1 mass%, has been shown to be advantageous for the sedimentation of photocatalyst particles in water [114]. This is probably because of the change in surface nature from hydrophilic to hydrophobic by carbon coating. For water purification, in addition to separation of the photocatalyst, removal

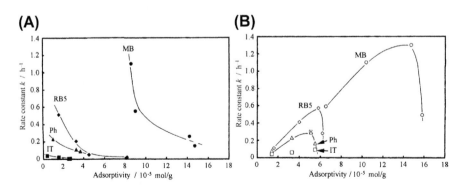

FIGURE 13.12 Dependence of k for the Decomposition of Four Model Pollutants—MB, IT, RB5, and Ph—on the Adsorptivity for Carbon-coated TiO_2

(A) TiO_2 carbon-coated at 700 °C, and (B) at 900 °C

of degradation products is also required. A new type of photocatalytic membrane reactor combining photocatalysis with direct-contact membrane distillation was constructed in the laboratory and tested using carbon-coated TiO_2, showing more efficient usage of the photocatalyst TiO_2 [115,116]. Photocatalytic membrane reactors for water treatment have been reviewed [117].

13.5 Synthesis of novel photocatalysts via carbon coating

13.5.1 Carbon-coated Ti_nO_{2n-1}

Carbon-coated Ti_nO_{2n-1}, which had visible-light sensitivity, was easily prepared from the mixture of rutile-type TiO_2 with PVA. The reduction of TiO_2 is reasonably supposed to occur from the surface of its particle, and so Ti_nO_{2n-1} was often formed in heterogeneous reduction states, i.e. a mixture of Ti_nO_{2n-1} with different n-values [118–122]. A temperature above 900 °C was required to reduce rutile-type TiO_2. At 1100 °C, the reducing reaction also depended on the mixing ratio of TiO_2/PVA, only a small amount of Ti_9O_{17} being formed at the ratio of 95/5, but Ti_4O_7 in a single phase at 50/50 through Ti_9O_{17}, Ti_6O_{11}, and Ti_5O_9 in intermediate mixing ratios between 90/10 and 60/40. For carbon-coated Ti_nO_{2n-1}, photocatalytic activity has been observed in a solution of IT under irradiation of visible light [120,121].

13.5.2 Carbon-coated $W_{18}O_{49}$

A new photocatalyst, carbon-coated $W_{18}O_{49}$, with visible-light activity has been successfully synthesized through a similar process: a mixture of para-ammonium tungstate $((NH_4)_{10}W_{12}O_{41}\cdot5H_2O)$ with PVA was prepared in aqueous solution and then heat-treated at 800 °C [123,124]. Carbon-coated $W_{18}O_{49}$ showed a marked decrease in Ph concentration under visible-light irradiation. For $W_{18}O_{49}$ prepared without carbon coating, however, photoactivity was not detected, probably because the particle size of $W_{18}O_{49}$ crystals was much larger than that of carbon-coated $W_{18}O_{49}$. The decomposition of Dimethyl sulfoxide (DMSO) was performed by detecting its oxidation product, methanesulfonic acid, CH_3SO_3H (MSA), revealing that the carbon-coated $W_{18}O_{49}$ produces active •OH radicals under visible light, like TiO_2 under UV. The main drawback of WO_3 as a visible-light-active photocatalyst has been pointed out to be its high solubility in water. However, $W_{18}O_{49}$ was confirmed to have much poorer solubility [124].

13.5.3 TiO_2 co-modified by carbon and iron

TiO_2 particles co-modified by carbon and iron have been prepared by heat treatment of mixtures of TiO_2/FeC_2O_4 in 1/10 mass ratio at different temperatures of 500–900 °C in Ar [125–131]. The resultant TiO_2 particles (TiO_2/C/Fe composites) showed a marked acceleration of Ph decomposition through the combination of two processes, photo-Fenton reaction and photocatalysis in solution. The composites contained small amounts of carbon and iron (3.3–0.2% carbon and 0.2–1.7% Fe), iron being in a form of ilmenite-type $FeTiO_3$ and TiO_2 in rutile phase above 700 °C. In Figure 13.13, changes in relative concentration of Ph with UV irradiation time are shown for TiO_2/C/Fe composites with 0.03 mol/L H_2O_2 added. Marked acceleration of Ph decomposition is observed for all composites after 1–3 h UV irradiation. The composite prepared at 900 °C does not show a rapid c/c_0 decrease, but still exhibits a higher decomposition rate than the pristine TiO_2 (ST-01) and the composite TiO_2/Fe prepared at 550 °C, both of which show a linear relationship and almost the same apparent rate constant. The trend observed for the TiO_2/C/Fe composites is very similar to that for H_2O_2 alone (no photocatalyst used), but the Ph decomposition occurs much faster with the TiO_2/C/Fe composites prepared at 400–700 °C, in which anatase-type TiO_2 is the predominant phase. These results suggest that the modification of TiO_2 by carbon together with Fe is effective, and only a small amount of carbon (less than 1 mass%) is necessary for photocatalytic activity.

13.6 Concluding remarks

The merits and drawbacks of carbon materials in photocatalysis have been reviewed. The principal role of carbon is the hybridization of adsorptivity and photoactivity, leading to an improvement in photocatalyst performance. The resulting composite materials were classified into four groups: TiO_2-loaded AC, mechanical mixtures of TiO_2 and AC, carbon-modified TiO_2, and carbon-coated TiO_2.

Enhancement of apparent photocatalytic activity has been reported in TiO_2-loaded ACs, which comes principally from the pollutant-rich environment at the interface

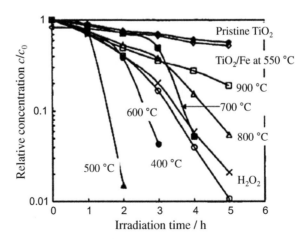

FIGURE 13.13 Changes in c/c_0 of Ph with UV Irradiation for TiO_2/C/Fe Composites Prepared at Different Temperatures, in Comparison with the Pristine TiO_2, TiO_2/Fe Composite, and H_2O_2

between the TiO_2 photocatalyst particle and the AC support. Optimum conditions for the loading amount of TiO_2 and the pore structure of AC have to be selected in order to achieve enhanced photocatalytic activity. Appropriate adsorptivity is required for the AC support: too low or too high an uptake of pollutant results in lower enhancement. More detailed studies on the effect of pore structure of AC, using different ACs, are needed to understand the mechanism of enhancement of photocatalytic performance and to develop practical applications of TiO_2-loaded ACs.

Powder mixtures of photocatalytic TiO_2 with AC exhibit enhanced activity for pollutant decomposition when suspended in solution. Their cyclic use has been confirmed in a pilot plant using sunlight. Properly selected AC mixed with TiO_2 was shown to achieve a high efficiency of photodecomposition.

Carbon doping, i.e. substitution of oxygen in the anatase lattice by carbon, has proved to be promising for preparation of visible-light-active photocatalysts. Carbon-modified TiO_2, in which some carbon atoms have substituted oxygen atoms in the lattice but others have precipitated as aggregates, also showed visible-light sensitivity. Carbon-modified TiO_2 achieved high practical efficiency, calculated based on the change in photocatalytic activity due to carbon modification and the turbidity in water under visible light.

A simple process for the preparation of carbon-coated TiO_2 has been developed: a mixture of carbon precursor (such as PVA or PET) and TiO_2 precursor (TiO_2 itself or, say, titanium tetraisopropoxide), either in powder or in solution, is heat-treated above 700 °C in an inert atmosphere. During heat treatment, the carbon precursor is pyrolyzed and carbonized, and the TiO_2 precursor decomposes to give anatase-type TiO_2 particles, all of which are coated by carbon. Carbon coating was found to have various advantages in photocatalysis. Carbon formed on TiO_2 particles was porous and thus high adsorptivity was obtained, without losing substrate TiO_2 photoactivity; hybridization of photoactivity and adsorptivity was thus achieved. When a pollutant is adsorbed in the coated carbon layer, its photodecomposition is accelerated because it reaches a higher concentration near the surface of TiO_2 particles than in solution. Phase transformation in TiO_2, from photoactive anatase to less active rutile, was strongly suppressed by the coated carbon layer by virtue of the inhibition of sintering and grain growth of TiO_2 particles. As a consequence, high crystallinity of anatase phase was obtained, which was preferable to the photodecomposition of pollutants in water. The carbon layer was also able interrupt direct contact between photoactive TiO_2 and organic binder, and so carbon-coated TiO_2 could be fixed either on an organic adhesive tape or in an organic film for cyclic use. Carbon-coating of TiO_2 led to much lower turbidity in solution: it settled out much faster in water than uncoated TiO_2, which is an advantage for practical water purification. High performance of a photocatalytic membrane reactor by coupling photocatalysis with direct membrane distillation has also been demonstrated using carbon-coated TiO_2 catalysts.

By applying the carbon-coating technique, two new photocatalysts have been developed, carbon-coated Ti_nO_{2n-1} and $W_{18}O_{49}$. During heat treatment of a mixture of rutile-type TiO_2 with PVA above 900 °C, reduced phases of TiO_2, Ti_nO_{2n-1} (Magnéli phases), were formed by reaction with carbon. The structure could be controlled from $n = 2$ to 9 by changing the TiO_2/PVA mixing ratio. These Ti_nO_{2n-1} phases, particularly Ti_2O_3 and Ti_3O_5, were found to be active under visible-light irradiation. Carbon-coated $W_{18}O_{49}$, which was prepared by the same procedure from a mixture of PVA and para-ammonium tungstate, was also active under visible light. In order to have appreciable visible-light sensitivity, these particles needed to be as small as possible and carbon-coated. Both requirements were satisfied by heat treatment at 800–900 °C in an inert atmosphere.

By applying the same procedure to a mixture containing FeC_2O_4, TiO_2 particles modified by both carbon and Fe were prepared, and they showed high photoactivity based on the photo-Fenton reaction.

In conclusion, carbon can contribute to photocatalysis not only by hybridization of adsorptivity and photoactivity for the cases of TiO_2-loaded AC and carbon-coated TiO_2, but also by the improved photocatalytic performance of carbon-modified TiO_2 and carbon-coated Ti_nO_{2n-1} due to higher visible-light sensitivity.

References

[1] Fujishima A, Honda K. Nature 1972;238:37.
[2] Toyoda M, Tsumura T, Tryba B, et al. Chem Phys Carbon 2012;31:171–267.
[3] Minero C, Carrozo F, Pelizzetti E. Langmuir 1992;8:481–6.
[4] Uchida H, Itoh S, Yoneyama H. Chem Lett 1993;22:1995–8.
[5] Ibusuki T, Takeuchi K. J Mol Catal 1994;88:93–102.
[6] Takeda N, Torimoto T, Sampath S, et al. J Phys Chem 1995;99:9986–91.
[7] Xu Y, Langford CH. J Phys Chem 1995;99:11501–7.
[8] Anderson C, Bard AJ. J Phys Chem 1995;99:9882–5.
[9] Dagan G, Sampath S, Lev O. Chem Mater 1995;7:446–53.
[10] Torimoto T, Ito S, Kuwabata S, et al. Environ Sci Technol 1996;30:1275–81.
[11] Takeda N, Ohtani M, Torimoto T, et al. J Phys Chem B 1997;101:2644–9.
[12] Anderson C, Bard AJ. J Phys Chem B 1997;101:2611–6.
[13] Torimoto T, Okawa Y, Takeda N, et al. J Photochem Photobiol A 1997;103:153–7.
[14] Takeda N, Iwata N, Torimoto T, et al. J Catal 1998;177:240–6.
[15] Zhanpeisov NU, Harada M, Anpo M. J Mol Struct THEOCHEM 2000;529:135–9.
[16] Ding Z, Hu X, Yue PL, et al. Catal Today 2001;68:173–82.
[17] Tsumura T, Kojitani N, Umemura H, et al. Appl Surf Sci 2002;196:429–36.
[18] Tryba B, Morawski AW, Inagaki M. Appl Catal B 2003;41:427–33.
[19] Tryba B, Morawski AW, Inagaki M. Appl Catal B 2003;46:203–8.
[20] Toyoda M, Nanbu Y, Kito T, et al. Desalination 2003;159:273–82.
[21] Tryba B, Morawski AW, Toyoda M, et al. Water Res 2003;4:35–42.
[22] Arana J, Dona-Rodriguez JM, Tello Rendon E, et al. Appl Catal B 2003;44:161–72.
[23] Inagaki M, Morishita T, Kuno A, et al. Carbon 2004;42:497–502.
[24] El-Sheikh AH, Newman AP, Al-Daffaee H, et al. Surf Coat Technol 2004;187:284–92.
[25] Duminica FD, Maury F, Senocq F. Surf Coat Technol 2004;188–189:255–9.
[26] Zhang X, Zhou M, Lei L. Carbon 2005;43:1700–8.
[27] Zhang X, Zhou M, Lei L. Mater Res Bull 2005;40:1899–904.
[28] Zhang X, Zhou M, Lei L. Carbon 2006;44:325–33.
[29] Li Y, Li X, Li J, et al. Water Res 2006;40:1119–26.

[30] Liu JH, Yang R, Li SM. J Environ Sci 2006;18:979–82.

[31] Kubo M, Fukuda H, Chua XJ, et al. Ind Eng Chem Res 2007;46:699–704.

[32] Liu Y, Yang S, Hong J, et al. J Hazard Mater 2007;142:208–15.

[33] Liu SX, Chen XY, Chen X. J Hazard Mater 2007;143:257–63.

[34] Chen ML, Lim CS, Oh WC. J Ceram Process Res 2007;8:119–24.

[35] Wang W, Gomez Silva C, Faria FL. Appl Catal B 2007;70:470–8.

[36] Puma GL, Bono A, Krishnaiah D, et al. J Hazard Mater 2008;157:209–19.

[37] Xu D, Huang ZH, Kang F, et al. Catal Today 2008;139:64–8.

[38] Akpan UG, Hameed BH. J Hazard Mater 2009;170:520–9.

[39] Zhu B, Zou L. J Environ Manage 2009;90:3217–25.

[40] Wang X, Liu Y, Hu Z, et al. J Hazard Mater 2009;169:1061–7.

[41] Lu Y, Wang D, Ma C, et al. Build Environ 2010;45:615–21.

[42] Tao Y, Wu CY, Mazyck DW. Chemosphere 2006;65:35–42.

[43] Wang X, Hu Z, Chen Y, et al. Appl Surf Sci 2009;255:3953–8.

[44] Li Y, Ma M, Sun S, et al. Appl Surf Sci 2008;254:4154–8.

[45] Li Y, Ma M, Sun S, et al. Catal Commun 2008;9:1583–7.

[46] Nagaoka S, Hamasaki Y, Ishihara S, et al. J Mol Catal A 2002;177:255–63.

[47] Tryba B, Tsumura T, Janus M, et al. Appl Catal B 2004;50:177–83.

[48] Eder D, Windle AH. Adv Mater 2008;20:1787–93.

[49] Yao Y, Li G, Ciston S, et al. Environ Sci Technol 2008;42:4952–7.

[50] Kim S, Lim SK. Appl Catal B 2008;84:16–20.

[51] Woan K, Pyrgiotakis G, Sigmund W. Adv Mater 2009;21:2233–9.

[52] Gao B, Chen GZ, Puma GL. Appl Catal B 2009;89:503–9.

[53] Teng DH, Yu YH, Liu HY, et al. Catal Commun 2009;10:442–6.

[54] Xu YJ, Zhuang YB, Fu XZ. J Phys Chem C 2010;114:2669–76.

[55] Leary R, Westwood A. Carbon 2011;49:741–72.

[56] Lee JJ, Suh JK, Hong JS, et al. Carbon 2008;46:1648–55.

[57] Wang S, Shi X, Shao G, et al. J Phys Chem Solids 2008;69:2396–400.

[58] Matos J, Laine J, Herrmann JM. Appl Catal B 1998;19:281–91.

[59] Herrmann JM, Matos J, Disdier J, et al. Catal Today 1999;54:255–65.

[60] Matos J, Laine J, Herrmenn J M. Carbon 1999;37:1870–2.

[61] Matos J, Laine J, Herrmann J M. J Catal 2001;200:10–20.

[62] Cordero T, Chovelon JM, Duchamp C, et al. Appl Catal B 2007;73:227–35.

[63] Matos J, Laine J, Herrmann JM, et al. Appl Catal B 2007;70:461–9.

[64] Cordero T, Duchamp C, Chovelon JM, et al. J Photochem Photobiol A 2007;191:
     122–31.

[65] Malato S, Blanco J, Fernadez-Ibanez P, et al. J Solar Energy Eng 2001;123:138–42.

[66] Di Valentin C, Pacchioni G, Sellon A. Chem Mater 2005;17:6656–65.

[67] Wang H, Lewis JP. J Phys Condens Mater 2006;18:421–34.

[68] Zabck P, Eberl J, Kisch H. Photochem Photobiol Sci 2009;8:264–9.

[69] Irie H, Watanabe Y, Hashimoto K. Chem Lett 2003;32:772–3.

[70] Choi Y, Umebayashi T, Yoshikawa M. J Mat Sci 2004;39:1837–9.

[71] Shen M, Wu Z, Huang H, et al. Mater Lett 2006;60:693–7.

[72] Sakthivel S, Kisch H. Ang Chem Int Ed 2003;42:4908–11.

[73] Ren W, Ai Z, Jia F, et al. Appl Catal B 2007;69:138–44.

[74] Dong F, Wang H, Wu Z. J Phys Chem C 2009;113:16717–23.

[75] Kahn SUM, Al-Shahry M, Ingler Jr WB. Science 2002;297:2243–5.

[76] Park Y, Kim W, Park H, et al. Appl Catal B 2009;91:355–61.

[77] Irie H, Washizuka S, Hashimoto K. Thin Solid Films 2006;510:21–5.

[78] Wong MS, Hsu SW, Rao KK, et al. J Mol Catal A 2008;279:20–6.

[79] Wang SH, Chen TK, Rao KK, et al. Appl Catal B 2007;76:328–34.

[80] Huang Y, Deng K, Ai Z, et al. Mater Chem Phys 2009;114:235–41.

[81] Yang X, Cao C, Erickson L, et al. J Catal 2008;260:128–33.

[82] Zhao L, Chen X, Wang X, et al. Adv Mater 2010;22:3317–21.

[83] Ohno T, Tsubota T, Nishijima K, et al. Chem Lett 2004;33:750–1.

[84] Ohno T, Tsubota T, Toyofuku M, et al. Catal Lett 2004;98:255–8.

[85] Lettmann C, Hildenbrand K, Kisch H, et al. Appl Catal B 2001;32:215–27.

[86] Treschev SY, Chou PW, Tseng YH, et al. Appl Catal B 2008;79:8–16.

[87] Kuo C, Tseng Y, Huang C, et al. J Mol Catal A 2007;270:93–100.

[88] Janus M, Tryba B, Inagaki M, et al. Appl Catal B 2004;52:61–7.

[89] Morawski AW, Janus M, Tryba B, et al. C R Chimie 2006;9:800–5.

[90] Janus M, Morawski AW. Appl Catal B 2007;75:118–23.

[91] Janus M, Inagaki M, Tryba B, et al. Appl Catal B 2006;63:272–6.

[92] Janus M, Kusiak K, Morawski AW. Catal Lett 2009;131:506–11.

[93] Janus M, Kusiak E, Choina J, et al. Catal Lett 2009;131:606–11.

[94] Janus M, Tryba B, Kusiak E, et al. Catal Lett 2009;128:36–9.

[95] Janus M, Kusiak E, Choina J, et al. Desalination 2009;249:358–63.

[96] Shao GS, Liu L, Ma TY, et al. Chem Eng J 2010;160:370–7.

[97] Geng J, Jiang Z, Wang Y, et al. Scr Mater 2008;59:352–5.

[98] Izumi I, Kuroda K, Ohnishi Y, et al. Mizushori Gizyutsu 2001;42:461–5 [in Japanese].

[99] Tsumura T, Kojitani N, Izumi I, et al. J Mater Chem 2002;12:1391–6.

[100] Inagaki M, Hirose Y, Matsunaga T, et al. Carbon 2003;41:2619–24.

[101] Tryba B, Tsumura T, Janus M, et al. Appl Catal B 2004;50:177–83.

[102] Tryba B, Morawski AW, Tsumura T, et al. J Photochem Photobiol A 2004;167: 127–35.

[103] Toyoda M, Yoshikawa Y, Tsumura T, et al. J Photochem Photobiol A 2005;171: 167–71.

[104] Inagaki M, Kojin F, Tryba B, et al. Carbon 2005;43:1652–9.

[105] Inagaki M, Matsunaga T, Tsumura T, et al. TANSO 2005; No. 219: 217–220 [in Japanese].

[106] Toyoda M, Tryba B, Kojin F, et al. TANSO 2005; No. 220: 289–299.

[107] Toyoda M, Tryba B, Ito E, et al. J Jpn Soc Water Environ 2006;29:9–14 [in Japanese].

[108] Matsunaga T, Inagaki M. Appl Catal B 2006;64:9–12.

[109] Toyoda M, Tryba B, Kojin F, et al. Adv Sci Technol 2006;46:180–7.

[110] Inagaki M, Nonaka M, Kojin F, et al. Environ Technol 2006;27:521–8.

[111] Inagaki M, Nonaka R, Tryba B, et al. Chemosphere 2006;64:437–45.

[112] Inagaki M, Imai T, Yoshikawa T, et al. Appl Catal B 2004;51:247–54.

[113] Toyoda M, Nanbu Y, Nakazawa Y, et al. Appl Catal B 2004;49:227–32.

[114] Janus M, Inagaki M, Tryba B, et al. Appl Catal B 2006;63:272–6.

[115] Mozia S, Toyoda M, Inagaki M, et al. Desalination 2007;212:141–51.

[116] Mozia S, Toyoda M, Inagaki M, et al. J Hazard Mater 2007;140:369–75.

[117] Mozia S. Sep Purif Technol 2010;73:71–91.

[118] Tsumura T, Hattori Y, Kaneko K, et al. Desalination 2004;169:269–75.

[119] Toyoda M, Yano T, Mozia S, et al. TANSO 2005; No. 220: 265–269.

[120] Toyoda M, Yano T, Tsumura T, et al. J. Adv Oxid Technol 2006;9:49–52.
[121] Toyoda M, Yano T, Tryba B, et al. Appl Catal B 2009;88:160–4.
[122] Toyoda M, Yano T, Ghafar HHA, et al. J Photocatal Sci 2011;12:79–85.
[123] Kojin F, Mori M, Morishita T, et al. Chem Lett 2006;35:388–9.
[124] Kojin F, Mori M, Noda Y, et al. Appl Catal B 2008;78:202–9.
[125] Tryba B, Toyoda M, Morawski AW, et al. Chemosphere 2005;60:477–84.
[126] Tryba B, Morawski AW, Inagaki M, et al. J Photochem Photobiol A 2006;179:224–8.
[127] Tryba B, Inagaki M, Toyoda M, et al. J Adv Oxid Technol 2007;10:25–30.
[128] Tryba B. J Adv Oxid Technol 2007;10:267–72.
[129] Tryba B. J Hazard Mater 2008;151:623–7.
[130] Tryba B, Morawski AW, Inagaki M, et al. Appl Catal B 2006;63:215–21.
[131] Tryba B, Morawski AW, Inagaki M, et al. Chemosphere 2006;64:1225–35.

Carbon Materials for Spilled-oil Recovery

14

There have been many oil-spill accidents across the world. A number of tanker accidents causing serious damage to the environment can be recalled: the heavy oil spill of c. 3.6×10^4 tons in Alaska in March 1989, the oil spill of c. 26×10^4 tons off Angola in May 1991, the wide spread of c. 9×10^4 tons of spilled oil in the North Sea in January 1993, the crash between two tankers and serious disturbance of ship transport with c. 2×10^4 tons of oil spilled in the Strait of Malacca in October 1997, the oil tanker Nakhodka spill of 4.5×10^3 tons of oil near the coast in the Sea of Japan in January 1997, the contamination by 70×10^4 liters of oil causing anxiety about rare species in the Galapagos Islands in January 2001, the stranding of a tanker on the coast of Spain with c. 4×10^4 tons of oil spilled in November 2002. In addition to these accidents that were mainly due to bad weather, the demolition of storage tanks in Kuwait during the 1991 Gulf War spilled a large amount of heavy oil into the sea. Although such disastrous accidents resulted in massive oil spills, statistics have been reported that tanker and rig accidents account for only 12.5% and 1.5% of the total amount of spilled oil, respectively. The principal loss of oil actually occurs during transportation and storage. Continuous leakage of oil through pipe joints, for example, may produce serious contamination of soil, river water, and sometime even subterranean water, and this has detrimental effects on the lives of human beings, as well as plants, fish, and animals. These oil spills have resulted not only in great damage to the environment, but also in a great loss of energy resources.

A common treatment technique used in oil-spill accidents on the sea is containment with large floating barriers (so-called "oil fences") followed by skimming with specialized ships that either vacuum the oil off the sea or soak it up with sorbing materials. So far, some porous polymers, such as polypropylene and poly(ethylene terephthalate), and some natural sorbents, such as cotton fiber, milkweed floss, and kenaf plant, have been used for the sorption of spilled oil. Their sorption capacity is in the range of 10–30 kg of heavy oil per 1 kg of polymer, although they sorb water as well as heavy oil.

In this chapter, the sorption capacity and kinetics for heavy oils are reviewed in various carbon materials, and the cycling performance of carbon sorbents in sorption and recovery are presented. Recycling of carbon sorbents is also considered,

---

Prerequisite for readers: Chapter 3.9 (Carbon materials for environment remediation) in *Carbon Materials Science and Engineering: From Fundamentals to Applications*, Tsinghua University Press.

in addition to reuse of the recovered heavy oils. Further, laboratory-scale experiments for practical recovery of spilled heavy oils are presented.

14.1 Sorption capacity for heavy oils

14.1.1 Exfoliated graphite

Exfoliated graphite (EG) has been found to be able to rapidly sorb a large amount of heavy oils [1–17]. As shown in Figures 14.2A and 14.2B, A-grade heavy oil floating on water was completely sorbed into the EG added. The characteristic brown color of A-grade heavy oil disappeared within 1 min after the addition of EG. The EG lost its luster with sorbing oil and appeared deep black (Figure 14.2B). Even after sorption of heavy oil, the EG remained floating on the water. After removing the EG, no contamination appeared in the water and no brown stain was observed after transferring oil-sorbed EG onto a white filter paper, as shown in Figure 14.2D. When the amount of heavy oil was a little larger than the sorption capacity of EG added, the periphery of the EG lump was rimmed by oil. When a large excess of oil was present, the entire EG appeared wetted and the brown color of oil remained on the water, as shown in Figure 14.2C. (The reader is referred to the web version of this book to see the figure in color.) The EG with the sorbed heavy oil could easily be separated from water by conventional filtration.

The sorption capacity was calculated by measuring the mass of carbon before and after oil sorption and expressed as kg of sorbed oil per 1 kg of carbon. Sorption capacity of EG depends strongly on the bulk density, decreasing rapidly with increasing bulk density for all oils. In the case of A-grade heavy oil, sorption capacity reaches up to 80 kg/kg for EG with a low bulk density of 6 kg/m^3, this sorption capacity being much higher than that of polypropylene nonwoven web, which is often used as a sorbent for heavy oils. The sorption rate of the EG for A-grade heavy oil was high: sorption

FIGURE 14.1

Manual Skimming of Spilled Heavy Oil on a Contaminated Beach (Nakhodka Accident in January, 1997)

completed within 1 min. The EG with a bulk density of 10 kg/m³ had a slightly lower sorption capacity (about 70 kg/kg). For crude oil, the sorption capacities of these two EGs were slightly smaller, at 75 and 65 kg/kg, respectively, and the sorption rate was high, reaching saturation within 2 min. In the case of C-grade heavy oil with a much higher viscosity, the sorption capacity was somewhat smaller, 67 and 60 kg/kg for EGs with bulk densities of 6 and 10 kg/m³, respectively, but sorption proceeded very slowly, about 8 h needed for saturation. For B-grade heavy oil, with high viscosity but lower than C-grade oil, sorption capacity was similar to that of crude oil, but sorption rate was similar to that for C-grade oil. In Figure 14.3, sorption capacity of EG is plotted against bulk density for heavy oils with different viscosities (A-grade 0.004, B-grade 0.27, C-grade 0.35, and crude oil 0.004 Pa·s). For the four grades of heavy oil, sorption capacity decreases rapidly with increasing bulk density of EG, viscous oils showing more rapid decrease. Over the whole range of bulk density, sorption capacity for viscous heavy oils (C- and B-grade) was smaller than for the less viscous oils (A-grade and

FIGURE 14.2 Appearance of Sorption and Recovery of Heavy Oil on Water Using EG

(A) A-grade heavy oil floating on water, (B) 2 minutes after the addition of EG with less oil than the sorption capacity of EG, (C) heavy oil with more than the sorption capacity of EG added, and (D) after transfer of heavy oil–sorbed EG onto a white filter paper

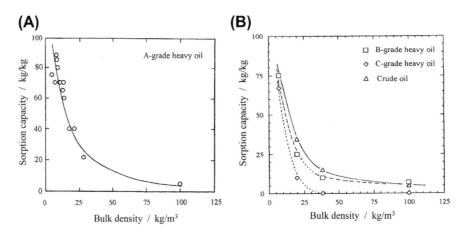

FIGURE 14.3 Dependence of Heavy Oil Sorption Capacity of EG on the Bulk Density at Room Temperature

(A) A-grade heavy oil, and (B) three different grades of heavy oils

crude). No sorption of viscous C-grade heavy oil was observed when the bulk density of EG was more than 40 kg/m^3.

In Figure 14.4, sorption capacity of EG can be seen to be linearly related to total pore volume for pores of 1–600 μm in radius, measured by mercury porosimetry using a special dilatometer [18]. Sorption of heavy oil occurs in this range of pore sizes, most of which are located among the worm-like particles; in other words, most of the heavy oil is sorbed into the void spaces formed by the entanglement of worm-like particles of EG (see Chapter 9 for worm-like particles in EG). Experimental points for the total pore volume above 0.1 m^3/kg tend to deviate from the linear relationship, which suggests that the larger void spaces cannot be completely filled with heavy oil.

Large void spaces among worm-like particles have been quantitatively evaluated by image analysis using thin slices prepared from EG after impregnation with paraffin oil [19]. By comparison with sorption capacity, the volume of these large spaces was found to be responsible for about 70% of total heavy-oil sorption capacity. However, crevice-like pores on the surface of particles and the ellipsoidal pores inside the particles also play an important role in heavy-oil sorption. Observation under an optical microscope showed that, at the beginning of sorption, oil was coming up at the edges of crevices formed on the particle surface to occupy whole crevice-like pores, and then filling large void spaces quickly. Such complicated pore structure is thought to cause strong holding of sorbed heavy oil in EG particles (see Figure 14.2D). The oleophilic (hydrophobic) nature of the carbon surface is also advantageous for such sorption and occlusion of oil.

In Figure 14.5A, it can be seen that sorption capacity of EG with a bulk density of about 6 kg/m^3 tends to decrease with lowering temperature [11]. In the case of viscous C-grade heavy oil, this temperature dependence is so pronounced that no sorption

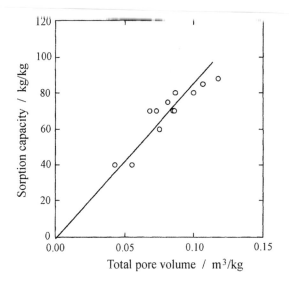

FIGURE 14.4

Relationship Between Sorption Capacity for A-grade Heavy Oil and the Total Pore Volume of EG

From [18]

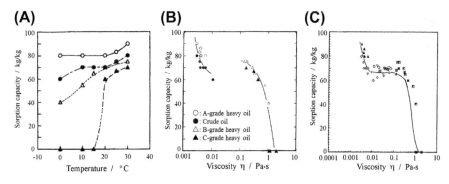

FIGURE 14.5 Dependence of Sorption Capacity of EG on the Temperature and Viscosity of Oil

(A) Temperature dependence for four grades of heavy oil, (B) viscosity dependence for the four heavy oils, and (C) viscosity dependence for various oils, including heavy oils, gasoline, and salad oil

From [11,17]

occurred below 15 °C. In the case of less viscous A-grade heavy oil, however, only a slight decrease in capacity is observed. The temperature dependence was reasonably supposed to be due to changes in viscosity of the oil with temperature. In Figure 14.5B, therefore, sorption capacity observed at different temperatures (Figure 14.5A) is re-plotted against viscosity, and pronounced dependence is observed. In Figure 14.5C, the sorption capacities for other oils with viscosities between 0.01 and 0.2 Pa·s, including kerosene, different cooking oils, and motor oils, exhibit general dependence of sorption capacity on the viscosity of the oil [17].

The sorption behavior of EG for several biomedical molecules, including ovalbumin, serum albumin, bovine serum albumin (BSA), lysine, and herring sperm DNA, has also been studied [20]. The sorption capacity and behavior for these molecules are very similar to those for heavy oils.

EG was prepared by microwave irradiation at 700 W for 60 s on the mixture of natural graphite with average flake size of 320 μm, nitric acid, and potassium permanganate. The bulk density was about 6 kg/m^3, and sorption capacity was about 56 kg/kg for engine oil and about 32 kg/kg for kerosene [21].

14.1.2 Carbonized fir fibers

The sorption capacity of a fibrous component extracted from fir trees (fir fibers) carbonized at 380 and 900 °C is shown as a function of its bulk density for A-grade and C-grade heavy oils in Figures 14.6A and 14.6B, respectively, in comparison with EG [22]. A strong bulk-density dependence is seen for both heavy oils, as with EG. For A-grade heavy oil, the capacity seems to be slightly lower than that of EG, particularly in the range of low bulk density, but it is still quite high, 60 to 80 kg per 1 kg of the fibers with bulk density of about 6 kg/m^3. Above 40 kg/m^3, the sorption capacity is comparable to that of EG, 10–20 kg/kg. For viscous C-grade heavy oil, however, sorption capacity of the fibers is much higher than that of EG, particularly those with bulk density above 10 kg/m^3. Although EG with bulk density more than 40 kg/m^3 could not sorb C-grade heavy oil, carbonized fir fibers with the same bulk density could sorb c. 15 kg/kg (Figure 14.6B). As shown in Figure 14.6, carbonization temperature seems not to have a marked effect on the sorption capacity for either A-grade or C-grade oils. A wide range of carbonization temperatures, 380–1200 °C, was employed but no marked effect on the sorption capacity for A-grade heavy oil was found [22].

14.1.3 Carbon fibers

The various carbon fibers have not shown high sorption capacities in comparison with the carbonized fir fibers and EG [23]. Even activated carbon fibers with high Brunauer-Emmett-Teller (BET) surface area did not have high sorption capacity for the low-viscosity A-grade heavy oil. Different granular activated carbons with high surface area (>1000 m^2/g) have also been used for heavy oil sorption, but their capacity was also very low, about 1 kg/kg or less.

Sorption capacity of carbon fiber felts with bulk density of 54–77 kg/m^3, which were prepared from polyacrylonitrile (PAN)-based and pitch-based carbon fibers, was 11–17 kg/kg for A-grade heavy oil, and the bulk-density dependence was almost the same as EG [23]. The sorption capacity of felts was not expected to be high because of their relatively high bulk densities, but cycle performance was excellent, much better than EG and better than fir fibers, as explained below.

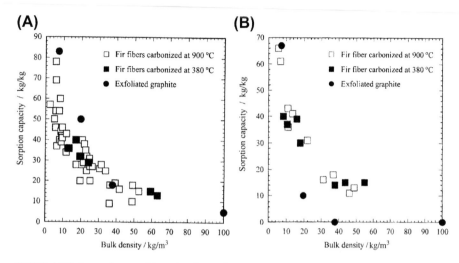

FIGURE 14.6 Dependence of Sorption Capacity of Carbonized Fir Fibers on Bulk Density for Heavy Oils

(A) A-grade and (B) C-grade heavy oils

From [22]

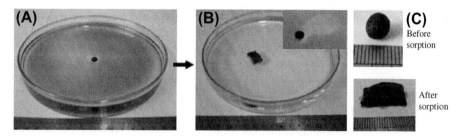

FIGURE 14.7 Sorption of Diesel Oil Spreading on Water by a CNT Sponge

(A) Just after placing the CNT sponge, (B) after the completion of sorption, and (C) appearance of the CNT sponge before and after sorption (The oil was colored by the blue by dye.)

From [24]

14.1.4 Carbon nanotube sponge

Sponge-like bulky material consisting of self-assembled interconnected multi-walled carbon nanotube (MWCNT) skeletons, carbon nanotube (CNT) sponge, has been reported to have a high sorption capacity for various organic solvents and oils [24]. CNT sponges have been prepared by chemical vapor deposition (CVD) of 1,2-dichlorobenzene with ferrocene at 850 °C, and had a bulk density of 5–25 kg/m^3 and a porosity of more than 99%. Snapshots during sorption are shown in Figures 14.7A and 14.7B; they were taken at the moment of the addition of densified CNT sponge in the center of diesel oil spread over a large area (227 cm^2) of water surface, and clear water surface was recovered by the completion of oil sorption after several minutes. The sponge had swollen from spherical to rectangular after sorption of oil (Figure 14.7C). Sorption capacity was 140 kg/kg for diesel oil and 110 kg/kg for gasoline.

A vertically aligned MWCNT membrane synthesized on stainless steel meshes in a flow of C$_2$H$_4$/H$_2$/Ar at 750 °C using an iron catalyst was tested as a filter for separating oil and water [25]. Diesel oil and engine oil with high viscosity could be easily separated from water, even after their emulsification.

14.1.5 Other carbon materials

Charcoals were prepared from three plants, balsa from Ecuador, giant ipil-ipil from the Philippines, and bamboo from Japan, under different conditions, and the sorption capacity for A-grade heavy oil was determined [26]. Charcoals prepared from balsa exhibited a relatively high sorption capacity of 30 kg/kg.

Sorption capacity of carbonized rice husk for oils has been tested [27,28]. Carbonized rice husk, containing residual SiO$_2$, had a sorption capacity of more than 6 kg/kg for B-grade heavy oil, and the reason was thought to be that oil adsorption capacity was closely related to oily components remaining in the rice husk even after carbonization, rather than its porosity [27]. Rice husks carbonized at 480 °C had a sorption capacity of 9.2 kg/kg for crude oil, 5.5 kg/kg for diesel oil, and 3.7 kg/kg for gasoline [28]. Sorption kinetics was studied by measuring penetration height in a column of carbonized rice husk, and a marked dependence on the density of oil was shown.

14.2 Selectivity of sorption

Although EG can sorb a small amount of water, about 1.8 kg/kg, it sorbs heavy oil by expelling water sorbed in advance. When A-grade heavy oil was dropped onto one end of a piece of water-saturated EG, water came out from the other end, as shown in Figure 14.8. With continuing the heavy-oil dropping, water came out continuously, and finally heavy oil appeared to come out. The sorption capacity of EG for A-grade heavy oil after saturation either with sea water or with tap water was measured to be 70 kg/kg, and the time for achieving this value was about 4 h; a slightly lower sorption capacity and much longer saturation time than EG without prior saturation by water (83 kg/kg and within 1 min, respectively). This experimental fact reveals that heavy oil displaces water sorbed into EG, but not completely. The reason for the lowering of capacity and increasing time for sorption is postulated to be that water pre-adsorbed in the small pores is not replaced completely by heavy oil.

In the case of carbonized fir fibers, water adsorption depended strongly on the purity of water: the fibers adsorbed a relatively large amount of distilled water, but only a small amount of sea water. After the fir fibers were saturated with distilled water, heavy oil was dropped onto it, but drainage of water, as in the case

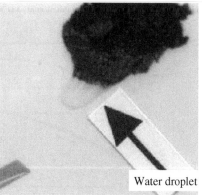

FIGURE 14.8

Water Coming Out of Water-Saturated Exfoliated Graphite by Dropping A-grade Heavy Oil onto the EG

of EG (Figure 14.8), was not observed. The sorption capacity of water-saturated 900 °C-carbonized fir fibers for A-grade heavy oil was very similar to that for the dry fibers.

14.3 Sorption kinetics

Sorption rate of heavy oils, A- and C-grade, into carbon materials has been evaluated by applying the so-called wicking method: mass increase by capillary suction of heavy oil into carbon sorbents (a column of EG, carbonized fir fibers or carbon fiber felt packed into a glass tube with different densities) was measured at room temperature as a function of time [13,14]. Some of the sorption curves for EG with different bulk densities are shown in Figure 14.9. The initial slope depends strongly on both the bulk density of EG and the viscosity of the heavy oil. For A-grade heavy oil, very rapid suction and saturation are observed for bulk densities more than 12 kg/m^3 (Figure 14.9A). A more gradual suction of heavy oil is observed for a bulk density of 7 kg/m^3, without reaching saturation even after 50 s. The initial slope increases with increasing bulk density, but the saturated mass increase shows a maximum at a bulk density of around 12 kg/m^3. A very similar dependence on the bulk density of EG is also observed for the much more viscous C-grade heavy oil, but saturation requires a much longer time, more than 3000 s (Figure 14.9B).

The sorption curve before saturation is well approximated by the equation:

$$m_s = K_s \, t^{1/2} + B \qquad (14.1)$$

where m_s is mass increase per cross-sectional area of the sorbent column, t is time, K_s is the sorptivity or liquid sorption coefficient, and B is a constant. Therefore, sorptivity, K_s (kg/m^{2}s$^{1/2}$), was determined as the slope of an m_s vs $t^{1/2}$ plot.

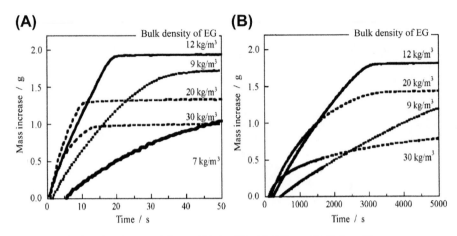

FIGURE 14.9 Sorption Curves of Heavy Oils for EG with Different Bulk Densities

(A) A-grade, and (B) C-grade heavy oils

From [13]

In Figure 14.10, K_s is plotted against bulk density of carbon sorbent. K_s of carbonized fir fibers drastically increases with increasing bulk density in the region of 8 to 20 kg/m³, and reaches a maximum at c. 5.5 kg/m²s$^{1/2}$ when the fibers are densified above 30 kg/m³. In the case of carbon fiber felt, which has high bulk density, the value of K_s is approximately constant at c. 5.5 kg/m²s$^{1/2}$, though experimental points are scattered somewhat. For EG, the maximum K_s is about half of that for the other two materials, achieved at 16 kg/m³.

For viscous C-grade heavy oil, the value of K_s was only about 0.2 kg/m²s$^{1/2}$. A slight dependence on bulk density of two carbon sorbents, carbonized fir fibers and carbon fiber felts, was observed, but it was difficult to analyze in detail.

The relationship between sorptivity, K_s, and bulk density, d_l, has been discussed [14] on the basis of the theoretical formula [29]:

$$K_s = \left[d_l \cdot \sqrt{\frac{\gamma}{\mu}} \right] \cdot \left[\frac{\varepsilon^*}{\lambda} \cdot \sqrt{r_0} \right] \cdot \left[\sqrt{\frac{\cos \theta}{2}} \right] \quad (14.2)$$

where the symbols are surface tension, γ, and viscosity, μ, of the sorbate, effective sorbent porosity, ε^*, average tortuosity factor of the capillaries, λ ($\lambda > 1$), average pore radius, r_0, of the sorbent, and the contact angle, θ, of the interface between the liquid sorbate and pore wall in the solid sorbent. The sorption rate of carbon sorbents for A-grade heavy oil, evaluated by K_s, depends on three factors, ε^*, λ, and r_0 of sorbent carbons, where ε^* increases rapidly but r_0 decreases with increasing bulk density, particularly in the low-density region. The tortuosity factor, λ, is mainly governed by the smoothness of the surface of sorbent particles. In order to have high K_s, therefore, a low value of λ, i.e. a smooth particle surface, and high d_l are desired. These conditions are best fulfilled in the carbon fiber felt among the three carbon

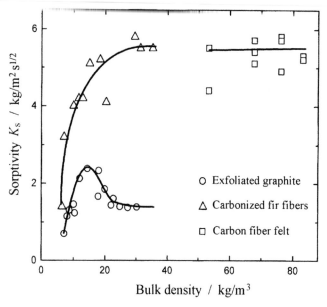

FIGURE 14.10

Dependence of Sorptivity, K_s, on the Bulk Density of Three Carbon Sorbents for A-grade Heavy Oil

From [14]

sorbents used. For EG, K_s is rather low, probably because of a high value of λ due to the complicated surface of worm-like particles. Carbonized fir fiber shows a wide range of sorption kinetics, from a low rate comparable to that of EG to a high rate comparable to carbon fiber felt.

The sorption kinetics for other oils, including kerosene, light oil, various cooking oils, different motor oils, and diesel oil, into a column of EG with a bulk density of about 7 kg/m³ was also studied and similar results were obtained [17].

14.4 Cycle performance of carbon sorbents and heavy oils

Recovery of sorbed heavy oil from carbon materials was carried out by filtering under suction (under 5–7 kPa pressure), washing with solvents, centrifuging at 3800 rpm, or by squeezing mechanically. The recovered sorbent was repeatedly used for sorption of the heavy oil. For filtration under suction, simple equipment using a paper filter was employed. In the process of washing out the sorbed heavy oil, *n*-hexane was used in the case of A-grade heavy oil, but A-grade heavy oil was employed, in addition to *n*-hexane, for viscous C-grade heavy oil.

As shown in Figure 14.11, sorption capacity of EG with a bulk density of c. 6 kg/m³ is as high as 80 kg/kg, and about 50% of sorbed oils are recovered in the first cycle

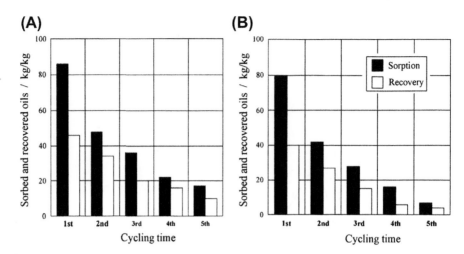

FIGURE 14.11 Changes in the Amount of Heavy Oil Sorbed and Recovered by Filtration Under Suction with Number of Cycles for EG

(A) A-grade heavy oil, and (B) crude oil. EG with a bulk density of about 6 kg/m^3

From [7]

of sorption/recovery [7]. However, sorbed and recovered amounts of the oil decrease rather markedly with increasing number of cycles. The reason for such rapid decrease with sorption/recovery cycles was shown by scanning electron microscopy (SEM) observations to be oil retention in the pores of the EG.

By insertion of a hexane washing step during the sorption/filtration cycle, sorption capacity dropped markedly. Shrinkage of EG was observed after *n*-hexane washing, presumably the cause of the reduction in sorption capacity. Washing with kerosene or xylene and heating in vacuum at 400 °C also resulted in marked reduction of sorption capacity. By squeezing EG after sorption, roughly 100% of sorbed oil was recovered, but this completely destroyed the characteristic bulky texture of the EG and so no further sorption of oil was possible.

Carbonized fir fibers showed better cyclability than EG for A-grade heavy oil [22], as shown in Figure 14.12. In the first cycle, the 5.5 kg/m^3 fibers (Figure 14.12A) sorb about 46 kg/kg, and then about 91% of sorbed oil (42 kg/kg) is recovered. After the second cycle, sorption capacity gradually decreases, but the recovery ratio of about 90% is maintained to at least the eighth cycle. For fibers with a high bulk density of 20 kg/m^3, the sorption capacity decreases gradually, maintaining 26 kg/kg even after the eighth cycle (about 84% of the first cycle); the recovery ratio is rather low, at 80%, in the first cycle, but it reaches more than 90% upon further cycling (Figure 14.12B). Carbonized fir fibers with lower bulk density have higher sorption capacity but poorer cycle performance, because the entangled texture of the fibers is more fragile.

Heavy oil sorbed into carbonized fir fibers was able to be recovered by washing with *n*-hexane at a recovery ratio more than 90%. Although C-grade heavy oil could

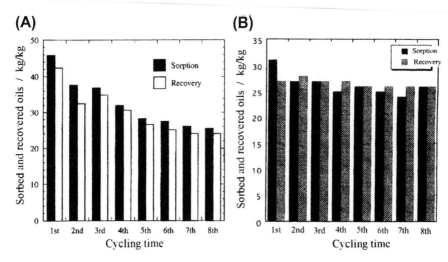

FIGURE 14.12 Changes in Sorbed and Recovered Amounts by Filtration Under Suction of A-grade Heavy Oil for Carbonized Fir Fibers with Different Bulk Densities

(A) Bulk density of 5.5 kg/m³, and (B) 20 kg/m³

From [22]

not be recovered by filtration, it was recovered with high efficiency by washing with *n*-hexane. After washing, the fibers were able to be reused for sorption of C-grade heavy oil at a rate depending on the bulk density, but the sorption capacity decreased by repeated use. With hexane washing for A-grade and C-grade heavy oils, carbonized fir fibers had a better cycle performance than EG.

PAN-based carbon fiber felts had much better cycle performance for A-grade heavy oil both by simple filtration under suction and by hexane washing [23], as shown in Figures 14.13A and 14.13B, respectively. More than 90% of sorbed oil was able to be recovered by simple filtration and almost 100% recovery was achieved by *n*-hexane washing, although the actual capacity for sorption is rather low, at 11 kg/kg.

For PAN-based carbon fiber felts, the recovery ratio for both A- and C-grade oils by centrifugation at 3800 rpm was almost 100% and the sorption capacity remained at almost the same level as in the first cycle even after eight cycles. Recovery by squeezing with twisting could be also applied for the felts sorbed with heavy oil. After the first cycle, the sorption capacity decreased a little, probably because of a slight change in the bulky network of carbon fibers in the felt. A small additional change in sorption capacity occurred in subsequent cycles, but recovery ratio was close to 100%. Pitch-based carbon fiber felt, on the other hand, was not possible to squeeze, because of its low mechanical strength.

Composition and structure analyses were performed on recovered heavy oils by different techniques, such as elemental analysis, proton nuclear magnetic resonance (^{1}H-NMR), flow-injection mass spectroscopy (FI-MS), field-desorption mass

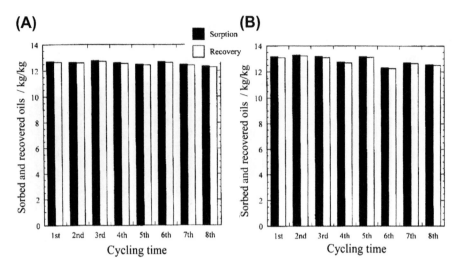

FIGURE 14.13 Cycle Performance of Carbon Fiber Felt with a Bulk Density of About 75 kg/m³ for A-grade Heavy Oil

(A) Filtration under suction of 0.5 kPa, and (B) *n*-hexane washing

From [22]

spectrometry (FD-MS), gel-permeation chromatography (GPC), and high-performance liquid chromatography (HPLC). The results demonstrated no noticeable differences from the original oils. In Table 14.1, the fraction of aromatic hydrocarbons, F_{arom}, measured by ^{1}H-NMR and average molecular weights—number-averaged, Mn, and weight-averaged, Mw—are listed. No pronounced differences between the original and recovered oils are detected in either Mn or Mw. Only a slight decrease in Mw/Mn is observed for C-grade heavy oil, suggesting a slight narrowing of the molecular-weight distribution. F_{arom} tends to decrease a little for the two oils recovered from EG, more markedly for the C-grade, though there was no detectable difference for the A-grade heavy oil.

14.5 Preliminary experiments for practical recovery of spilled heavy oils

14.5.1 Exfoliated graphite packed into a plastic bag

EG was packed into bags of polyethylene and polypropylene with different mesh openings (characterized by the unit of kg/m²) and heavy oil sorption was measured [30]. The bags packed with EG ($200 \times 200 \times 50$ mm³) were placed on different amounts of A-grade heavy oil floating on 500 cm³ water in a $373 \times 309 \times 43$ mm³ tray. The appearance of the bag placed on the heavy-oil layer is shown in Figure 14.14.

Table 14.1 Averaged Molecular Weight Values and Fractions of Aromatic Hydrocarbons of Heavy Oils, Original and Recovered

Heavy Oil Sample		F_{arom}	Mn (%)	Mw (%)	Mw/Mn
A-grade heavy oil	Original	4.0	258.1	273.5	1.06
	Recovered	4.2	258.1	274.4	1.06
Crude oil	Original	4.9	645	869	1.35
	Recovered	4.5	672	915	1.36
C-grade heavy oil	Original	5.4	1071	1768	1.65
	Recovered	4.6	1207	1839	1.52

F_{arom}, fraction of aromatic hydrocarbon; Mn, number-averaged molecular weight; Mw, weight-averaged molecular weight

FIGURE 14.14 Plastic Bags Packed with EG Dipped Into Heavy Oil

(A) Thin layer of heavy oil, and (B) thick layer of heavy oil on the surface of water

From [30]

When the oil layer was as thin as 0.9 mm (100 cm³ heavy oil used), contact between the EG inside the bag and the heavy oil was not sufficient, as shown in Figure 14.14A. When the oil layer was sufficiently thick, thicker than about 4 mm (more than 500 cm³ of heavy oil used), the EG was wetted completely, as shown in Figure 14.14B.

When the bag containing EG was immersed into sufficient A-grade heavy oil for 1 h and lifted up using a stainless steel mesh, the sorption capacity measured was the same as that measured for EG itself and was governed by bulk density, irrespective of bag material and water-proofing treatment.

14.5.2 Formed exfoliated graphite

Instead of packing into a plastic bag, EG was formed into compacts with a size of 40×40×10 mm³ by compression in a mold without using any binder. The compacts thus prepared were easily handled, but their bulk density became more than 30 kg/m³.

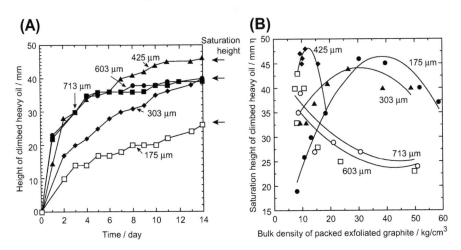

FIGURE 14.15 Heavy Oil Climbing into EG from Contaminated Sand

(A) Changes over time in the height of heavy oil climbing from contaminated sands with different particle sizes, and (B) saturation height of climbed heavy oil as a function of particle size of the sand and bulk density of EG

From [31]

Sorption capacities of these EG compacts were comparable to that of EG without compaction, although the actual value of sorptivity cannot be high because of the high bulk density [30].

14.5.3 Heavy oil sorption from contaminated sand

Sorption of heavy oil into EG was measured by placing EG with different packed densities onto contaminated sands (mixture of α-alumina powder with different particle sizes with A-grade heavy oil) [31]. As the heavy oil climbed, the EG appeared wet and changed color to a much darker black than the original. The change in height of this darker black color was measured as a function of time.

Figure 14.15A shows the changes in height of heavy oil with time for model sands with different particle sizes: the bulk density of packed EG was kept at about 10 kg/m³. With increasing time, the height increases gradually and appears to level out. After 14 days it hits a maximum, except for sand with the smallest particles. With a particle size of sand of 425 μm, heavy oil reached the highest position in the EG; in other words, the largest amount of heavy oil could be pumped into the EG. With a smaller particle size of 175 μm, sorption into EG was very slow, so the height attained was rather low, still increasing gradually even after 14 days. With a larger particle size of 713 μm, the climbing rate was almost the same as for 425-μm particles, but the height of the oil hits a maximum in a shorter time. In Figure 14.15B, the saturation height is plotted against the bulk density of EG using particle size of sand as a parameter. The results suggest that there is an appropriate combination of particle size of sand contaminated by heavy oil and bulk density of EG to attain the maximum adsorption; in other words, a balance in pore size between sand and EG. When a sand with very fine size, e.g. 175 μm, is contaminated by A-grade heavy oil, EG with a bulk density of about 35 kg/m^3 gives the greatest height, i.e. the highest efficiency of pumping. When the sand particle size is large, e.g. 425 μm, EG of about 10 kg/m^3 is more suitable for heavy oil pumping.

For real sea sand (average particle size 236 μm), the same experiment was performed using EG with a bulk density of 10 kg/m^3, and the results were consistent with the relationship observed using model sand (Figure 14.15B) [31].

14.5.4 Sorption of heavy-oil mousse

When heavy oil is mixed with water, it is known to change to a mousse-like material. Sorption of such heavy-oil mousse, prepared by mixing 50 cm^3 of C-grade heavy oil with 25 cm^3 of water, into EG and carbonized fir fibers was studied. Sorption capacity reached a maximum of about 61 kg/kg after 10 h for EG with a bulk density of about 6 kg/m^3; a lower capacity and longer time than for pure C-grade heavy oil (83 kg/kg and 8 h, respectively). For carbonized fir fibers with a bulk density of about 7.2 kg/m^3, the sorption capacity for the mousse was slightly lower and saturation was achieved after c. 20 h.

Even though the sorption performance (both capacity and rate) for heavy-oil mousse was slightly inferior to that for pure oil, it was experimentally demonstrated that heavy oil can be recovered from the mousse by using carbon materials.

14.5.5 TiO$_2$-loaded exfoliated graphite

EG loaded with photoactive anatase-type TiO$_2$ particles was prepared from the mixture of titanium isopropoxide and expandable graphite (precursor of EG) by abrupt heating to 1000 °C. Under ultra-violet irradiation, A-grade heavy oil sorbed into the EG was decomposed much faster than the case where TiO$_2$ particles were mixed with heavy oil [32]. Acceleration of heavy-oil decomposition by loading TiO$_2$ onto EG was thought to be due to a high dispersion of fine TiO$_2$ particles and also to sorption of heavy oil into large pores in the EG.

14.6 Concluding remarks

14.6.1 Comparison among carbon materials

In Table 14.2, the heavy-oil sorption performance is compared for the three carbon sorbents by listing their sorption capacity, sorptivity (K_s) as a measure of sorption rate, and possible process for the recovery of heavy oil. Both sorption capacity and K_s depend strongly on sorbent bulk density, as shown in Figures 14.16A and 14.16B, respectively.

Table 14.2 Comparison of Heavy Oil Sorption Performance Among Various Materials

Sorbing Material	Heavy Oil	Sorption Capacity (kg/kg)	Sorptivity, K_s (kg/m²s^{1/2})	Recovery Process
Exfoliated graphite	A	87	c. 0.7	Filtration only
	C	67	–	No recovery
Carbonized fir fibers	A	78	c. 1.4	Filtration and washing
	C	66	c. 0.1	Washing
Carbon fiber felt	A	20	c. 5.5	Filtration, washing, centrifuging, and squeezing
	C	22	c. 0.2	Washing, centrifuging, and squeezing
Polypropylene nonwoven web	Crude	11.3	–	Squeezing
Cotton fibers	Crude	33.2	–	Squeezing
Milkweed fibers	Crude	38.2	–	Squeezing
Kenaf heated at 400 °C	Salad oil	33	–	–

EG has a very high sorption capacity for heavy oil, but its sorption rate is rather low. By increasing its bulk density, the sorption rate can be improved slightly, but sorption capacity decreases markedly at the same time. Carbonized fir fibers have a sorption capacity and rate comparable to EG for both A- and C-grade heavy oils. By densification of carbonized fir fibers, sorption rate increases rather rapidly, but sorption capacity decreases. On the other hand, carbon fiber felts have a relatively low capacity but a high rate, because of their high bulk density, and their sorptivity seems to be at the upper limit for carbon sorbents, as shown in Figure 14.16B.

For practical application to heavy-oil spill accidents, prompt action is so crucial that both high sorption capacity and high sorption rate are required. With currently available carbon sorbents, however, these two factors cannot be satisfied by using single material, as illustrated in Figure 14.16. Therefore, an adequate balance between capacity and rate has to be considered by carefully selecting the bulk density of the sorbent. From Figure 14.16, a bulk density of about 20 kg/m³ seems to be the most appropriate, leading to a capacity of about 40 kg/kg and sorptivity close to the highest value of 5.5 kg/m²s^{1/2}.

Sorbent recovery is also an important point for practical applications. Thus, bulky EG and carbonized fir fibers have some drawbacks, and there may also be problems with handling and storage of these materials in large amounts. In order to solve such problems, EG was packed into plastic bags and examined for heavy-oil sorption and

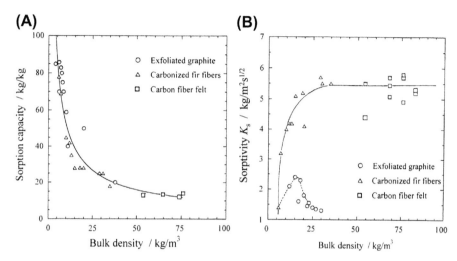

FIGURE 14.16 Dependence of Sorption Performance on Bulk Density of Carbon Adsorbents: EG, Carbonized Fir Fibers, and Carbon Fiber Felts

(A) Sorption capacity, and (B) sorptivity, K_s (sorption rate)

From [14]

recovery. If sufficient attention is paid to maintaining contact between the EG and the heavy oil, and to recovery of the bag after sorption by dipping, a high capacity can be achieved. No problems with handling or storage are expected for carbon felt, but sorption capacity is not so high. When fine carbon powder produced from waste wood is added to an oil layer on the surface of water and kept for several hours, the carbon/oil mixture becomes so intimate that part of the mixture can easily be picked out without any oil ooze from neighboring parts [33], suggesting that such a carbon/oil mixture can be handled easily and separation of carbon powder from oil can be done without difficulty.

Reuse of recovered heavy oils is also necessary from the viewpoint of energy resources. The recovery process for EG was very limited: only filtration under suction with less viscous A-grade and crude oil is possible, but cycle efficiency is poor. Carbonized fir fibers have somewhat better cycle performance than EG. The carbon fibers were able to withstand washing using organic solvents, with a fairly high recovery ratio. For the felt, not only filtration and washing but also centrifugation and even squeezing processes were possible to be applied, with high cycle efficiency.

14.6.2 Mechanism of heavy oil sorption

The strong dependence of sorption capacity on bulk density of EG and carbonized fir fibers suggests that the spaces formed by entangled worm-like particles and fibrous particles with irregular surfaces are primarily responsible for the heavy-oil sorption. It has to be pointed out that these void spaces are flexible (flexible pores). EG and carbonized fir fibers show different cycle performances, which are mainly due to the strength of particles. Carbonized fir fibers are much stronger than worm-like particles of EG and so the bulky network of fir fibers is somewhat stronger (more rigid) than that of EG; in other words, spaces formed by worm-like particles of EG are deformed more easily than those of carbonized fir fibers. Carbon fiber felts, particularly from PAN-based carbon fibers, have excellent cycle performance for filtration under suction, washing with solvents, centrifuging, and even squeezing, although the sorption capacities are not very high. This is because of the high mechanical strength and high elastic modulus of carbon fibers, which makes the spaces among the fibers deformable but recoverable.

The hydrophobic and oleophilic nature of the carbon surface is also an important factor for achieving high sorption capacity, in addition to large void spaces among sorbent particles. Fir fibers with low bulk density without carbonization showed a relatively low capacity for A-grade oil, 17–24 kg/kg, but after carbonization the fibers showed a high sorption capacity of 80 kg/kg. In the case of natural sorbents, such as milkweed, the presence of wax has been pointed out to be important. However, the results for fir fibers clearly indicated that fibers carbonized up to 900 °C, in which no organic materials such as wax remained, showed a high sorption capacity, even higher than the original fibers [22]. This is also the case with rice husks [27,28].

14.6.3 Comparison with other materials

Oil sorbents used and studied so far can be classified into three groups: inorganic minerals such as perlite and vermiculite; organic synthetic materials such as polypropylene and polyurethane; and biomass such as peat moss, kenaf, straw, and wood fibers. Synthesis and properties of materials for oil sorption have been reviewed, focusing on hydrophobic silica aerogels, zeolites, organoclays, and natural sorbents [34]. In Table 14.2 the sorption performance of carbon materials is compared to that of some other sorbents. Since no quantitative rate measurements have been reported on these other sorbents, thorough comparison cannot be made. Although the heavy oils used in these experiments are not exactly the same, two carbon sorbents, EG and carbonized fir fibers, have markedly higher sorption capacity even for viscous C-grade heavy oil, more than twice that of conventional polypropylene foam and most natural fibrous sorbents.

Organic synthetic materials have already been commercialized as sorbents for oils in the form of nonwoven web and foam. Both size and shape of pores formed in these materials were shown to be important by using polypropylene filaments and rectangular shape was favorable rather than circular [13]. These organic synthetic sorbents have been used as reference materials for other sorbents in most research papers. Research on the development of biomass-derived sorbents is very active, mainly because most of them are low-cost waste materials, and also many of them are biodegradable, even though the sorption capacity of these materials is not so high, typically about 10 kg/kg [35–37]. For milkweed and cotton fibers, however, relatively high sorption capacities have been reported, c. 33 and 50 kg/kg, respectively, for light oils (a diesel oil and a crude oil) [37]. Similar values have been reported for unstructured cotton fibers. Silk-floss fibers showed a high sorption capacity (85 kg/kg) for crude oil, comparable to that of carbon materials. For biomass sorbents, water uptake must also be taken into account. After soaking the sorbent in water, its capacity for oil was reported to decrease slightly. Therefore, the surface of many of such sorbents must be changed from hydrophilic to hydrophobic. For this purpose, acetylation of free hydroxyl groups was carried out on various biomass materials, e.g. rice straw, raw cotton, sugarcane bagasse, and wheat straw; sorption capacity for oil was reported to be doubled, from c. 10 to 20 kg/kg [38].

For the process of heavy oil recovery from these sorbents, only squeezing has been reported and most of them showed very high performance. The experimental result that oils taken up by sorbents can be recovered upon squeezing suggests that milder processes for recovery, i.e. filtration under suction and washing with organic solvents, could easily be applied. From the viewpoint of recovery performance, therefore, EG and carbonized fir fibers are inferior to polypropylene nonwoven webs and natural fibrous sorbents.

Oil sorption, recovery of sorbed oils, and recycling of carbon materials as sorbents have been reviewed [39–41].

References

[1] Shen WC, Cao NZ, Wen SZ, et al. Carbon'96 (European Carbon Conference). Newcastle upon Tyne, UK: The British Carbon Group, 1996, 256.
[2] Toyoda M, Aizawa J, Inagaki M. Desalination 1998;115:199–201.
[3] Toyoda M, Moriya K, Aizawa J, et al. Nihon Kagaku Kaishi 1999:193–8, [in Japanese].
[4] Toyoda M, Moriya K, Inagaki M. TANSO 1999; No.187: 96–100 [in Japanese].
[5] Toyoda M, Moriya K, Inagaki M. Nihon Kagaku Kaishi 2000:217–20, [in Japanese].
[6] Toyoda M, Moriya K, Aizawa J, et al. Desalination 2000;128:205–11.
[7] Inagaki M, Konno H, Toyoda M, et al. Desalination 2000;128:213–8.
[8] Inagaki M, Shibata K, Setou S, et al. Desalination 2000;128:219–22.
[9] Tryba B, Kalenczuk RJ, Kang F, et al. Mol Cryst Liq Cryst 2000;340:113–9.
[10] Toyoda M, Inagaki M. Carbon 2000;38:199–210.
[11] Toyoda M, Moriya K, Inagaki M. Sekiyu Gakkaishi 2001;44:169–72, [in Japanese].
[12] Toyoda M, Nishi Y, Iwashita N, et al. Desalination 2002;151:139–44.
[13] Nishi Y, Dai G, Iwashita N, et al. Mater Sci Res Int 2002;8:243–8.
[14] Nishi Y, Iwashita N, Sawada Y, et al. Water Res 2002;36:5029–36.
[15] Tryba B, Morawski AW, Kalenczuk RJ, et al. Spill Sci Technol Bull 2003;8:569–71.
[16] Toyoda M, Inagaki M. Spill Sci Technol Bull 2003;8:467–74.
[17] Inagaki M, Nagata T, Suwa T, et al. New Carbon Mater 2006;21:97–101.
[18] Nishi Y, Iwashita N, Inagaki M. TANSO 2002; No. 201: 31–34 [in Japanese].
[19] Zheng YP, Wang HN, Kang F, et al. Carbon 2004;42:2603–7.
[20] Kang F, Zheng YP, Zhao H, et al. New Carbon Mater 2003;18:161–73.
[21] Wei T, Fan Z, Luo G, et al. Carbon 2008;47:337–9.
[22] Inagaki M, Kawahara A, Konno H. Carbon 2002;40:105–11.
[23] Inagaki M, Kawahara A, Iwashita N, et al. Carbon 2002;40:1487–92.

[24] Gui X, Wei J, Wang K, et al. Adv Mater 2010;22:617–21.

[25] Lee C, Baik S. Carbon 2010;48:2192–7.

[26] Iwashita N, Nishi Y, Sawada Y, et al. Zairyou 2004;53:818–25, [in Japanese].

[27] Kumagai S, Noguchi Y, Kurimoto Y, et al. Water Manage 2007;27:554–61.

[28] Angelova D, Uzunov I, Uzunova S, et al. Chem Eng J 2011;172:306–11.

[29] Beltran V, Escardino A, Feliu C, et al. Br Ceram Trans J 1988;87:64–9.

[30] Toyoda M, Dogawa N, Seki T, et al. TANSO 2001; No.199: 166–169 [in Japanese].

[31] Inagaki M, Kawahara A, Konno H. Desalination 2004;17:77–82.

[32] Tsumura T, Kojitani N, Umemura H, et al. Appl Surf Sci 2002;196:429–36.

[33] Samoilov NA, Khlestkin RN, Osipov MI, et al. Russ J Appl Chem 2004;77:327–32.

[34] Adebajo MO, Frost RL, Kloprogge JT, et al. J Porous Mater 2003;10:159–70.

[35] Zahid MA, Halligan JE, Johnson RF. Ind Eng Chem Process Des Develop 1972; 11:550–5.

[36] Johnson RF, Manjrekar TG, Halligan JE. Environ Sci Technol 1973;7:439–43.

[37] Chol HM, Cloud RM. Environ Sci Technol 1992;26:772–6.

[38] Sun RC, Sun XF, Sun JX, et al. C R Chim 2004;7:125–34.

[39] Inagaki M, Nishi Y, Iwashita N, et al. Fresenius Environ Bull 2004;13:183–9.

[40] Toyoda M, Iwashita N, Inagaki M. Chem Phys Carbon 2007;30:177–234.

[41] Inagaki M, Toyoda M, Iwashita N, et al. Adsorption by Carbon. In: Botani EJ, Tascon JD, editors. Elsevier. ; 2008. p. 711–34.

Carbon Materials for Adsorption of Molecules and Ions

15

Carbon materials can have large amounts and various sizes of pores. Pores in carbon materials have been introduced and controlled mostly via mild oxidation, the process being called "activation," and the resultant porous carbon materials (activated carbons, ACs) have played an important role as adsorbents for various gases and liquids since the prehistoric era. Thanks to a dramatic revolution in modern technology, however, the importance of porous carbons has increased and, as a consequence, pore structure in ACs is required to be controlled more and more precisely. Various processes to control pore structure in carbon materials without activation, such as template carbonization (Chapter 7), have been proposed, in addition of the development of new activating agents [1].

Pores in carbon materials have been classified into macropores, mesopores, and micropores, as shown in Figure 15.1. The proportions of macro-, meso-, and micropores and the distribution of their sizes can be controlled by changing the carbonization conditions (precursor, temperature, process, etc.) and activation conditions (temperature, duration, activating agent, etc.). Pores formed in three carbon materials are illustrated in Figure 15.1, as examples. In granular ACs, most micropores are formed inside mesopores, and the mesopores are formed inside macropores, such that adsorbate molecules have to diffuse through macropores and then mesopores to be adsorbed into micropores. In activated carbon fibers (ACFs), micropores can be formed on the surface of the fiber directly, where the adsorbate can easily reach the entrance of the micropore, which is preferable for the kinetics of the adsorption. In MgO-templated carbons, a large number of mesopores are formed, of which the number and size are controlled by the content of the MgO template and its size, and micropores whose characteristics depend on the carbon precursor are formed on the mesopore walls. In most ACs, mesopores are created by widening micropores, in other words by sacrificing the existed micropores, and carbon yield is reduced. However, in most porous carbons prepared via template carbonization, including MgO-templated carbon, micropores and mesopores are formed directly without any activation process; in other words, without losing carbon atoms for pore formation.

Prerequisite for readers: Chapters 3.5 (Porous carbons) and 3.8 (Carbon materials for energy storage) in *Carbon Materials Science and Engineering: From Fundamentals to Applications*, Tsinghua University Press.

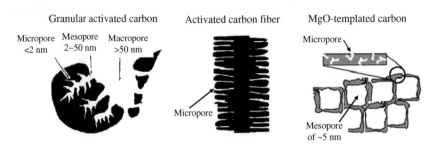

FIGURE 15.1 Pores Formed in Carbon Materials

In Chapter 10, nanoporous carbons were reviewed, with an emphasis on preparation into membranes. In the present chapter, nanoporous carbons are reviewed focusing on their use for adsorption and storage of molecules and ions, or adsorbates. The adsorbates selected are hydrogen; methane; carbon dioxide; organic molecules, including volatile organic compounds (VOCs); and heavy metals. Capacitive deionization of brackish water is also reviewed briefly. Storage and/or removal of these gases and ions via adsorption are very important for global energy and environmental problems.

15.1 Adsorption and storage of hydrogen

Hydrogen adsorption into carbon materials has attracted attention since papers were published that reported an enormous amount, more than 60 mass%, of adsorption into carbon nanofibers [2,3], because this reported H_2 capacity was much higher than the benchmark for application to automobile fuel-cell systems proposed by US Department of Energy (DOE), i.e. 6.0 mass% (45 g/L) for the year 2010 and 9.0 mass% (81 g/L) for 2015 [4]. Different carbon materials, not only carbon nanofibers but also microporous ACs, including ACFs and carbon nanotubes, and also those doped by alkali metals have been studied for hydrogen storage [5–10]. Most of the results, however, were very low in adsorptivity and also very poor in reproducibility. Theoretical calculations for hydrogen storage in carbons have also been reported, but the results varied over a wide range [11–14]. Milling of natural graphite in a hydrogen atmosphere resulted in high uptake of hydrogen, but it was necessary to heat to above 350 °C for hydrogen desorption [15]. Hydrogen storage during water electrolysis has also been proposed, using carbon nanotubes [16], ACs [17–19], and ACF cloth [20]. The majority of this electrochemical storage of hydrogen was the intercalation of atomic hydrogen into the interlayer space of carbons, and a minority was electric double-layer adsorption [19].

Hydrogen-adsorption isotherms have been carefully measured on various ACs and ACFs [21–37], but the capacities were not so high as reported before. Hydrogen adsorption of carbon materials, including single-walled carbon nanotubes

(A)

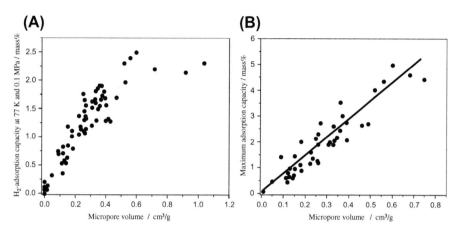

(B)

FIGURE 15.2 Dependence of Hydrogen Adsorption Capacity at 77 K and 0.1 MPa on Micropore Volume of Various Carbons

(A) As a function of micropore volume measured by N_2 adsorption at 77 K, and (B) as a function of micropore volume measured by CO_2 adsorption at 273 K

From [28]

(SWCNTs) and templated carbons, was closely related to micropore volume, V_{micro}, particularly that measured from CO_2-adsorption isotherms at 273 K [22,24–26,32]. The data on hydrogen adsorption capacities of various ACs are summarized as a function of V_{micro} in Figure 15.2A, and suggest that H_2 adsorption is mainly governed by micropores [28]. By selecting the data on the maximum capacity, which are either measured directly as a plateau or extrapolated from the isotherm using an approximated equation and V_{micro} measured by CO_2 adsorption at 273 K, the linearity of the relationship is improved, as shown in Figure 15.2B, suggesting the importance of ultramicropores (<0.7 nm diameter). The importance of micropores, particularly narrow micropores, for hydrogen adsorption was concluded from detailed studies on various carbon materials [29]. Adsorption capacity at 77 and 303 K under 0–3.5 MPa of the phenolic-resin-derived ACFs, which had Brunauer-Emmett-Teller (BET) surface area (S_{BET}) up to 2350 m²/g, gave a relationship with V_{micro} similar to zeolites having S_{BET} of 710 and 330 m²/g [22]. SWCNTs after oxidation in 4 mol/L HNO_3 gave higher S_{BET} and V_{micro} (710 m²/g and 0.25 cm³/g, respectively) and their capacity at a given V_{micro} was higher than that of the zeolite and ACFs, suggesting that acid treatment of SWCNTs may increase the number of sites with high interaction potential for hydrogen adsorption [22]. The effect of the activation reagent on hydrogen storage was compared between CO_2 and KOH by additional activation of a commercially available AC (S_{BET} = 1585 m²/g and V_{micro} = 0.59 cm³/g), showing that KOH activation was more effective, increasing V_{micro} to 1.09 cm³/g and hydrogen adsorption capacity to 7.08 mass% at 77 K and 2 MPa [33].

AC monoliths prepared by compression with a polymeric binder, followed by carbonization of the binder at 750 °C, gave a volumetric hydrogen capacity of

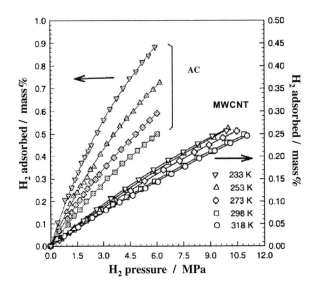

FIGURE 15.3 Amount of H₂ Adsorbed into an AC and a MWCNTs at Different Temperatures

From [23]

29.7 g/L at 77 K and 4 MPa [30]. Monoliths have been prepared from amorphous carbon nanofibers (CNFs) produced by the polymer blend technique from a novolac-type phenolic resin, followed by CO_2 activation, and gave 1.3 mass% hydrogen storage at 298 K [37]. The activated CNFs had enough compressibility to give a high packing density. Electrospun polyacrylonitrile (PAN)-based carbon nanofibers have also been studied after activation with NaOH and K_2CO_3 at 750 °C [34].

In Figure 15.3, the amount of H_2 adsorbed at different temperatures into an AC with S_{BET} of 3000 m²/g (AX-21) and into multi-walled carbon nanotubes (MWCNTs) prepared by catalytic chemical vapor deposition (CVD) of acetylene are compared [23]. The AC has a much higher adsorption capacity than the MWCNTs, but has much stronger dependence of capacity on temperature. Double-walled carbon nanotubes (DWCNTs) were shown to adsorb relatively large amounts of H_2 at 77 K under pressure of less than 0.1 MPa, higher than SWCNTs, even though DWCNTs have much smaller S_{BET} and V_{micro} [38]. Hydrogen storage in carbon nanotubes (CNTs) was reported to be less than 1.7 mass% under pressure up to 12 MPa at room temperature [39].

Hydrogen adsorption of porous carbons prepared via template methods has also been studied [27,40–47]. Mesoporous carbons prepared using mesoporous silicas (MCM-48, SBA-15, etc.) have been modified by coupling with other techniques: electrolysis in KOH [40], activation [41,43], treatment in NH_3 [42,44], and enriching nitrogen [42]. By CO_2 activation at 950 °C, S_{BET} increased to 2749 from 984 m²/g for the pristine carbon, and V_{micro} increased markedly to 0.96 from 0.37 cm³/g; consequently, maximum hydrogen uptake increased to 2.24 from 0.98 mass% at 77 K and

0.1 MPa [41]. Nitrogen-doped mesoporous carbons prepared from ethylenediamine and CCl_4 using SBA-15 as a template gave a hydrogen capacity of 6.84 mass% at 77 K and 2 MPa after KOH activation [43]. Templated microporous carbons using zeolite Y with different S_{BET} of 1610–3800 m^2/g and $V_{micro}(N_2)$ of 0.6–1.58 cm^3/g exhibited hydrogen uptake as high as 2.2 mass% at 30 °C and 34 MPa. This hydrogen uptake was higher than that of commercial ACs (S_{BET} of 1700–2680 m^2/g and $V_{micro}(N_2)$ of 0.74–1.20 cm^3/g), which is mainly owing to a marked contribution of uniform micropores with a diameter of 1.2 nm [45]. A zeolite-templated carbon was also used after KOH activation, reporting increases in V_{micro} from 0.74 to 1.01 cm^3/g and in hydrogen uptake from 4.81 to 6.30 cm^3/g [46]. Nanoporous carbon with S_{BET} of 3405 m^2/g prepared from the mixture of a zeolite-type metal-organic framework (MOF-5; as a template and a carbon precursor) and furfuryl alcohol gave a H_2-storage capacity of 2.77 mass% at 77 K and 0.1 MPa [47].

Microporous carbons prepared from metal carbides, TiC and B_4C, by chlorination gave H_2 storage of 3.0 and 1.06 mass%, respectively, at 77 K and 0.1 MPa after annealing in H_2 to remove trapped Cl_2 [48–50]. Microporous carbon nanofibers prepared by electrospinning of polycarbosilane-THF (tetrahydrofuran) solution followed by chlorination to decompose SiC gave hydrogen storage of 3.86 mass% at 77 K and 1.7 MPa [51].

Loading of metals, mostly Ni, on nanoporous carbons has been used to try to increase the H_2-adsorption capacity [52–59]. The hydrogen uptake increased a little but heating of the adsorbent was required for hydrogen desorption because of loose chemisorption of hydrogen atoms on the carbon surface as spilt-over species. The spillover effect was reported to play a significant role at 25 °C and 20 MPa, although the storage capacity at 77 K and 4 MPa seems to be mainly influenced by the textural properties of the carbon support used [59].

Hydrogen storage capacity was compared between two ACs (one commercially available and one laboratory-made from an anthracite by KOH activation) and MOF-5 [60]. Although the three samples showed similar hydrogen storage on a gravimetric basis, the two ACs showed higher hydrogen storage than MOF-5 on a volumetric basis. MOFs were reported to have a gravimetric adsorption of 4.3–5.5 mass% and volumetric capacity of 33.2–37.8 g/L at 77 K and around 2 MPa [61].

15.2 Adsorption and storage of methane and methane hydrate

Much research work has been carried out on the adsorption of methane (CH_4), the principal component of natural gases, by various adsorbents under different conditions, similar to that on the adsorption and storage of hydrogen described in the previous section. The US DOE set a target figure of 180 v/v for application in vehicles, at 25 °C and 3.5 MPa [62]. Theoretical calculations of the storage capacity for CH_4 of porous carbons predicted 200–270 v/v. Other microporous materials, such as various zeolites, MOFs, and other compounds containing micropores in their lattice, have also been tested for

CH_4 adsorption and storage. On some MOFs, CH_4-storage capacity was reported to be 230 v/v at 290 °C and 3.5 MPa [63], and 171 v/v at 25 °C and 3.5 MPa [61].

The adsorption performance of various adsorbents, including ACFs, was reviewed in 1998 [64]. In the review, the published data on the amount of CH_4 adsorbed by various microporous adsorbents were discussed mainly in relation to the surface area of the adsorbents, even though the presence of micropores was emphasized to be an important factor for CH_4 adsorption. Adsorption capacity per mass of adsorbent can be approximated by a linear relationship to the surface area, as shown in Figure 15.4A. By expressing the capacity per volume of the adsorbent, however, monoliths of ACs and ACFs give relatively high values, higher than granular ACs and mesoporous silica MCM-48 (Figure 15.4B). In another review published in 2002 [65], linear dependence of gravimetric capacity for CH_4 adsorption on micropore volume for various activated carbon materials was presented, although all carbon materials had been prepared in their laboratory. Practical application of adsorbed natural gas for vehicles has been reviewed based on the literature published before 1999 [66].

Carbon monolith was produced by the compression of carbon particles produced by CO_2 laser ablation of carbon at room temperature without a metal catalyst (called single-walled carbon nanohorns, SWNHs), and was found to have a relatively high CH_4-storage capacity, 160 v/v at 30 °C and 3.5 MPa [67]. SWNHs were oxidized either in H_2O_2 or O_2 after dispersion in ethanol, followed by repeated compression and crushing, in order to produce a high enough micropore volume. After nine cycles of compressing under 50 MPa and crushing, a thin disk with a bulk density of 0.97 g/cm^3 was obtained, which had S_{BET} of 1097 m^2/g and V_{micro} of 0.55 cm^3/g, although SWNH before compression had 1030 m^2/g and 0.50 cm^3/g, respectively.

CH_4 adsorption by carbide-derived microporous carbons has also been studied [68–70]. AN adsorption capacity of 18.5 mass% at 25 °C and 6 MPa was obtained in

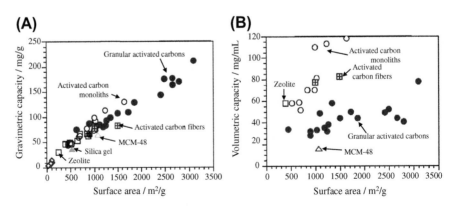

FIGURE 15.4 Adsorption Capacity of Various Adsorbents for CH_4 at 298 K and 3.5 MPa, as a Function of Surface Area

(A) Gravimetric capacity, and (B) volumetric capacity

From [64]

TiC-derived carbon after activation with CO_2 at 975 °C, with S_{BET} of 3360 m²/g [68]. Application of template techniques using mesoporous silica (SBA-15) and the surfactant P123 on SiC precursors, followed by carbonization and chlorination, resulted in a microporous carbon with ordered mesopores that gave a high CH_4 adsorption, 19 mass% at 25 °C and 10 MPa [69,70]. Further treatment of SiC-derived carbon in a flow of hydrogen at 600 °C was effective to remove remaining metal particles and to enhance CH_4 adsorption [69].

A noticeable enhancement of CH_4 adsorption by the pre-adsorbed water in micropores was found using steam-activated pitch-based ACFs with different S_{BET} values (900–1800 m²/g) and KOH-activated pitch-based carbon with a high surface area (2290 m²/g) [71]. Although CH_4 adsorption by the microporous carbons at 30 °C was less than 9.4 mg/g at 101 kPa, the presence of the pre-adsorbed water noticeably enhanced methane adsorption at 303 K, even under sub-atmospheric pressure. The increase in CH_4 adsorption depends on pressure and reaches saturation after 20–50 h, with a maximum at 1–2 h, as shown in Figure 15.5A. It also depends on the fractional filling, ϕ_w, of micropores by the pre-adsorbed water, as shown in Figure 15.5B for an ACF with S_{BET} of 1800 m²/g. CH_4-adsorption capacity of the ACF increased linearly with increasing ϕ_w up to 0.35, indicating the formation of stable methane-water clathrate (methane hydrate). The proportion of methane to water (CH_4/H_2O) of the hydrate formed in micropores was thought to be 1/2.

The effect of pre-adsorbed water on the adsorption and desorption of CH_4 has been studied in different ACs under various conditions [72–78]. Adsorption

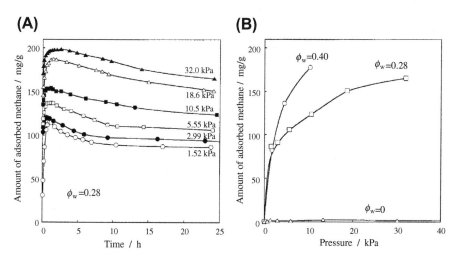

(A)

(B)

FIGURE 15.5 Enhancement of CH_4 Adsorption at 30 °C by Pre-adsorbed Water in an ACF with S_{BET} of 1800 m²/g

(A) Dependence on adsorption time under different pressures, and (B) that on CH_4 pressure with different ϕ_w (fractional filling of micropores by pre-adsorbed water)

From [71]

isotherms depended strongly on the water content, R_w (mass%), of the AC, R_w being expressed as g of water adsorbed into 100 g of carbon [73]. In Figure 15.6A, isotherms observed at 2 °C on an AC prepared from coconut shells are shown. The saturated amount of CH_4 adsorbed increases with increasing R_w to 1.4 mass%, but too much water (i.e. 3.0 mass%) disturbs CH_4 adsorption. It is characteristic for the CH_4 adsorption into AC containing pre-adsorbed water to show a marked inflection point. This inflection was reasonably thought to occur owing to the formation of methane hydrate in the micropores, because this inflection pressure around 4 MPa at 2 °C is a little higher than the formation pressure of the hydrate. In Figure 15.6B, a marked hysteresis in adsorption/desorption isotherms is shown for an AC with R_w of 1.4 mass%. The volumetric storage capacity reached 152 v/v at charging pressure of 8 MPa [75]. Methane hydrate was formed quickly in the micropore space at a pressure above 4.12 MPa at 2 °C, and stored methane could be continuously released at a constant flow rate. On a series of ACs prepared from corncob particulates activated by KOH with pore volume controlled in the range of 1.29–1.77 cm^3/g by heating with pre-adsorbed water at 750 °C, pronounced inflections in CH_4-adsorption isotherms at a little less than 5 MPa were observed, and hysteresis in the desorption isotherm was also observed on ACs with pre-adsorbed water in an R_w range of 1.5–3.75 mass% [78]. Gravimetric CH_4-storage capacity showed a maximum of about 60 mass% (based on dry AC) at R_w of around 3 mass%, and experimental volumetric capacity of 204 v/v was obtained at 2 °C and 9 MPa.

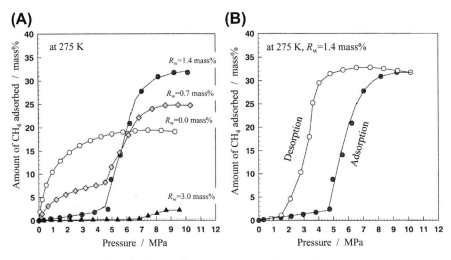

FIGURE 15.6 CH₄-Adsorption/Desorption Isotherms of ACs at 2 °C

(A) Adsorption isotherms with different water content, R_w, and (B) an example of the marked hysteresis observed

From [73]

15.3 **Adsorption and storage of CO_2**

Global warming is now understood to be one of the world's major environmental issues to be solved, and to be mainly due to carbon dioxide gas, CO_2. Its capture with amine solvents was suggested to be the most applicable technology [79], but physical adsorption into some adsorbents, including nanoporous carbon materials, has been proposed as an alternative process for CO_2 capture [80,81]. It was originally planned that adsorbents, after the adsorption of CO_2, would be sequestrated underground [82], but now the possibility of regeneration of the adsorbents is also being studied. The selective adsorption of CO_2 from the gaseous mixture with either CH_4 or N_2 is now understood to be an important process for its removal from syngas and natural gas.

Carbon fiber composites with a bulk density of 0.35 g/cm^3 were prepared from the mixture of isotropic-pitch-based carbon fibers and phenol resin by carbonization at 650 °C, followed by activation with either steam or CO_2 at 800–950 °C [83,84]. Their adsorption of CO_2 and CH_4 was studied as a function of burn-off during activation. Adsorption of CO_2 depended on the temperature; adsorption capacity of CO_2 reached 55 cm^3/g at 30 °C and reduced to 15 cm^3/g at 100 °C for the composite with 9% burn-off. Breakthrough experiments have shown selective separation of CO_2 from CO_2/CH_4 gas mixtures [84]. On cylindrical blocks of the same kind of composite, in which there were 17 evenly distributed longitudinal channels with a diameter of 0.03 cm, CO_2 adsorption was measured at 0 and 25 °C [85]. As shown in Figure 15.7, adsorption capacity at different temperatures corresponded reasonably to micropore volume and size as determined from the adsorption isotherms of CO_2 at 0 °C. The CO_2-adsorption capacity reaches 4.0–4.5 mmol/g at 0 °C and

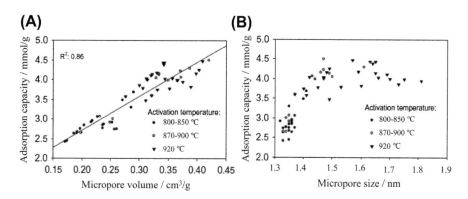

FIGURE 15.7 Adsorption Capacity for CO_2 at 0 °C of Carbonized Carbon Fiber/Phenol Resin Composites

(A) As a function of micropore volume, and (B) as a function of micropore size, determined from CO_2 adsorption

From [85]

2.6–2.9 mmol/g at 25 °C. Porous carbons prepared from resorcinol-formaldehyde by using MOF-5 at 950 °C gave a CO_2-adsorption capacity of 2.9 mmol/g at 300 K [86].

Nitrogen-doped nanoporous carbons prepared from mixtures of melamine-formaldehyde resins through silica-template carbonization gave 2.25 mmol/g at 25 °C and 0.86 mmol/g at 75 °C [87], and those from resorcinol and formaldehyde in the presence of L-lysine as the catalyst gave 3.13 mmol/g at room temperature [88]. The adsorption capacity and selectivity for CO_2 from the mixed gases with CH_4 and N_2 reduced with increasing temperature [88]. MgO/CaO-loaded porous carbons have also been employed for the adsorption of CO_2 and SO_2 [89,90].

For the regeneration of the adsorbents, heating by passing a low voltage DC current through the composite adsorbents has been proposed (electrical swing adsorption, ESA) [84], and CO_2 desorption has also been accomplished by heating the ACF adsorbent up to 60 °C [91]. CO_2 desorption from MgO/CaO-loaded porous carbon has been tried up to 850 °C [90].

Irreversible adsorption of CO_2 has been observed in phenol-based carbon spheres with glass-like carbon nanotexture, which were carbonized up to 1000 °C in a CO_2 atmosphere (named APT) [92–95]. APT showed a marked irreversibility in volumetric adsorption/desorption isotherms of CO_2 at different temperatures, although the carbon spheres prepared at the same temperature in a N_2 atmosphere (named APS) showed completely reversible adsorption/desorption of CO_2. The rate of adsorption of CO_2 on APT was very slow, so that a long time was necessary to reach equilibrium. The gravimetric adsorption/desorption behavior of CO_2 on APT is compared with APS in Figure 15.8 [94]. After the saturation of adsorption, changes in mass were measured by evacuating to 670 kPa pressure at 0 °C and also by heating under evacuation to 250 and 500 °C. On APS, adsorption occurs immediately with the introduction of CO_2 gas and saturation is achieved within 10 min; desorption also occurred rapidly and completely under vacuum at 0 °C. On APT, however, a certain

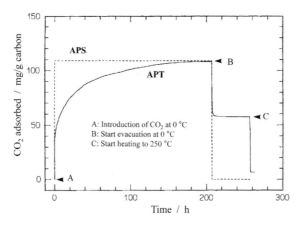

FIGURE 15.8 Gravimetric Adsorption/Desorption of CO_2 for APT and APS

From [94]

amount of CO_2, about 40% of the saturated amount, is adsorbed quickly during the first 10 min, and then CO_2 adsorption continues very slowly, taking more than 200 h to reach saturation. On evacuation at 0 °C, about 40% of the adsorbed CO_2 is desorbed immediately and another 10% within 2 h, but approximately half of the adsorbed CO_2 is not able to be desorbed even after 20 h evacuation at 0 °C. In order to remove this strongly adsorbed CO_2, it is necessary to heat up to 250 °C under evacuation, but c. 5% of the adsorbed CO_2 still remains. A second exposure of APT to CO_2 led to the same behavior as observed at the first cycle, i.e. a rapid adsorption to about 40% and then a gradual adsorption to reach saturation.

The effect of outgassing at 0, 250, and 500 °C under vacuum on adsorption behavior of CO_2 was studied in APT. The mass loss by outgassing at 0, 250, and 500 °C was 4.4, 6.7, and 8 mass%, respectively. The higher the outgassing temperature, the larger the adsorbed amount of CO_2 obtained at saturation and the longer the time needed to reach saturation. The proportion of CO_2 that was rapidly adsorbed during the first 10 min was almost the same, and not dependent on outgassing temperature.

These gravimetric studies on the adsorption/desorption of CO_2 suggest the presence of three types of micropores in the sample of APT. The first type of micropore adsorbs CO_2 molecules immediately and also desorbs quickly just by evacuation at 0 °C, of which the volume is supposedly about 40% in APT. The second type of micropore adsorbs CO_2 very slowly, taking about 200 h to reach saturation, and requires heating to 250 °C under vacuum for desorption, being about 45% in APT. From the third type of micropore, the adsorbed CO_2 is not released even by heating up to 250 °C under vacuum (strongly trapped CO_2). The main characteristics of pores in the samples of APS and APT were studied by applying the Dubinin-Astakhov equation and through immersion calorimetry using molecular probes of different sizes: CH_2Cl_2 (0.33 nm), C_6H_6 (0.41 nm), and CCl_4 (0.63 nm) [94]. From this characterization, the presence of constriction of pore walls around CO_2 molecules was considered to explain their trapping in the micropores of APT; a weak constriction from the first type of micropores, a strong constriction from the second type, and the strongest constriction from the third. Detailed study was carried out on ultramicropores in APS and APT by supercritical gas adsorption analysis assisted by grand canonical Monte Carlo (GCMC) simulation, together with other molecular sieve carbons and ACFs [96]. From comparison of the pore width determined from adsorption isotherms of N_2 and Ar at 303 K to that from an α_s plot of the N_2-adsorption isotherm at 77 K, APT was found to have a special pore morphology, in which pore entrance size (pore mouth width) was narrower than inner pore size. Such micropores have been observed in other molecular sieve carbons, and called "ink-bottle-type" micropores. In APT, a similar irreversible adsorption/desorption behavior of N_2, which has a slightly smaller cross-section than CO_2, was observed only at very low temperature, i.e. -196 °C. Irradiation of oxygen plasma has been found to modify the surface of glass-like carbon spheres to improve selective adsorption of CO_2 [97]. Similar irreversible adsorption for CO_2, as well as for H_2O, has been observed in graphitic nanoribbons [98].

Adsorption of NO_x gases into porous carbons prepared from different precursors under different conditions has been studied [99,100]. MgO-loaded porous carbon was able to remove a trace of SO_2 from ambient air [90].

15.4 Adsorption of organic molecules

15.4.1 Organic gases (including VOCs)

Volatile organic compounds (VOCs) are known to be harmful to human beings even at low concentration. Adsorption has attracted a great deal of attention as an effective technique to remove VOC gases at low concentration, and much research has been carried out.

The effects of pore size distribution and surface nature on the adsorption of some of VOCs, such as benzene, toluene, acetone, and trichloromethane, has been studied using commercially available ACs derived from different carbon precursors with different activation processes [101–107]. Micropores with a size less than 0.7 nm, the volume of which was determined by CO_2 adsorption at 273 K, were shown to be responsible for the adsorption of benzene and toluene vapors (200 ppmv) by commercially available ACs, and oxygen-containing functional groups on the surface of the ACs seemed to disturb the adsorption of VOC gases [107]. Since organic molecules have a wide variety of structures, sizes, and polarities, not only pore structure in carbon adsorbents but also their surface nature has been pointed out to be important for the interaction with organic molecules. In order to understand this interaction, ACs were subjected to oxidation in HNO_3 and H_2O_2, which disturbed the adsorption of phenanthrene [106], and to ozonation, which increased the adsorption of methyle-thylketone and benzene [104].

The effects of carbonization atmosphere and subsequent activation of glass-like carbon spheres were investigated, focusing on the adsorption behavior of chlorinated methanes [108,109]. The pristine glass-like carbon spheres, with a diameter of about 50 μm, were prepared from phenol spheres by carbonization in an atmosphere of either N_2 or CO_2 at 700 and 1000 °C. After carbonization, spheres were subsequently oxidized in boiling nitric acid at 13 vol%. The adsorption capacity, measured by a gravimetric method, of these carbon spheres is shown in Figure 15.9, plotting relative mass increase with time. There is little relationship between the adsorption capacities for chlorinated methanes and pore characteristics. HNO_3 treatment gives different surface groups to the carbon spheres carbonized at different temperatures: acidic groups, both carboxylic and lactone types, increase on 700 °C-carbonized spheres, but basic groups become predominant on 1000 °C-carbonized spheres. This surface nature of carbon spheres has a certain effect on the adsorption capacity: spheres containing a larger number of basic groups show higher adsorption capacity for $CHCl_3$ and CH_2Cl_2 than spheres with a predominance of acidic groups, though no adsorption of CCl_4. Considering the fact that CCl_4, $CHCl_2$, and CH_2Cl_2 molecules have very similar cross-sectional areas of 0.32, 0.28, and 0.25 nm^2, but very different dipole moments of 0, 1.1, and 1.8 Debye, respectively, the adsorption of polar molecules

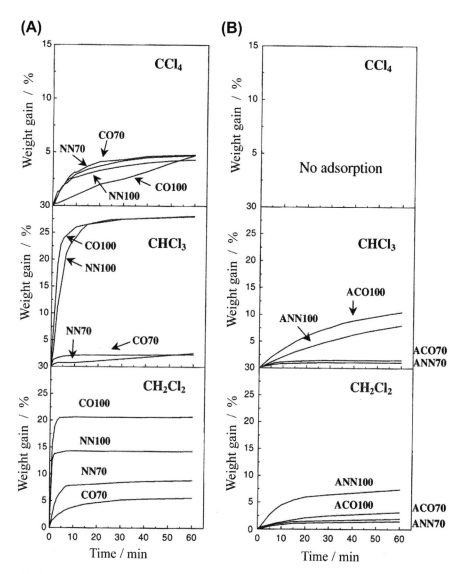

FIGURE 15.9 Adsorption Capacities of Chlorinated Methanes for Carbon Spheres Before and After HNO₃ Treatment

Samples CO70 and CO100 are the carbon spheres carbonized in CO_2 at 700 and 1000 °C, and NN70 and NN100 are those carbonized in N_2 at 700 and 1000 °C, respectively: (A) before and (B) after HNO₃ treatment

From [109]

may be driven by dipole-dipole interaction between the adsorbate molecules and the basic functional groups on the surface of the adsorbent. Adsorption of CCl_4 vapor has been studied on aligned activated CNFs (ACNFs) prepared using epoxy/toluene solution with an anodic aluminum oxide (AAO) template, followed by KOH activation [110]. ACNFs prepared using AAO film with a nominal pore size of 20 nm showed high S_{BET} and high V_{micro}, and consequently high adsorptivity of CCl_4.

ACFs were thought to have more potential in the abatement of VOCs than ACs, because of the smaller diameter (sub-micrometer) and larger number of micropores in ACFs, and much research has concentrated on their use for VOC adsorption [111–123]. An ACF cloth (novolac-derived, S_{BET} of 1600 m²/g) showed abrupt adsorption and higher adsorption capacities for acetone and benzene than an AC (coal-derived, S_{BET} of 965 m²/g) [121]. Breakthrough curves for polar methylethylketone and nonpolar benzene were studied using ACFs with different surface areas [119]. The effects of S_{BET}, adsorbent-bed temperature, and relative humidity are shown in Figure 15.10. For the ACF with S_{BET} of 600 m²/g, the breakthrough time for benzene becomes short, and adsorption proceeds gradually (Figure 15.10A). Increase in adsorption temperature (adsorbent-bed temperature) shortens the

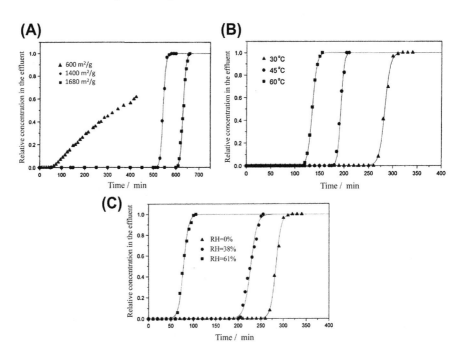

FIGURE 15.10 Breakthrough Curves of Benzene Adsorption for ACFs (250 ppm in N₂ at 30 °C)

(A) ACFs with different S_{BET}; (B) effects of adsorption temperature, and (C) of humidity for an ACF with $S_{BET}=1400$ m²/g. RH, relative humidity

From [119]

breakthrough time (Figure 15.10B), and higher relative humidity makes the break-through time markedly shorter (Figure 15.10C) for an ACF with S_{BET} Of 1400 m²/g.

The micropore surface of ACF cloths was chemically modified by treating with ammonia, mixed acid HNO_3/H_2SO_4, or chlorine to be basic, acidic, or polar, respectively, and their adsorptivity was then evaluated for removal of VOCs (acetalde-hyde, acetone, and benzene) at 10–1000 ppmv concentration [112]. The oxidized surface displayed greatly enhanced adsorption of acetone and acetaldehyde at 25 and 50 ppmv, respectively. The effect of relative humidity (RH) in the atmosphere was studied on the adsorption of acetone and benzene (350–1000 ppmv) with ACF cloths [113]; RH had little effect on the adsorption of acetone (miscible with water) even at 90% RH, and RH above 65% made the adsorption of benzene (immiscible with water) unattainable. Vacuum treatment of adsorbent ACF increased its adsorp-tion capacity for benzene [118]. Modification of ACF surfaces by $CuSO_4$ has been reported to increase adsorption capacities for benzene, toluene, methanol, ethanol, and acetone [122]. Regeneration of the absorbent ACFs has also been studied; elec-trical heating by passing of a DC current was effective for complete regeneration [120,123].

The removal of VOCs by adsorption using ACFs was found to be effective if VOC concentrations in the pollutant-laden gas were in the levels of ppm, whereas removal by cryogenic condensation was found to be effective at relatively high VOC concentrations (>1%) [120].

Carbon nanofibers prepared by different techniques, such as electrospinning, catalytic CVD, and the template method, with diameters thinner than commercially available ACFs, were found to have highly developed microporosity after activation [124,125]. These ACNFs have been studied as an adsorbent for VOCs in many papers [126–135]. In Figure 15.11, two PAN-based fibrous carbons, ACF and ACNF, are

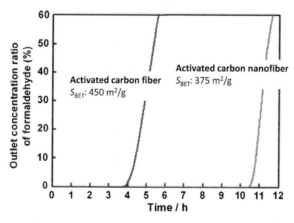

FIGURE 15.11 Breakthrough Curves of Formaldehyde Adsorption for PAN-based ACFs and ACNFs

From [134]

compared using the breakthrough curve for formaldehyde adsorption [134]. Breakthrough time for the ACNF is much longer than for the ACF, even though S_{BET} of the ACNF is lower than that of the ACF, probably owing to more effective usage of micropores for adsorption in the ACNF. On the basis of the isosteric enthalpy and energy distribution for adsorption, the electrospun PAN-based ACNFs were thought to have a more homogeneous surface than ACFs for adsorption [126]. Carbon nanofibers, prepared with an AAO template from an epoxy/toluene carbon precursor, had a relatively high surface area and high adsorption capacities for acetone, n-hexane, and CCl_4 [127,128].

15.4.2 Organic molecules in water

The adsorption behavior of trihalomethanes (THMs; $CHCl_3$, $CHBrCl_2$, $CHBr_2Cl$, and $CHBr_3$) from their diluted aqueous solutions was studied using glass-like carbon spheres oxidized under different conditions [136,137]. The pristine carbon spheres were prepared from phenol-resin spheres by carbonization at 1000 °C in CO_2 (referred to as APT in the previous section) and oxidized either in boiling HNO_3 or in air. Adsorption isotherms of THMs for these carbon spheres were determined at room temperature using their aqueous solutions in different concentrations from 0.1 to 1.7 mg/L. In Figure 15.12A, adsorption isotherms for $CHCl_3$ are shown for five APTs oxidized under different conditions, consequently having different S_{BET}. Although the pristine APT and the spheres after oxidation in nitric acid (APT-N) have a very small capacity for adsorption of $CHCl_3$, oxidation in air at 400 °C seems to be effective to increase the capacity for $CHCl_3$ adsorption. APT-44, which was oxidized in air at 400 °C for 4 h, shows the highest adsorption capacity, 18.4 μmol/g, in 1.0 mg/L $CHCl_3$ solution. No clear correspondence was found between the adsorption capacity of $CHCl_3$ and S_{BET}. Adsorption of asymmetrical molecules like THMs has been discussed in relation to the size and the shape of pore entrances located on the surface of spherical particles, determined from scanning tunneling microscope (STM) [137]. In order to determine the selectivity of adsorption for different THM molecules, adsorption was examined with APT-44 by measuring the adsorption isotherm for each THM in an equimolar mixture. These adsorption isotherms are shown in Figure 15.12B. The largest molecule, $CHBr_3$, was the most preferentially adsorbed onto APT-44, followed by $CHBr_2Cl$, $CHBrCl_2$, and then $CHCl_3$. The adsorption of $CHCl_3$ from the solution mixed with other THMs was strongly depressed, reaching about 40% of the adsorption capacity from an aqueous solution of $CHCl_3$ alone, and the total adsorbed amount of all four THM molecules is about twice of that for pure $CHCl_3$.

Adsorption of a large number of organic compounds (aromatics and aliphatics) in water was carried out onto rayon-based ACF cloths (which were microporous by CO_2 activation and mesoporous by H_2O activation) and coconut-derived AC granules (microporous by CO_2-H_2O activation). Kinetic and equilibrium data were discussed on the basis of the approximation of adsorption isotherms by the Freundlich equation [138].

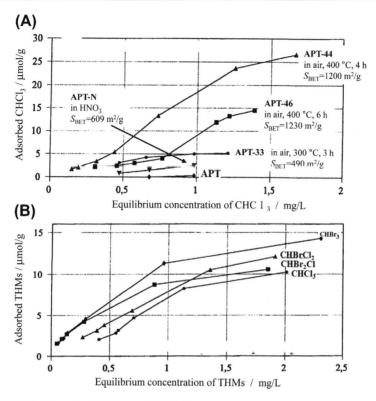

FIGURE 15.12 Adsorption Isotherms of AC Spheres

(A) Isotherms of carbon spheres activated under different conditions for $CHCl_3$ and
(B) those of APT-44 for different THMs from their equimolar mixture

From [137]

The adsorption behavior of various organic molecules from their aqueous solution has been studied in glass-like carbon spheres as a function of activation conditions, and analyzed through a so-called master curve [139–141]. Activation of carbon spheres was carried out in a flow of ambient air at 330, 350, and 370 °C for various times from 1 to 100 h. Master curves of methylene blue, diquat, and phenol are shown in Figure 15.13, which were constructed by shifting the relationship between relative concentration of each pollutant and activation time observed at 330 and 370 °C to that at 350 °C along the time-axis, to get the best overlap. The results demonstrate that the adsorption of organic molecules from aqueous solutions depends strongly on the activation conditions, in other words pore structure, of the adsorbent carbon, and also that the conversion between activation temperature and time is possible for the adsorption of organic molecules in water.

Adsorption of organic molecules onto carbon nanotubes was studied on a SWCNT (2 nm diameter) and two MWCNTs (10–30 nm and 40–60 nm in

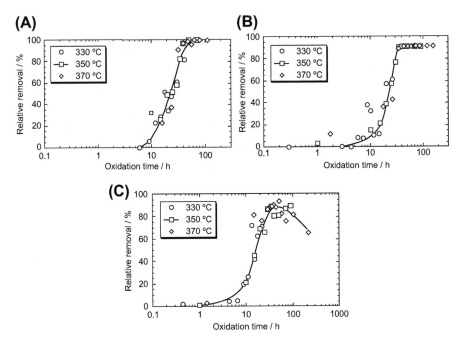

FIGURE 15.13 Master Curves of Adsorption of Three Organic Molecules from their Aqueous Solutions

(A) Methylene blue, (B) diquat, and (C) phenol

From [140,141]

diameter) [142]. The strong adsorptive interaction between the carbon nanotubes and nitroaromatics was due to the π-π electron donor–acceptor interaction between nitroaromatic molecules (acceptors) and the highly polarizable surfaces (donors) of CNTs. For a given CNT, the adsorption affinity increased in the order of nonpolar aliphatic < nonpolar aromatics < nitroaromatics, and increased with increasing number of nitrogen-containing functional groups in the nitroaromatics.

MgO-loaded porous carbon prepared from the mixture of poly(ethylene terephthalate) (PET) and $MgCO_3$ by carbonization at 850 °C showed high adsorption capacity for an anionic dye (reactive red 198) in water, higher than the porous carbon after removing MgO by acid washing [143]. For a cationic dye (basic red 18), however, the former had lower adsorption capacity than the latter. Porous carbons prepared from the mixtures of PET and different inorganic magnesium compounds (carbonate, hydroxide, and oxide) have been tested as adsorbents for model pollutants in water [144].

BC027 bacteria (a kind of *Pseudomonas*) were inoculated onto an activated bamboo charcoal substrate, and the resultant biologically activated carbon showed high performance for the removal of quinoline in culture medium [145].

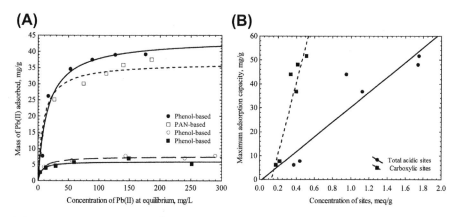

FIGURE 15.14 Adsorption of Pb²⁺ Ions in Water at 298 K and pH=4 by Different ACFs

(A) Adsorption equilibrium for different ACFs, and (B) maximum adsorption capacity against the concentration of total acidic sites and carboxylic sites in ACFs

From [156]

15.5 Adsorption and removal of heavy-metal ions in water

Removal of heavy metals (Pb, Cu, Cr, Cd, Ni, etc.) in industrial waste using an appropriate adsorbent is an important process to prevent damage to our environment. Since the adsorption of heavy metals to carbon is understood to be mostly due to the complexation with functional groups on the carbon surface, ACs are used for this purpose after surface treatment in various acids, such as HNO_3 and malonic acid. Different biomasses, such as silk cotton hull, tree sawdust, maize cob, and banana pith, have been selected as carbon precursors for ACs because of their high oxygen content and for ecological reasons [146–156]. In Figure 15.14A, adsorption isotherms for Pb^{2+} ions in $Pb(NO_3)_2$ aqueous solution at 298 K and pH=4 are shown for commercially available PAN- and phenol-based ACFs, revealing quite different performances of the ACFs even though they were prepared from the same precursor, phenol resin [156]. This difference was explained by a difference in the number of acidic sites (functional groups), particularly carboxylic sites, which were mainly introduced onto the ACF surface by the treatment in 15 vol% HNO_3, and were quantitatively determined by the Boehm titration method, as shown in Figure 15.14B by plotting maximum adsorption capacity against the concentration of total acidic sites and carboxylic sites. The contribution of carboxylic sites is shown to be very important. The pH value of the aqueous solution containing the heavy-metal ions had to be selected in order to interact with functional groups on the surface of the adsorbent carbon. In the most cases, heavy metals adsorbed on the carbon surface were easily desorbed by lowering the pH. The modification of ACs with ethylenediamine after the conversion of carboxylic groups to acyl chloride functional groups by thionyl chloride in toluene was reported to increase the adsorption capacity for Pb^{2+} ions [154].

Carbon nanotubes (CNTs) have been found to be able to adsorb a large amount of heavy metals [157–170], owing to chemical interactions between the metal ions and the surface functional groups of CNTs [168]. However, it has to be pointed out that most CNTs are used after treatment in concentrated HNO_3, for example, to exclude any remaining metal nanoparticles from the catalyst used during CVD; the resultant CNTs being metal free, but also oxidized to a certain extent. Effectiveness of surface oxidation of MWCNTs prepared by catalytic CVD has been demonstrated on Cd^{2+}-ion adsorption [160]. Adsorption isotherms are shown for CNTs oxidized in H_2O_2, $KMnO_4$, and HNO_3 in Figure 15.15. The oxidation, particularly that with $KMnO_4$, increases adsorption capacity although surface area and pore volume did not show marked increases. The maximum sorption capacity was 11.0 mg/g for $KMnO_4$-oxidized MWCNTs, compared to 1.1 mg/g for as-grown ones. SWCNTs had a slightly higher adsorption capacity for Zn^{2+} ions than MWCNTs after oxidation in sodium hypochlorite solution [165]. Efficient adsorption and removal of Pb^{2+} ions in water has been performed using CNTs formed on bamboo charcoal [170].

15.6 Capacitive deionization

Capacitive deionization (CDI) is a method for removing dissolved salts from brackish water by adsorbing salt ions onto electrode surfaces to form electric double layers, and is a potential method for desalination of drinking water [171,172]. Since the adsorption mechanism of CDI is the same as that of electric double-layer capacitors for energy storage (Chapter 11), the devices are called flow-through capacitors [172].

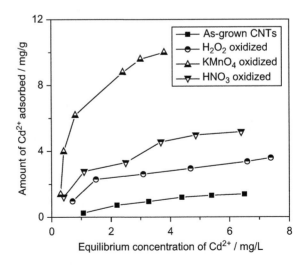

FIGURE 15.15

Adsorption Isotherms of Cd^{2+} Ions in Water for MWCNTs Oxidized Under Different Conditions

From [160]

After the electrodes become saturated, they can easily be regenerated by reversing the potential between the electrodes. The following advantages have been noted: no chemicals are required during the cycles of desalination and regeneration, and the applied voltages are fairly low to avoid the electrolysis of water, which make this method environmentally friendly and energy efficient for water purification, in comparison with traditional techniques, such as thermal distillation, reverse osmosis, and electrodialysis [173]. Thus, the CDI process could play an important role in providing potable water and agricultural water at low cost and without pollution. This process has been reviewed, focusing on the use of ACs prepared through electrospinning as the electrodes [174].

Various carbon electrodes have been tested for CDI: ACs [175,176], ACFs [177–180], templated nanoporous carbons [175,181], carbon aerogels [182,183], MWCNTs [184–188], and graphene [189]. Mesoporous carbon prepared from tetraethyl orthosilicate using triblock copolymer P123 as the template had a much higher adsorption capacity for NaCl than a microporous AC [175]. By Langmuir approximation of the adsorption isotherm measured on MWCNTs in 0.5 mol/L NaCl aqueous solution, the equilibrium adsorption capacity at terminal voltage of 1.2 V was estimated to be 9.35 mg/g [186]. Treatment of an ACF cloth in 1mol/L KOH solution at 90–100 °C improved the kinetics of CDI response [178]. PAN-based electrospun carbon nanofibers have been tested for CDI after steam activation [179]. MWCNT sponge, fabricated by CVD at 860 °C using 1,2-dichlorobenzene as carbon precursor and ferrocene as catalyst precursor, showed a high CDI capacity in NaCl solution of 40 mg/g [188]. In Figure 15.16, adsorption capacities of different carbon materials for Na$^+$ in water are compared, the capacity values being determined under similar conditions.

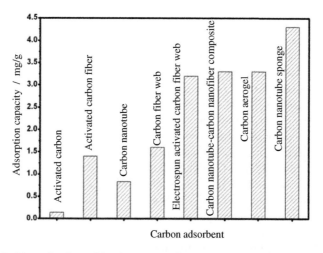

FIGURE 15.16 Adsorption Capacities for Na$^+$ Ions in Aqueous Solution of Various Carbon Materials

TiO_2 nanoparticles have been loaded onto an ACF cloth to improve CDI performance, and this was found to improve adsorption capacity and adsorption/desorption rate [177].

Electrochemical polarization is known to enhance adsorption capacity and rate of AC electrodes (electrosorption). This has been reviewed with emphasis on its application for the conservation and cleaning of the environment, not limiting its application to desalination [174].

15.7 Concluding remarks

Hydrogen storage will be important in efforts to combat the predicted global energy problem, and the US Department of Energy has announced the 2015 benchmark for application to automobile fuel-cell systems as 9.0 mass% at 25 °C under 3.5 MPa. Many papers have reported a storage capacity higher than this benchmark at 77 K, but not at room temperature. The amount of hydrogen that can be adsorbed on porous materials, including carbon materials, at room temperature under high pressure is much lower (<0.5 mass%) than the benchmark. The strong temperature dependence of hydrogen physisorption on porous materials, in addition to a low storage capacity, becomes a high barrier to application of this physisorption for hydrogen storage [28].

For methane, however, porous carbon materials could surpass the benchmark set by the DOE, i.e. 180 v/v at 25 °C under 3.5 MPa. A marked enhancement of CH_4 adsorption into microporous carbons by pre-adsorbed water has been observed. For practical application of porous carbons to CH_4 storage, some technological investigation is required, in addition to the development of low-cost raw materials and production procedures.

CO_2 is known to be adsorbed into micropores, as is CH_4, and the adsorbed amount of CO_2 at room temperature is known to be a measure to evaluate micropores in nanoporous carbons. Work has been done to improve CO_2 adsorptivity of porous carbons, either by controlling carbonization conditions or loading metallic species. Irreversible adsorption of CO_2 into glass-like carbon spheres carbonized in a CO_2 atmosphere [92,94], which is thought to be caused by the formation of ink-bottle-type micropores, is interesting for the application to reduce CO_2 concentration in waste gases by fixation in carbon adsorbents. For practical fixation of CO_2 gas, development of low-cost carbon adsorbents is strongly desired; for example, biomass-derived nanoporous carbon with high fixation ability for CO_2.

Adsorptive fixation of organics in gas phase and in water is becoming important for the remediation of our environment. Since the size of organics and their interactions with the carbon surface vary widely, however, systematic studies on adsorption performance are required, taking into account the pore structure and surface nature of carbon adsorbents. Processes for the treatment of adsorbed organics, particularly hazardous organics, have to be established, and the cyclic performance of the adsorbent carbon has to be studied on the basis of resources conservation.

For the adsorption and removal of heavy metal ions in water, CNTs have been shown to have high performance. It has to be pointed out that the MWCNTs used in most papers are produced by catalytic CVD of different hydrocarbon gases. Their nanotexture is not the same as that of SWCNTs, consisting of small carbon layers aligned preferentially along the tube axis, so that they may be called carbon nanofibers. Since heavy-metal adsorption onto the carbon surface is known to occur mainly owing to interaction between metal ions and functional groups on the carbon surface, the introduction of certain functional groups, particularly acidic groups, on the surface of carbon nanofibers can be effective. Considering the cost of adsorbent CNTs, their repeated use has to be seriously investigated, although only a few papers have reported on cycle performance of CNTs for heavy-metal removal. The use of composites of CNTs with bamboo charcoal is interesting; these can be prepared at low cost and have some ability for heavy metal adsorption [170,190].

Application of carbon materials to capacitive deionization (capacitive desalination) is not yet established. Optimum pore structure of adsorbent carbon has to be explored in more detail with reference to the data on various porous carbon materials for capacitive energy storage, which is reviewed in Chapter 11. For practical application, cycle performance of capacitive deionization also needs to be studied.

For the control of adsorption performance, pore structure in carbon materials has to be controlled, which has been done either by activation or templating. Template carbonization has advantages in controlling pore structure in carbon, giving large numbers of pores that are homogeneous in size, as explained in Chapter 7. However, activation remains an important industrially employed process for fabricating activated carbon materials, as explained in Chapter 5. The activation process can be classified into two, physical activation with steam, CO_2, or air, and chemical activation with chemical agents such as $ZnCl_2$ and KOH. Chemical activation can give higher porosity, 1 cm^3/g in pore volume and 3000 m^2/g in surface area, than physical activation, and presents a much higher yield and less damaged AC surface [191]. Functionalities formed on the surface of ACs are also expected to be different. Therefore, the activation process has to be studied in more detail, including investigation of surface functionalities, to improve the adsorption performance for target adsorbates.

References

[1] Inagaki M. New Carbon Mater 2009;24:193–222.
[2] Chambers A, Park C, Baker RTK, et al. J Phys Chem B 1998;102:4253–6.
[3] Park C, Anderson PE, Chambers A, et al. J Phys Chem B 1999;103:10572–81.
[4] US Deparment of Energy, 2006 Annual Progress Report for the DOE Hydrogen Program, November 2006, available online at: http://www.hydrogen.energy.gov/annual_progress.html
[5] Dillon AC, Jones KM, Bekkedahl TA, et al. Nature 1997;386:377–9.
[6] Liu C, Fan YY, Liu M, et al. Science 1999;286:1127–9.
[7] Ye Y, Ahn CC, Witham C, et al. Appl Phys Lett 1999;74:2307–9.
[8] Fan YY, Liao B, Liu M, et al. Carbon 1999;37:1649–52.

[9] Chen P, Wu X, Lin J, et al. Science 1999;285:91–3.

[10] Yang RT. Carbon 2000;38:623–6.

[11] Rzepka M, Lamp P, de la Casa-Lillo MA. J Phys Chem B 1998;102:10894–8.

[12] Dresselhaus MS, Williams KA, Eklund PC. MRS Bull 1999;24:45–50.

[13] Darkrim F, Levesque D. J Phys Chem B 2000;104:6773–6.

[14] Lee SM, Lee TH. Appl Phys Lett 2000;76:2877–9.

[15] Orimo S, Matsushima T, Fujii H, et al. J Appl Phys 2001;90:1545–9.

[16] Lee SM, Park KS, Choi YC, et al. Synth Met 2000;113:209–16.

[17] Jurewicz K, Frackowiak E, Beguin F. Electrochem Solid State Lett 2001;4:A27–9.

[18] Jurewicz K, Frackowiak E, Beguin F. Appl Phys A 2004;78:981–7.

[19] Qu DY. J Power Sources 2008;179:310–6.

[20] Beguin F, Friebe M, Jurewicz K, et al. Carbon 2006;44:2392–8.

[21] de la Casa-Lillo MA, Lamari-Darkrim F, Cazorla-Amoros D, et al. J Phys Chem B 2002;106:10930–4.

[22] Takagi H, Hatori H, Soneda Y, et al. Mater Sci Eng B 2004;108:143–7.

[23] Zhou L, Zhou Y, Sun Y. Int J Hydrogen Energy 2004;29:475–9.

[24] Zhao XB, Xiao B, Fletcher AJ, et al. J Phys Chem B 2005;109:8880–8.

[25] Texier-Mandoki N, Dentzer J, Piquero T, et al. Carbon 2004;42:2744–7.

[26] Panella B, Hirscher M, Roth S. Carbon 2005;43:2209–14.

[27] Gadiou R, Saadallah S, Piquero T, et al. Microp Mesop Mater 2005;79:121–8.

[28] Thomas KM. Catal Today 2007;120:389–98.

[29] Xu WC, Takahashi K, Matsuo Y, et al. Int J Hydrogen Energy 2007;32:2504–12.

[30] Jorda-Beneyto M, Lozano-Castello D, Suarez-Garcia F, et al. Microp Mesop Mater 2008;112:235–42.

[31] Kunowsky M, Weinberger B, Lamari Darkrim F, et al. Int J Hydrogen Energy 2008;33:3091–5.

[32] Tian HY, Buckley CE, Wang SB, et al. Carbon 2009;47:2128–30.

[33] Wang H, Gao Q, Hu J. J Am Chem Soc 2009;131:7016–22.

[34] Im JS, Park S, Lee Y. Mater Res Bull 2009;44:1871–8.

[35] Fierro V, Szczurek A, Zlotea C, et al. Carbon 2010;48:1902–11.

[36] Kunowsky M, Marco-Lozar JP, Cazorla-Amoros D, et al. Int J Hydrogen Energy 2010;35:2393–402.

[37] Kunowsky M, Marco-Lozar JP, Oya A, et al. Carbon 2012;50:1407–16.

[38] Miyamoto J, Hattori Y, Noguchi D, et al. J Am Chem Soc 2006;128:12636–7.

[39] Liu C, Chen Y, Wu CZ, et al. Carbon 2010;48:452–5.

[40] Vix-Guterl C, Frackowiak E, Jurewicz K, et al. Carbon 2005;43:1293–302.

[41] Xia K, Gao Q, Wu C, et al. Carbon 2007;45:1989–96.

[42] Giraudet S, Zhu Z, Yao X, et al. J Phys Chem C 2010;114:8639–45.

[43] Zheng Z, Gao Q, Jiang J. Carbon 2010;48:2968–73.

[44] Giraudet S, Zhu Z. Carbon 2011;49:398–405.

[45] Nishihara H, Hou PX, Li LX, et al. J Phys Chem C 2009;113:3189–96.

[46] Wang H, Gao Q, Hu J, et al. Carbon 2009;47:2259–68.

[47] Jiang HL, Liu B, Lan YQ, et al. J Am Chem Soc 2011;133:11854–7.

[48] Gogotsi Y, Dash RK, Yushin G, et al. J Am Chem Soc 2005;127:16006–7.

[49] Yushin G, Dash R, Jagiello J, et al. Adv Funct Mater 2006;16:2288–93.

[50] Wang H, Gao Q. Carbon 2009;47:820–8.

[51] Rose M, Kockrick E, Senkovska I, et al. Carbon 2010;48:403–7.

[52] Zielinski M, Wojcieszak R, Monteverdi S, et al. Catal Commun 2005;6:777–83.

[53] Kim HS, Lee H, Han KS, et al. J Phys Chem B 2005;109:8983–6.

[54] Lee YS, Kim YH, Hong JS, et al. Catal Today 2007;120:420–5.

[55] Zielinski M, Wojcieszak R, Monteverdi S, et al. Int J Hydrogen Energy 2007; 32:1024–32.

[56] Li YW, Yang RT. J Phys Chem C 2007;111:11086–94.

[57] Im JS, Kwon O, Kim YH, et al. Microp Mesop Mater 2008;115:514–21.

[58] Zubizarreta L, Menendez JA, Pis JJ, et al. Int J Hydrogen Energy 2009;34:3070–6.

[59] Zubizarreta L, Menendez JA, Job N, et al. Carbon 2010;48:2722–33.

[60] Juan-Juan J, Marco-Lozar JP, Suarez-Garcıa F, et al. Carbon 2010;48:2906–9.

[61] Wang XS, Ma S, Rauch K, et al. Chem Mater 2008;20:3145–52.

[62] Burchell T, Rogers M. SAE Tech Pap Ser 2000:01–2205.

[63] Ma S, Sun D, Simmons JM, et al. J Am Chem Soc 2008;130:1012–6.

[64] Menon VC, Komarneni S. J Porous Mater 1998;5:43–58.

[65] Lozano-Castello D, Alcaniz-Monge J, de la Casa-Lillo MA, et al. Fuel 2002; 81:1777–803.

[66] Cook TL, Komodromos C, Quinn DF, et al. Carbon Materials for Advanced Technologies. In: Burchell TD, editor. Pergamon; 1999. p. 269–302.

[67] Bekyarova E, Murata K, Yudasaka M, et al. J Phys Chem B 2003;107:4681–4.

[68] Yeon SH, Osswald S, Gogotsi Y, et al. J Power Sources 2009;191:560–7.

[69] Kockrick E, Schrage C, Borchardt L, et al. Carbon 2010;48:1707–17.

[70] Oschatz M, Kockrick E, Rose M, et al. Carbon 2010;48:3987–92.

[71] Miyawaki J, Kanda T, Suzuki T, et al. J Phys Chem 1998;102:2187–92.

[72] Zhou L, Li M, Sun Y, et al. Carbon 2001;39:773–6.

[73] Zhou L, Sun Y, Zhou YP. AICHE J 2002;48:2412–6.

[74] Zhou YP, Dai M, Zhou L, et al. Carbon 2004;42:1855–8.

[75] Zhou YP, Wang YX, Chen HH, et al. Carbon 2005;43:2007–12.

[76] Liang MY, Chen GJ, Sun CY, et al. J Phys Chem B 2005;109:19034–41.

[77] Yan LJ, Chen GJ, Pang WX, et al. J Phys Chem B 2005;109:6025–30.

[78] Liu J, Zhou Y, Sun Y, et al. Carbon 2011;49:3731–6.

[79] Khatri RA, Chuang SSC, Soong Y, et al. Energy Fuels 2006;20:1514–20.

[80] Powell CE, Qiao GG. J Membr Sci 2006;279:1–49.

[81] Yang H, Xu Z, Fan M, et al. J Environ Sci 2008;20:14–27.

[82] White CM, Smith DH, Jones KL, et al. Energy Fuels 2005;19:659–724.

[83] Burchell TD, Judkins RR. Energy Convers Manage 1996;37:947–54.

[84] Burchell TD, Judkins RR, Rogers MR, et al. Carbon 1997;35:1279–94.

[85] An H, Feng B, Su S. Carbon 2009;47:2396–405.

[86] Deng HG, Jin SL, Zhan L, et al. New Carbon Mater 2012;27:194–9.

[87] Pevida C, Drage TC, Snape CE. Carbon 2008;46:1464–74.

[88] Hao GP, Li WC, Qian D, et al. Adv Mater 2010;22:853–7.

[89] Przepiorski J, Czyewski A, Pietrzak R, et al. J Therm Anal Calorim 2013;111:357–64.

[90] Przepiorski J, Czyewski A, Kapica J, et al. Chem Eng J 2012;191:147–53.

[91] Moon SH, Shim J W. J Colloid Interface Sci 2006;298:523–8.

[92] Inagaki M, Nakashima M. Carbon 1992;30:1135–6.

[93] Inagaki M, Nakashima M. TANSO 1994; No. 162: 61–65 [in Japanese].

[94] Nakashima M, Shimada S, Inagaki M, et al. Carbon 1995;33:1301–6.

[95] Przepiorski J, Tryba B, Morawski AW. Appl Surf Sci 2002;196:296–300.

[96] Kobori R, Ohab T, Suzuki T, et al. Adsorption 2009;15:114–22.

[97] Inagaki M, Sunahara M, Shindo A, et al. J Mater Res 1999;14:3208–10.

[98] Asai M, Ohba T, Iwanaga T, et al. J Am Chem Soc 2011;133:14880–3.
[99] Okuni T, Sahashi Y, Satsuma A, et al. TANSO 2005; No. 217: 95–98.
[100] Nowicki P, Supłat M, Przepiórski J, et al. Chem Eng J 2012;195–196:7–14.
[101] Benkhedda J, Jaubert JN, Barth D, et al. J Chem Eng Data 2000;45:650–3.
[102] Chiang YC, Chiang PC, Chang EE. J Environ Eng 2001;127:54–62.
[103] Chiang YC, Chiang PC, Huang CP. Carbon 2001;39:523–34.
[104] Chiang HL, Chiang PC, Huang CP. Chemosphere 2002;47:267–75.
[105] Ryu YK, Lee HJ, Yoo HK, et al. J Chem Eng Data 2002;47:1222–5.
[106] Garcia T, Murillo R, Cazorla-Amoros D, et al. Carbon 2004;42:1683–9.
[107] Lillo-Ródenas MA, Cazorla-Amorós D, Linares-Solano A. Carbon 2005;43:1758–67.
[108] Kim MI, Yun CH, Kim YJ, et al. Carbon 2002;40:2003–12.
[109] Kim YJ, Kim MI, Yun CH, et al. J Coll Interface Sci 2004;274:555–62.
[110] Hsieh CT, Chen W Y. Dia Relat Mater 2007;16:1945–9.
[111] Foster KL, Fuerman RG, Economy J, et al. Chem Mater 1992;4:1068–73.
[112] Dimotakis ED, Cal MP, Economy J, et al. Environ Sci Technol 1995;29:1876–80.
[113] Cal MP, Rood MJ, Larson SM. Gas Sep Purif 1996;10:117–21.
[114] Yun JH, Hwang KY, Choi DK. J Chem Eng Data 1998;43:843–5.
[115] Huang ZH, Kang F, Zheng YP, et al. Adsorpt Sci Technol 2002;20(5):495–500.
[116] Park JW, Lee SS, Choi DK, et al. J Chem Eng Data 2002;47(4):980–3.
[117] Singh KP, Mohan D, Tandon GS, et al. Ind Eng Chem Res 2002;41:2480–6.
[118] Huang ZH, Kang F, Zheng YP, et al. Carbon 2002;40:1363–7.
[119] Huang ZH, Kang F, Liang KM, et al. J Hazard Mater 2003;98:107–15.
[120] Dwivedi P, Gaur V, Sharma A, et al. Sep Purif Technol 2004;39:23–37.
[121] Ramirez D, Qi S, Rood MJ. Environ Sci Technol 2005;39:5864–71.
[122] Yi FY, Lin XD, Chen SX, et al. J Porous Mater 2009;16:521–6.
[123] Ramos ME, Bonelli PR, Cukierman AL, et al. J Hazard Mater 2010;177:175–82.
[124] Tavanai H, Jalili R, Morshed M. Surf Interface Anal 2009;41:814–9.
[125] Jimenez V, Diaz JA, Sanchez P, et al. Chem Eng J 2009;155:931–40.
[126] Shim WG, Kim C, Lee JW, et al. J Appl Polym Sci 2006;102:2454–62.
[127] Hsieh CT, Chou YW. Sep Sci Technol 2006;41:3155–68.
[128] Hsieh CT, Chen WY. Diamond Relat Mater 2007;16:1945–9.
[129] Oh GY, Ju YW, Jung HR, et al. J Anal Appl Pyrolysis 2008;81:211–7.
[130] Cuervo MR, Asedegbega-Nieto E, Diaz E, et al. J Chromatogr A 2008;1188:264–73.
[131] Oh GY, Ju YW, Kim MY, et al. Sci Total Environ 2008;393:341–7.
[132] Tsai JH, Chiang HM, Huang GY, et al. J Hazard Mater 2008;154:1183–91.
[133] Liu CH, Li JJ, Zhang HL, et al. Colloid Surf A 2008;313:9–12.
[134] Lee KJ, Shiratori N, Lee GH, et al. Carbon 2010;48:4248–55.
[135] Wang MX, Huang ZH, Kang F, et al. Mater Lett 2011;65:1875–7.
[136] Morawski AW, Inagaki M. Desalination 1997;114:23–7.
[137] Morawski AW, Kalenczuk R, Inagaki M. Desalination 2000;130:107–12.
[138] Brasquet C, Le Cloirec P. Langmuir 1999;15:5906–12.
[139] Tanaike O, Fukuoka M, Inagaki M. Synth Met 2002;125:255–7.
[140] Inagaki M, Sakanishi M. Adsorpt Sci Technol 2003;21:587–95.
[141] Toyoda M, Nanbu Y, Kito T, et al. Desalination 2003;159:273–82.
[142] Chen W, Duan L, Zhu DQ. Environ Sci Technol 2007;41:8295–300.
[143] Czyzewski A, Karolczyk J, Usarek A, et al. Bull Mater Sci 2012;35:211–9.
[144] Karolczyk J, Janus M, Przepiorski J. Pol J Chem Technol 2012:14: 95–99.
[145] Zhu L, Huang ZH, Wen D, et al. J Phys Chem Solids 2010;71:704–7.

[146] Monser L, Adhoum N. Sep Purif Technol 2002;26:137–46.

[147] Faur-Brasquet C, Reddad Z, Kadirvelu K, et al. Appl Surf Sci 2002;196:356–65.

[148] Kadirvelu K, Kavipriya M, Karthika C, et al. Bioresour Technol 2003;87:129–32.

[149] Goel J, Kadirvelu K, Rajagopal C, et al. J Hazard Mater 2005;125:211–20.

[150] Issabayeva G, Aroua MK, Sulaiman NMN. Bioresour Technol 2006;97:2350–5.

[151] Issabayeva G, Aroua MK, Sulaiman NM. J Hazard Mater 2008;155:109–13.

[152] Imamoglu M, Tekir O. Desalination 2008;228:108–13.

[153] Rao MM, Ramana DK, Seshaiah K, et al. J Hazard Mater 2009;166:1006–13.

[154] Zhu J, Yang J, Deng B. Environ Chem Lett 2010;8:277–82.

[155] Momcilovic M, Purenovic M, Bojic A, et al. Desalination 2011;276:53–9.

[156] Leyva-Ramos R, Berber-Mendoza MS, Salazar-Rabago J, et al. Adsorption 2011;
17:515–26.

[157] Li YH, Wang SG, Wei JQ, et al. Chem Phys Lett 2002;357:263–6.

[158] Chen JP, Wu SN, Chong KH. Carbon 2003;41:1979–86.

[159] Li YH, Wang S, Luan Z, et al. Carbon 2003;41:1057–62.

[160] Li YH, Ding J, Luan ZK, et al. Carbon 2003;41:2787–92.

[161] Liang P, Liu Y, Guo L, et al. J Anal At Spectrom 2004;19:1489–92.

[162] Li YH, Di ZC, Ding J, et al. Water Res 2005;39:605–9.

[163] Li YH, Zhu YQ, Zhao YM, et al. Diamond Relat Mater 2006;15:90–4.

[164] Lu C, Liu C. J Chem Technol Biotechnol 2006;81:1932–40.

[165] Lu C, Chiu H. Chem Eng Sci 2006;61:1138–45.

[166] Lu C, Chiu H, Liu C. Ind Eng Chem Res 2006;45:2850–5.

[167] Chen C, Wang X. Ind Eng Chem Res 2006;45:9144–9.

[168] Rao GP, Lu C, Su F. Sep Purif Technol 2007;58:224–31.

[169] Stafiej A, Pyrzynska K. Sep Purif Technol 2007;58:9–52.

[170] Huang ZH, Zhang F, Wang MX, et al. Chem Eng J 2012;184:193–7.

[171] Welgemoed TJ, Schutte CF. Desalination 2005;183:327–40.

[172] Oren Y. Desalination 2008;228:10–29.

[173] Anderson MA, Cudero AL, Palma J. Electrochim Acta 2010;55:3845–56.

[174] Foo KY, Hameed BH. J Hazard Mater 2009;170:552–9.

[175] Zou LD, Li LX, Song HH, et al. Water Res 2008;42:2340–8.

[176] Zou L, Morris G, Qi D. Desalination 2008;225:329–40.

[177] Ryoo MW, Seo G. Water Res 2003;37:1527–34.

[178] Ahn HJ, Lee JH, Jeong Y, et al. Mater Sci Eng A 2007;449:841–5.

[179] Wang M, Huang ZH, Wang L, et al. New J Chem 2010;34:1843–5.

[180] Wang G, Pan C, Wang L, et al. Electrochim Acta 2012;69:65–70.

[181] Mayes RT, Tsouris C, Kiggans JO, et al. J Mater Chem 2010;20:8674–8.

[182] Farmer JC, Fix DV, Mack GV, et al. J Electrochem Soc 1996;143:159–69.

[183] Gabelich CJ, Tran TD, Suffet IH. Environ Sci Technol 2002;36:3010–9.

[184] Zhang DS, Shi LY, Fang JH, et al. Mater Chem Phys 2006;97:415–9.

[185] Wang XZ, Li MG, Chen YW, et al. Appl Phys Lett 2006;89:53127.

[186] Wang S, Wang DZ, Ji LJ, et al. Sep Purif Technol 2007;58:12–6.

[187] Li HB, Gao Y, Pan LK. Water Res 2008;42:4923–8.

[188] Wang L, Wang M, Huang ZH, et al. J Mater Chem 2011;21:18295–9.

[189] Li HB, Lu T, Pan LK, et al. J Mater Chem 2009;19:6773–9.

[190] Zhang JN, Huang ZH, Lv R, et al. Langmuir 2009;25:269–74.

[191] Macia-Agullo JA, Moore BC, Cazorla-Amoros D, et al. Carbon 2004;42:1367–70.

Highly Oriented Graphite with High Thermal Conductivity

Graphite has a very high thermal conductivity along the layer plane and, coupled with its light weight, this is expected to lead to promising applications in heat dissipation for electronic micro-devices. In Table 16.1, thermal conductivity at room temperature, κ_{RT}, is listed for various materials. Graphite has exceptionally high κ_{RT} along the a-axis (along the layer plane), much higher than metals, like Cu and Al. On the other hand, graphite has very small κ_{RT} perpendicular to the film, as shown in Table 16.1, almost comparable to ceramics like glass, and it makes these films possible to use as a heat-insulating materials, particularly at high temperatures. This pronounced anisotropy is due to the graphite structure: graphite layers consisting of strong σ bonding, and parallel stacking of these layers by van der Waals bonding between π electron clouds. Natural diamond is isotropic and has the highest value of κ_{RT} in solids at about 2500 W/m·K.

Highly graphitized carbon materials exhibit very high thermal conductivities; a highly oriented pyrolytic graphite (HOPG) is reported to have κ_{RT} of 1950 W/m·K and a single crystal of graphite with a density of 2.265×10^3 kg/m^3 has about 2000 W/m·K [1–3]. Graphite films with high thermal conductivity are now commercially available for radiating heat from electronic devices containing integrated circuits, such as personal computers and mobile phones. The κ_{RT} of the commercially available graphite films reaches 1600 W/m·K, which is about four times higher than the values for Ag and Cu. In order to develop practical materials with sufficiently high κ, the preparation of highly oriented and highly crystallized graphite is essential. Highly oriented and highly crystallized graphite films insulate electromagnetic waves effectively because the electrical conductivity along the longitudinal direction of the films reaches 1–2×10^6 S/m at room temperature.

In this chapter, the preparation and characterization of highly oriented and highly crystallized graphite films from pyrolytic carbon and aromatic polyimides are briefly reviewed. Carbon materials with high thermal conductivity are discussed: highly oriented and highly crystallized graphite films, graphitized mesophase-pitch-based carbon fibers, and sintered diamond and diamond-like carbon, with reference to single crystals of graphite and diamond.

Prerequisite for readers: Chapter 3.2 (Highly oriented graphite) in *Carbon Materials Science and Engineering: From Fundamentals to Applications*, Tsinghua University Press.

Table 16.1 Thermal Conductivity at Room Temperature, κ_{room}, for Various Materials

Material	κ_{room} (W/m K)	Material	κ_{room} (W/m K)
Cu	398	Glass	~1
Al	236	Polyethylene	0.41
Graphite along a-axis	1950	Wood	0.15–0.25
Graphite along c-axis	6.9	Air	0.0241

16.1 Preparation

Highly oriented graphite flakes with irregular shape have been found in natural graphite and kish graphite [4,5]. However, it is very difficult to find a flake that is a single crystal, and so they are unavailable as a high-thermal-conductivity material. Highly oriented graphite films and plates with controlled size and thickness have been prepared from natural graphite via its exfoliation and compression, from pyrolytic carbons annealed under high temperature and pressure, and from various organic precursor films via controlled carbonization and graphitization [4,5].

Graphite films are prepared from natural graphite powder through its conversion either to graphite oxide or residue compounds of sulfuric acid, exfoliation by rapid heating to a high temperature, and then compression without any binding material. The graphite films thus prepared are widely used as seals and packings in various industries [6]. However, thermal conductivity of these graphite films is usually not satisfactorily high because the grains of constituent natural graphite are irregular and in contact only mechanically with each other.

Highly oriented and highly crystallized graphite plates have been prepared by heat treatment of pyrolytic carbons at high temperature, above 3000 °C, under compression, and commercialized as HOPG [7]. Their characteristics, including structural perfection, orientation of crystallites, electronic properties, and thermal conductivity, have been studied in detail and reviewed [7].

Graphite films with high crystallinity have been fabricated by heat treatment of various organic precursors, such as poly(p-phenylene vinylene) (PPV) [8], poly(p-phenylene-1,3,4-oxadiazole) (POD) [9–11], aromatic polyimides [12–14], and benzimidazobenzophenanthroline ladder (BBL) polymer [15]. Carbonization and graphitization of aromatic polyimides have been studied, focusing on molecular structure of starting polyimides, forming conditions of polyimide films, and carbonization and graphitization conditions, and preparation of high-quality graphite films from polyimides has been reviewed [16]. In order to achieve high orientation degree and high crystallinity in the resultant graphite films, heat treatment above 3000 °C under a simple mechanical constraint between artificial graphite blocks has been reported [3,17,18]. Heating to high temperatures with a step at around 2200 °C has been recommended for avoiding deformation of the film due to pore formation caused by the departure of nitrogen [19].

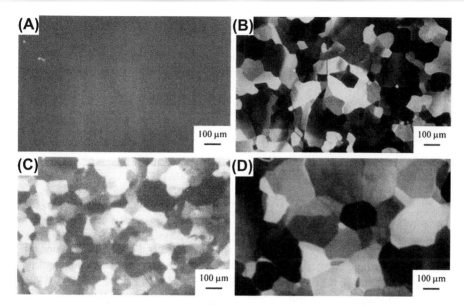

FIGURE 16.1 Channeling Contrast Images of Different Highly Oriented and Highly Crystallized Graphite Films

(A) Kish graphite, (B) HOPG, (C) Kapton graphitized at 3100 °C, and (D) PTT graphitized at 3200 °C

Courtesy of Prof. A. Yoshida of Tokyo City University

In Figure 16.1, electron channeling contrast images of polyimide films Kapton (pyromellitic dianhydride (PMDA) with 4,4′-oxydianiline (ODA): PMDA/ODA) and PPT (PMDA with *p*-phenylenediamine (PPD) and 3,30,4,40-tetraaminobiphenyl (TAB): PMDA/(PPD + TAB)), heat-treated at 3100 or 3200 °C are compared with those of kish graphite and HOPG [17,18]. Kish graphite has no crystallographic contrast because it consists of large grains, larger than the observation field (Figure 16.1A). Films derived from polyimides, Kapton and PTT, consist of grains with different orientations of *a*-axes (Figures 16.1C and 16.1D), like HOPG (Figure 16.1B). These channeling contrast images suggest that polyimide films such as Kapton and PTT can provide films comparable to HOPG by heat treatment at high temperatures above 3100 °C under mechanical constraint.

Graphite plates as large as $150 \times 50 \times 13$ mm^3 have been fabricated by heat treatment of a sheaf of 1500 films of Kapton of 25 μm thick at 2800–3000 °C under 10–30 MPa, as shown in Figure 16.2 [20]. The mosaic spread of the block, which was a measure of preferred orientation of crystallites along the plane of the block, was improved to 0.45° and thermal conductivity at room temperature along the plate reached more than 1000 W/m·K.

FIGURE 16.2 Graphite Blocks Prepared from Polyimide Kapton Under Hot-pressing

From [20]

16.2 Characterization

The development of graphitic structure is usually evaluated by interlayer spacing, d_{002}, determined by X-ray diffraction. Raman spectroscopy is also used, by calculating the intensity ratio of D-band to G-band, I_D/I_G. In Figure 16.3A, I_D/I_G is plotted against d_{002} for Kapton heat-treated at 2200–3400 °C [21]. With increasing heat-treatment temperature (HTT), d_{002} decreases, approaching the value for graphite (0.3354 nm), and I_D/I_G decreases rapidly (I_D/I_G is on a logarithmic scale) in direct relation to d_{002}. However, the D-band in the Raman spectrum disappears below d_{002} of about 0.336 nm, suggesting I_D/I_G is not a useful parameter to evaluate highly crystalline graphite. Full width at half maximum intensity of the G-band (G-FWHM) also decreases rapidly with increasing HTT and decreasing d_{002}, as shown in Figure 16.3B. These results suggest that there is a step for the further improvement in structural perfection below d_{002} of 0.337 nm, and also that even G-FWHM is not an appropriate parameter to evaluate structural perfection in highly crystallized graphite, as change in G-FWHM is limited to 13 to 15 cm^{-1}.

Crystallite orientation and crystallinity were characterized by measurement of X-ray diffraction and galvanomagnetic properties on various highly oriented and highly crystallized graphite films—kish graphite, HOPG, films prepared from polyimides, PPT, Kapton and Novax (PMDA/(ODA + PTD)), and pyrolytic graphite (PG-3200)—for which heat-treatment conditions are described in Table 16.2, together with various structural parameters [18].

Ratio of the in-plane electrical resistivity at room temperature to that at 4.2 K ($\rho_{RT}/\rho_{4.2K}$ or residual resistivity ratio, RRR) is used as a measure of the crystallinity of these highly crystallized graphite materials [22]. Two films derived from PTT and Kapton by heat treatment at 3200 °C under simple mechanical constraint gave high crystallinity as determined by $\rho_{RT}/\rho_{4.2K}$, comparable to that of kish graphite and HOPG, which were subjected to more severe thermal conditions, as explained in Table 16.2. Mosaic spread, MS, maximum magnetoresistance, $(\Delta\rho/\rho)_{max}$, and mean free path of carriers, λ, for these films proved high crystallinity. Even for PTT and Kapton, however, heat treatment at 3200 °C for just 10 min or at lower temperatures

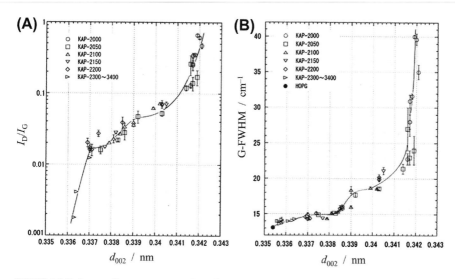

FIGURE 16.3 Raman Parameters as a Function of Interlayer Spacing, d_{002}, for Kapton Films Heat-treated at 2000–3400 °C

(A) Intensity ratio, I_D/I_G, and (B) full width at half maximum intensity of G-band, G-FWHM

From [21]

could not give high crystallinity. For another polyimide, Novax, and pyrolytic carbon, high crystallinity comparable to HOPG was not attained by heat treatment at 3200 °C.

The peak intensity of the 002 X-ray diffraction is measured as a function of rotation angle, φ, of the film (rocking curve), and is shown in Figure 16.4. The MS, defined as the full width at half maximum intensity of these rocking curves, reflects directly the preferred orientation of the c-axes of the crystal grains in the film. The rocking curves observed for HOPG and two films derived from PPT are smooth and symmetrical at $\varphi = 0$, where the maximum intensity is observed. However, the rocking curve for kish graphite is irregular and unsymmetrical, probably owing to the fact that it is an aggregation of large single-crystal domains, larger than the area seen in Figure 16.1A, with different thicknesses. Therefore, it gives a relatively high value of MS, higher than that of HOPG, as listed in Table 16.2. The measurement of a cleaved kish graphite flake with a thickness of about 40 μm still gave an irregular and unsymmetrical rocking curve, but a much smaller MS value of 0.5°, smaller than HOPG, was measured.

Maximum magnetoresistance, $(\Delta\rho/\rho)_{max}$, is plotted against magnetic field at 77 and 4.2 K in Figure 16.5, and the values at a field of 1 T are listed in Table 16.2. From the measurement of $(\Delta\rho/\rho)_{max}$ it is concluded that high crystallinity can be obtained by heat treatment of PPT and Kapton films under mechanical constraint. In Figure 16.5B, the magnetic-field dependence measured on another kish graphite flake with higher crystallinity ($\rho_{RT}/\rho_{4.2K} = 11.9$; KG-2) is also shown for comparison, which has slightly better crystallinity than PPT-3100–3200 (the $(\Delta\rho/\rho)_{max}$ scale for KG-2 is

Table 16.2 Structural and Galvanomagnetic Parameters for Highly Oriented and Highly Crystallized Graphite Films

Sample	Heat-treatment Conditions	$\rho_{RT}/\rho_{4.2K}$	MS (degrees)	$(\Delta\rho/\rho)_{max}$ at 1 T		λ (μm)
				77 K	4.2 K	
Kish graphite	Residue after vaporization of Fe	4.65	2.2 (0.5)	12.25	75.32	3.2
HOPG	3600 °C under 30 MPa	5.50	0.9	13.94	112.52	5.4
PPT-3200	3200 °C for 10 min	3.45	5.7	12.06	57.91	2.6
PPT-3100–3200	3100 °C and 3200 °C for 23 min	4.90	2.3	16.21	158.3	6.1
Kapton-3100	3100 °C for 17 min	3.32	6.7	12.54	71.80	3.5
Kapton-3100–3200	3100 °C and 3200 °C for 23 min	4.79	1.8	17.26	–	6.5
Novax-3100	3100 °C for 17 min	2.67	6.9	8.72	–	–
PG-3200	3200 °C for 7 min	1.60	8.6	3.38	15.75	1.5

$(\Delta\rho/\rho)_{max}$, maximum magnetoresistance; λ, mean free path; $\rho_{RT}/\rho_{4.2K}$, residual resistivity ratio; MS, mosaic spread; PG, pyrolytic carbon
From [18]

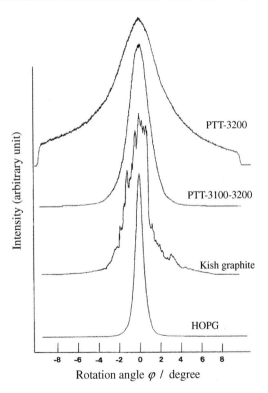

FIGURE 16.4 Rocking Curves Determined from OO2 Diffraction Intensity as a Function of Rotation Angle, φ

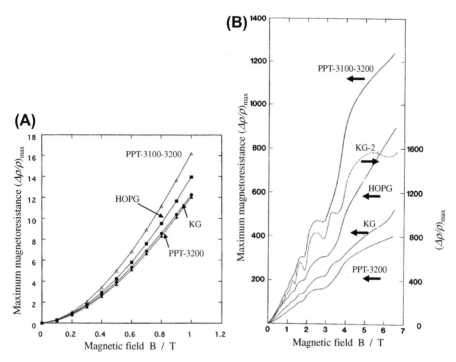

FIGURE 16.5 Magnetic-field Dependence of Maximum Magnetoresistance, $(\Delta\rho/\rho)_{max}$

(A) Low magnetic-field range at 77 K, and (B) high magnetic-field range at 4.2 K. KG, kish graphite

From [18]

different from the other films). The dependence at 4.2 K shows Shubnikov–de Haas oscillation, of which the amplitude decays considerably from KG-2 to PPT-3200, suggesting a decrease in crystal perfection.

The λ for electrical conduction is estimated from $(\Delta\rho/\rho)_{max}$ at 4.2 K in a field of 0.1 T and listed in Table 16.2. The λ values are consistent with those of $\rho_{RT}/\rho_{4.2K}$ and $(\Delta\rho/\rho)_{max}$.

In Table 16.2, the parameters for the films derived from three different poly-imides, PPT, Kapton, and Novax, and also pyrolytic carbon by the same heat treatment, are listed [3]. The results show that, in order to have highly oriented and highly crystallized graphite films, the molecular structure of the precursor polyimide has to be selected; under mild mechanical constraint at a high temperature of 3200 °C, PPT and Kapton gave films comparable to, and/or better orientation and crystal perfection than HOPG, but Novax did not. Even for PPT and Kapton, severe heat-treatment conditions, such as at 3200 °C for more than 20 min, are necessary. Pyrolytic carbon could not give marked improvement in orientation and crystallinity by heat treatment at 3200 °C under mechanical constraint. For pyrolytic carbon, heat treatment under much more severe conditions is needed to improve crystallinity.

In Table 16.3, parameters including electrical conductivity are compared for the films prepared from PPT with different thicknesses, 18 and 45 μm [17]. PPT films less than 45 μm thick can give almost the same characteristics, preferred orientation of crystallite, structural perfection, and electrical properties, but heat-treatment conditions govern the resultant characteristics, heat treatment for a long time at a high temperature of 3200 °C being preferable. Detailed parameters including phonon and carrier mean free paths were determined for the highly oriented graphite films by raising HTT to 3400 °C [23]. Among the different parameters of structural, galvanomagnetic, and thermal properties, clear relationships were obtained for various highly oriented graphite films. Electrical conductivity along the film, σ_{RT}, is related to d_{002} and ρ_{RT}/ρ_{77K}, as shown in Figure 16.6.

Properties of the highly oriented and highly crystallized graphite plates shown in Figure 16.2 are compared with those of HOPG and single-crystal graphite in Table 16.4 [20]. The performance of this graphite plate in applications for radiation optics, as an X-ray monochrometer and neutron filter, is almost comparable to the highest grade HOPG.

16.3 Carbon materials with high thermal conductivity
16.3.1 Pyrolytic graphite

Thermal conduction in graphite is due to lattice vibration, phonon, in contrast to electric conduction which is due to free electrons in metals. Thermal conductivity, κ, is known to be proportional to mean free path, l, of lattice vibration, l being governed by the crystallite boundaries and structural defects, and also by temperature, T. κ of well-crystallized graphite is known to depend on T^2 at low temperatures and on $1/T$ at high temperatures. For various graphitized materials, κ has been measured at low temperatures below 100 K [24,25]. Theoretically, κ is expected to give a maximum at a temperature where l governed by the scattering due to crystallite imperfections is comparable to l governed by temperature. For well-graphitized carbon block, κ values measured by different methods along and perpendicular to the extrusion direction showed maxima at around 400 K [26]. For carbon materials with low crystallinity, however, κ is governed by the scattering at the crystallite boundaries and defects, and accordingly depends strongly on the preparation conditions. Crystallite size, La, and κ of a coke increased with increasing HTT. Only a block graphitized at 3100 °C, of which La became more than 100 nm, showed a broad maximum [27].

In Figure 16.7A, temperature dependence of κ for various pyrolytic graphites, which are more or less oriented and less graphitized, is reproduced and compared with that of natural graphite [1,2]. κ increases quickly with increasing temperature and tends to show a maximum. The temperature giving a maximum in κ depends strongly on the pyrolytic graphite sample, which is considered to be related to crystallite size [1]. Temperature dependence of κ for highly oriented and highly crystallized graphite prepared from pyrolytic carbon by heat treatment at 3250 °C is shown in Figure 16.7B [2]. Above 10 K, κ exhibits $T^{2.7}$-type dependence, and reaches a

Table 16.3 Structural and Galvanomagnetic Parameters for Graphite Films Prepared from PPT with Different Thicknesses

Thickness of PPT (μm)	Heat-treatment Conditions	$\sigma \times 10^{-6}$(S/m)		$\rho_{RT}/\rho_{4.2K}$	MS (degrees)	$(\Delta\rho/\rho)_{max}$ at 1 T		λ (μm)
		300 K	77 K			77 K	4.2 K	
18	3100 °C for 40 min and 3200 °C for 8 and 15 min	1.8	2.5	4.24	1.7	14.12	–	5.7
45	3200 °C for 10 min	–	–	3.45	5.7	12.06	57.91	2.6
	3100 °C for 40 min and 3200 °C for 8 and 15 min	1.9	2.7	4.90	2.3	16.21	158.3	6.1

$(\Delta\rho/\rho)_{max}$ maximum magnetoresistance; λ, mean free path; $\rho_{RT}/\rho_{4.2K}$, residual resistivity ratio; σ, electrical conductivity; MS, mosaic spread
From [17]

Table 16.4 Properties of Graphite Plates Prepared from Kapton at 3000 °C for 1 h Under 10–30 MPa, in Comparison with HOPG and Single-crystal Graphite

	d_{002} (nm)	Density (g/cm³)	σ_{RT} (Ωcm)$^{-1}$		κ_{RT} (W/m·K)		CTE$\times 10^{-6}$	
			//	$\perp$	//	$\perp$	//	$\perp$
Graphite plate prepared from Kapton	0.3354–0.3358	2.24–2.25	20,000–23,000	5–6	>1000	5	–1.0	27
HOPG	0.3354–0.3359	2.250–2.266	22,000–28,600	4.0–6.6	1600–2000	7–9	–1.0	25
Single crystal	0.3354	2.267	25,000	1–100	200–500	40–80	–1.5	27

// and $\perp$, parallel and perpendicular to the plane of the sheets and plates, respectively; κ_{RT}, thermal conductivity at room temperature; σ_{RT}, electrical conductivity at room temperature; d_{002}, interlayer spacing; CTE, coefficient of thermal expansion
From [20]

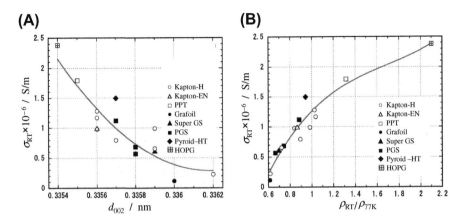

FIGURE 16.6 Relationship Between Structural Parameters and Galvanomagnetic Properties

(A) σ_{RT} against d_{002}, and (B) σ_{RT} against ρ_{RT}/ρ_{77K}. Most of samples are commercially available and expressed by using their trade names, refer the text (16.3.2)

Courtesy of Prof. Y. Hishiyama of Tokyo City University

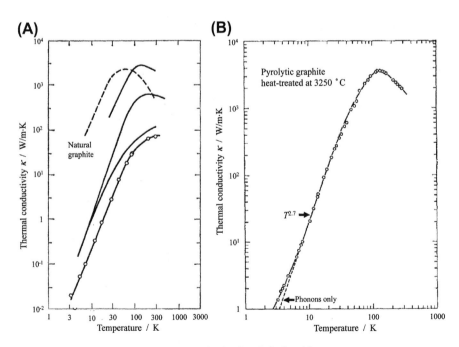

FIGURE 16.7 Temperature Dependence of κ for Pyrolytic Graphite

(A) Various pyrolytic graphite plates [1], and (B) pyrolytic graphite heat-treated at 3250 °C

(A) from [1], (B) from [2]

maximum of about 3000 W/m·K at around 120 K. These results suggest that, in order to realize high thermal conductivity along the layer plane of graphite, highly oriented and highly crystallized graphite has to be developed.

Commercially available pyrolytic graphite sheet with high thermal conductivity has been used in a fuel cell to reduce the weight, volume, and costs [28].

16.3.2 Polyimide-derived graphite

Thermal conductivity of various highly oriented and highly crystallized graphite films prepared from polyimides was determined at room temperature in air by a steady-state method and discussed in comparison with commercially available graphite films [29]. The graphite films used were commercially available Grafoil of 127 μm thickness (eGRAF SS300-0.13; Graf Tech Inc.), Super Graphite Sheet (Super GS) of 100 μm, and PGS graphite films of 25, 70, and 100 μm (Super GS, PGS25, PGS70, and PGS100; Panasonic Electronic Devices Co., Ltd), and Pyroid-HT of 1.65 mm (MINTEQ Co., Ltd), in addition to laboratory-made films prepared from polyimides, Kapton of 25 μm (Kapton-H and Kapton-EN; Toray-DuPont Co., Ltd) and PPT of 45 μm (Toho Rayon Co., Ltd), at high temperatures above 3000 °C. Observed κ was calibrated by referring to the κ-values measured for Cu and Al.

The κ_{RT} was reported to be related to σ_{RT}, although the former is governed by phonons but the latter by electrons. In Figure 16.8, κ_{RT} for the graphite films mentioned above is plotted against σ_{RT}, adding the experimental points for HOPG and a single crystal of graphite [29]. In Figure 16.8, the relationship reported for various artificial graphite materials with La larger than 20 nm [30] is also shown, by extrapolating to 900 W/m·K, for comparison. The values of κ_{RT} are much higher than the values reported for polycrystalline artificial graphite materials, because in the former the contribution of layer plane with high thermal conductivity is almost 100% owing to the highly oriented texture, but in the latter it is limited depending on the orientation of crystallites. The mean free path of phonons at room temperature becomes insensitive to the mean crystallite size, because the crystallite size exceeds the intrinsic mean free path determined by phonon-phonon interaction: this is the reason why the value of κ_{RT} tends to saturate in the extremely high crystallinity region, that is the high κ_{RT} and σ_{RT} region in Figure 16.8.

By intercalation of $FeCl_4^-$ ions into the graphite films derived from Kapton, electrical conductivity increased along the film surface by about one order of magnitude, but κ_{RT} decreased from 950 to 340 W/m·K [31]. The intercalation reaction inevitably introduces some structural defects, depending on the intercalate and the intercalation condition. The $FeCl_4^-$ can be intercalated into the graphite gallery by electrolysis at room temperature and the resultant intercalation compounds are stable in air, but the intercalate is a large ion and so a large amount of structural defects are thought to be introduced in graphite layers; as a consequence, a large decrease in κ_{RT} is observed.

In Figure 16.9, κ_{RT} is plotted against d_{002} and resistivity ratio ρ_{RT}/ρ_{77K} [23]. The relationship of κ_{RT} to d_{002} shows scattering but κ_{RT} shows definite dependence on ρ_{RT}/ρ_{77K} (Figure 16.9A). Between κ_{RT} and maximum magnetoresistance,

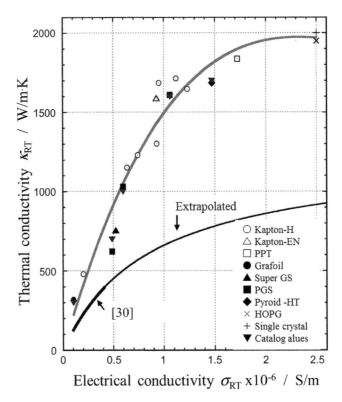

FIGURE 16.8 Relationship Between κ_{RT} and σ_{RT} for Highly Oriented and Highly Crystallized Graphite Films

Courtesy of Prof. Y. Kaburagi of Tokyo City University

$(\Delta\rho/\rho)_{max}$, at 77 K and 1 T, a similar relationship was observed [29]. Since both ρ_{RT}/ρ_{77K} and $(\Delta\rho/\rho)_{max}$ are known to be parameters representing the degree of graphitization or crystallinity, the dependence of κ_{RT} on these two parameters shows the same trend as shown in Figure16.9B and Figure 16.8: κ_{RT} increases rapidly with increasing crystallinity and tends to saturate in the high crystallinity region. For single-crystal graphite, ρ_{RT}/ρ_{77K} was reported to be 2.1 [32], and so the saturating κ_{RT} in the high-crystallinity region is confirmed in Figure 16.9B.

16.3.3 Natural graphite and its composites

Highly crystalline natural graphite flakes are an important raw material for making materials with high thermal conductivity. However, processing of the flakes to films and plates without loss of high thermal conductivity along the layer plane—in other words, how to orient the flakes in the film or plate—is an engineering challenge.

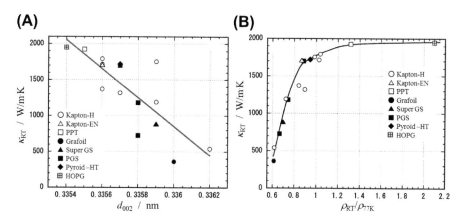

(A) ... **(B)**

FIGURE 16.9 Thermal Conductivity, κ_{RT}, of Highly Oriented Graphites

As a function of (A) d_{002} and (B) resistivity ratio ρ_{RT}/ρ_{77K}

From [23]

Natural graphite is formed into thin sheets after conversion to exfoliated graphite, and the sheets are widely used as packing and sealing materials [6]. Graphite films prepared by uniaxial compression of exfoliated graphite without any binder showed marked anisotropy in κ_{RT}, 500 W/m·K along the sheet and 6.5 W/m·K perpendicular to the sheet, with bulk density of 1.8 g/cm³ [33]. At present, graphite films with κ_{RT} values of about 320 W/m·K are commercially available.

Natural graphite with a flake size of more than 495 μm mixed with 14 mass% mesophase pitch was hot-pressed at 500 °C under 10 MPa to form into a graphite block with a size of 80 × 40 × 20 mm³ [34]. In order to have better orientation of graphite flakes in the block, natural graphite and pitch powder were mixed in a ball-milling jar by adding water and isopropyl alcohol to form slurry. To achieve low electrical resistivity and high thermal conductivity, heat treatment above 2800 °C was needed, as shown in Figure 16.10. The block after heat treatment at 2800 °C had low ρ_{RT} of 1.45 μΩm and high κ_{RT} of 522 W/m·K in the direction perpendicular to the hot-pressing, as well as high thermal diffusivity, 396 mm²/s. On a consolidated block of exfoliated graphite with a bulk density of 0.83 g/cm³, κ_{RT} was reported to be 337 W/m·K [35].

Natural graphite flakes with different flake sizes were hot-pressed under 30 MPa at 2700 °C with mesophase pitch containing 4 mass% Si and 12 mass% Ti: the flake size of 246 μm gave the highest bulk density of 2.26 g/cm³ and the highest κ_{RT} value of 654 W/m·K [36]. The increase in hot-pressing temperature to 3000 °C resulted in an increase in κ_{RT} to 704 W/m·K. The addition of Ti and Si into petroleum coke powder with coal-tar pitch followed by calcination and graphitization was found to be effective to increase κ_{RT} of the graphitized blocks [37,38].

Composites of natural graphite flakes with polymer and metal were prepared and their thermal conductivity measured [39–44]. Slurry of natural graphite

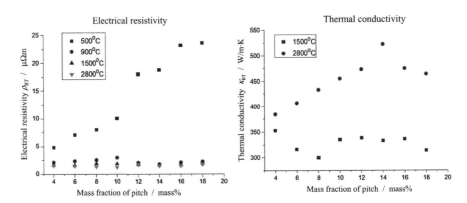

FIGURE 16.10 Dependence of ρ_{RT} and κ_{RT} of Natural Graphite/Mesophase Pitch Composites as a Function of HTT

From [34]

flakes (mean size 3.57 μm) and an ethanol solution of poly(vinyl butyral) was cast on tapes, revealing that blade height for casting governed preferred orientation of graphite flakes, and consequently κ_{RT} of the tape [42]. Addition of a small amount of graphene (10 mass%) into natural graphite/polymer composites was shown to improve thermal conductivity to about 500 W/m·K [44]. Graphite films, which were prepared from exfoliated graphite by further alcohol and oxidative acid treatment and had thickness of 2–8 nm, were mixed with epoxy in hot *N,N*-dimethylformaldehyde (DMF) via suspension, and then cast on an etched glass to obtain nanocomposites [45].

16.3.4 Carbon fibers

Mesophase-pitch-based carbon fibers (MPCFs) have been studied as a highly thermally conductive material, because of the high degree of crystallite orientation along the fiber axis and also because of graphitization. In Figure 16.11, κ_{RT} is plotted against ρ_{RT} for commercially available MPCFs and laboratory-made ones [46]. A well-defined relationship between κ_{RT} and ρ_{RT} is obtained, as seen with other carbon materials (Figure 16.8). κ_{RT} values higher than 1000 W/m·K have been reported for MPCFs [47].

Ribbon-shaped MPCFs are particularly interesting for obtaining high thermal conductivity, because of their specific nanotexture in cross-section, as illustrated in Figure 16.12A [48]. A high graphitization degree, large crystallite size, and high degree of orientation of crystallites perpendicular to the ribbon surface were realized by using naphthalene-based mesophase pitch with a specially designed nozzle for melt spinning and by high-temperature treatment [49]. Spinning temperature was shown to be important to achieve a high graphitization degree and high thermal conductivity along the fiber axis at room temperature, as shown in Figure 16.12B [50].

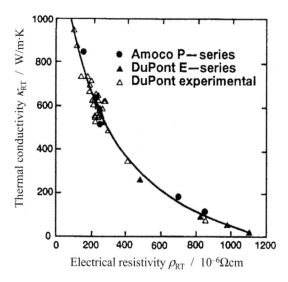

FIGURE 16.11 Relationship Between κ_{RT} and ρ_{RT} for MPCFs

From [46]

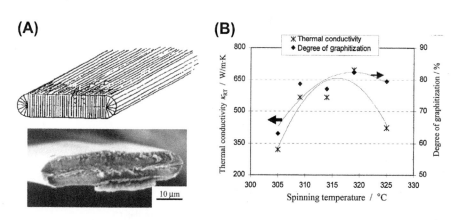

FIGURE 16.12 Ribbon-shaped MPCFs

(A) Illustration of nanotexture and scanning electron microscopy image, and (B) changes in κ_{RT} and graphitization degree with spinning temperature

From [50]

The ribbon-shaped fibers gave higher κ than ones with a round cross-section [51]. Ribbon-shaped mesophase-pitch fibers after stabilization were aligned in one direction and hot-pressed at 2400 °C under 30 MPa, followed by 3000 °C treatment, to form plates 27 μm wide and 5.7 μm thick [52]. The plates obtained showed marked anisotropy in thermal conductivity, along the fiber axis in the plate about 840 W/m·K, but perpendicular to the fiber axis in the plate only 11 W/m·K, and perpendicular to the plate 68 W/m·K [53]. In composites prepared from mesophase-pitch-based carbon fibers with high κ_{RT} of about 500 W/m·K and a resol-type phenol resin via a conventional process, the addition of 7 mass% multi-walled carbon nanotubes (MWCNTs) heat-treated at 2800 °C was found to be effective to improve κ_{RT} to 393 W/m·K [54].

On vapor-grown carbon-fiber mats (sheets), the changes in κ_{RT} with volume fraction of fibers and by heat treatment at 3000 °C were measured [55], as shown in Figure 16.13. The sheets (buckypaper) after graphitization at 3000 °C showed the highest κ_{RT} of about 150 W/m·K along the sheet, but 4 W/m·K across the sheet, at a fiber fraction of about 65 vol%.

16.3.5 Carbon nanotubes and graphene

Extremely high κ values of 6600 W/m·K were predicted for single-walled carbon nanotubes (SWCNTs) from theoretical calculations [56]. Measurement of single SWCNTs at 300–800 K showed a κ-value of 3500 W/m·K at room temperature [57]. Thermal and electrical transport properties were measured on aligned thin films of SWCNT ropes deposited from suspension in a high magnetic field [58,59]. In Figure 16.14, temperature dependence of κ along the direction parallel to the film is shown for two films prepared with and without the magnetic field (aligned and unaligned). The κ value reaches about 240 W/m·K in the aligned film, but that in the unaligned film is low, revealing that the orientation of SWCNT bundles in the film governs the κ value. The temperature dependence of κ and thermal conductance

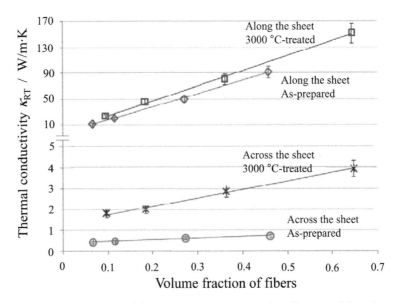

FIGURE 16.13 Thermal conductivity, κ_{RT}, for vapor-grown carbon-fiber mats (sheets)

From [55]

were also measured on a single MWCNT with a diameter of 14 nm [60]. Temperature dependence is shown in Figure 16.15, together with the results measured for MWCNT bundles of different diameters. The single MWCNT shows a high κ value of about 3000 W/m·K, much closer to the predicted value than the bundles of MWCNTs, which show much smaller values.

A vertically aligned MWCNT array (forest) had κ_{RT} of 8.3 W/m·K [61], and values of 0.5–1.2 W/m·K have been reported for arrays with bulk density of 0.06 g/cm^3 [62].

The κ value for a single-layer graphene prepared by cleavage of graphite was estimated to be 3080–5150 W/m·K from the temperature dependence of the position of the G-peak in Raman spectra [63,64], which was comparable to SWCNTs [59].

16.3.6 Diamond and diamond-like carbons

The very high and isotropic κ of diamond makes it very attractive as a high thermal conductivity material. Measurements of κ at 320 K on a number of natural and synthetic diamond crystals have been performed and κ values in the range of 500–2000 W/m·K were obtained [65]. The κ correlated well with the nitrogen content of the crystal. The value of 3320 W/m·K has been reported for ^{12}C-enriched diamond

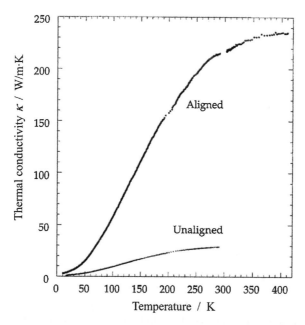

FIGURE 16.14 Temperature Dependence of κ Along the Film Surface for SWCNT Films Aligned Under Magnetic Field and Those Without Alignment

From [59]

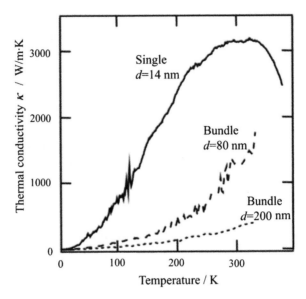

FIGURE 16.15 Temperature Dependence of κ for a Single MWCNT (Diameter, d, of 14 nm) and Bundles (d of 80 and 200 nm)

From [60]

synthesized by chemical vapor deposition (CVD) [66]. A diamond single crystal synthesized from ^{12}C-enriched pyrolytic carbon at 1430 °C and 6.1 GPa (99.97% ^{12}C) showed κ_{RT} of 2990 W/m·K, higher than the κ_{RT} of 2260 W/m·K for a single crystal synthesized from natural graphite under the same conditions (99.0% ^{12}C) [67]. Polycrystalline diamond films about 10 μm thick, which were grown in a DC arc discharge in different CH_4/H_2 mixtures, gave κ_{RT} values of 200–500 W/m·K, inversely proportional to the content of sp^2 carbon [68]. Polycrystalline diamond films prepared by microwave plasma CVD in CH_4/H_2 gave κ_{RT} of 800–2000 W/m·K, depending strongly on hydrogen content [69]. A small anisotropy in κ was observed. Epitaxially textured diamond films grown on the (100) plane of a silicon substrate showed κ_{RT} of 1120 W/m·K, but non-epitaxial diamond gave 550 W/m·K, revealing that κ was influenced by grain boundary defects [70]. A diamond/carbon nanocomposite was prepared by filling pyrolytic carbon into the spaces between nanodiamond particles (about 6 nm in size), of which the κ_{RT} value was in the range of 0.3–1.7 W/m·K [71]. Sintering of a natural microdiamond (10–14 μm in size) and a nanodiamond synthesized by detonation (about 6 nm in size) was performed at 1200–2300 °C under 5–7 GPa [72,73]. The κ_{RT} of sintered diamond block increases with increasing sintering temperature under pressure, as shown in Figure 16.16. Around diamond-graphite equilibrium conditions, κ becomes a maximum and then decreases, mainly owing to the graphitization of diamond at high temperatures. Sintered microdiamond showed κ_{RT} of 500 W/m·K, but nanodiamond gave a much smaller κ_{RT} of 10–50 W/m·K.

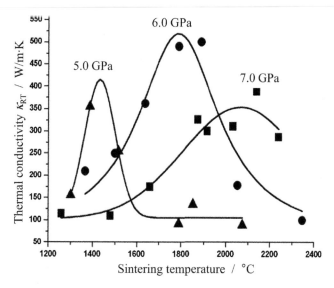

FIGURE 16.16 Dependence of Thermal Conductivity, κ_{RT}, of Natural Diamond on Sintering Temperature and Pressure

From [73]

The results suggested the importance of the grain boundary between the particles, at which the sp^2 bond is formed (graphitization) [74]. Diamond fibers prepared by CVD of CH_4 at 900 °C on tungsten wire, with a diameter of 25–250 μm, gave κ_{RT} of 750–1088 W/m·K along the fiber [75].

Diamond-like carbon (DLC) is amorphous, consisting of a significant fraction of sp^3 bonds. It is classified into tetrahedral amorphous carbons (*ta*-C) that are hydrogen-free and have the highest sp^3 content, and hydrogenated amorphous carbons (*a*-C:H). The latter are classified into four: polymer-like *a*-C:H (PLCH), containing 30–60 at% H and up to 70% sp^3 bonds; diamond-like *a*-C:H (DLCH), containing 20–35 at% H and 20–60% sp^3; hydrogenated tetrahedral amorphous carbon (*ta*-C:H), having 25–30 at% H and about 70% sp^3; and graphite-like *a*-C:H (GLCH), having less than 20% sp^3. The κ of diamonds, including various types of DLCs and synthesized nanodiamond, has been discussed as a function of various structural parameters including density, Young's modulus (E), and the G-band width of the Raman spectrum [76–78], these parameters depending strongly on the sp^3 content. In Figure 16.17, a plot of κ_{RT} vs density of various sintered diamonds is reproduced [77]. With the decrease in density, mainly due to the decrease in sp^3 content, κ_{RT} decreases abruptly by two orders of magnitude and then decreases gradually.

Composites of diamond with different materials have been prepared: with SiC [79], SiC-Si [80], Ag [81], Cu [82], and epoxy [75]. To make a composite of diamond (about 180 μm in size) with metallic Cu by a pulse plasma sintering method

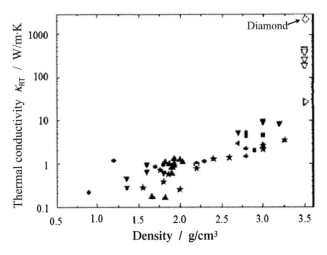

FIGURE 16.17 κ_{RT} **as a Function of Density for Various Sintered Diamonds**

Open marks correspond to the κ_{RT} for sintered diamonds and closed marks to diamond-like carbon films

From [78]

at 900 °C under 80 MPa, mixing Cr (0.8 Cr-Cu) was effective to avoid delamination at the interface between diamond and Cu due to the difference in thermal expansion coefficients [82].

16.4 Concluding remarks

Diamond is a fascinating material with high thermal conductivity, but it is difficult to use in practice, not only because of high cost but also difficulty in managing the interface with other materials. The presence of grain boundaries in diamond films disturbs thermal conductance, as shown by the fact that a sintered nano-sized diamond has a much lower κ_{RT} than a micron-sized one [73]. Structural defects and imperfects make the κ_{RT} value smaller, which is particularly pronounced by bonding with hydrogen and by the formation of sp^2 bonds, as shown in Figure 16.16 [78]. The interface between diamond crystal and matrix materials has to be controlled, as shown for composites with Cu [81] and also the interface of deposited-DLC and Al$_2$O$_3$ [83]. Coating of cutting tools, 6 mass% Co-WC substrates, and Si$_3$N$_4$-SiC composites with diamond film has been reported to be successful to reduce temperature rise at the tips [84]. So far, making better use of the high thermal conductivity of diamond is difficult for practical composites with various materials, including metals, as has been pointed out in the literature [85], and there are many further challenges in this area.

High thermal conductivity can be achieved in highly oriented and highly crystallized graphite; in other words, high orientation of perfect graphite layers has to be realized in carbon materials. There are four routes to prepare highly oriented and highly crystallized graphite materials:

1. Forming of either natural graphite flakes or mesophase-pitch-based carbon fibers by using a carbonaceous binder.
2. Formation of pyrolytic carbon via the CVD process.
3. Carbonization of polyimide films.
4. Preparation of oriented graphite films from natural graphite flakes by exfoliation and compression.

In the first three routes, graphitization at high temperature, as high as 3400 °C, is essential.

The last route, (4), has an advantage in that no high-temperature treatment is needed, and industrial experience has accumulated to allow production of so-called flexible graphite films [32]. However, much more careful processing to prepare highly oriented films is required. Complete exfoliation and recovery of well-separated and thinner flakes are needed, and the introduction of structural defects in graphite flakes has to be avoided, so that the procedure has to be modified, similar to the case of graphene synthesis described in Chapter 3. Formed natural graphite is commercially available as thermal interface materials [86,87].

In route (1), the forming process of natural graphite flakes is an important factor, as well as the selection of carbonaceous binder. To enhance the preferred orientation of flakes, it was experimentally shown that blade height for tape casting has to be carefully controlled [43]. In this route, orientation of flakes perpendicular to the film surface might be possible; in other words, high thermal conductivity could be achieved along the direction perpendicular to the sheet, this situation being difficult to realize in other routes. When mesophase-pitch-based carbon fibers with radial nanotexture in cross-section are used after graphitization, instead of natural graphite flakes, it might be much easier to achieve high thermal conductivity perpendicular to the formed films. In route (2), synthesis, characterization, and applications of pyrolytic carbons have been studied, and HOPG was developed by the treatment of specific pyrolytic carbon under high-temperature and high-pressure conditions [7]. Pyrolytic-carbon-derived materials with different grades of orientation and crystallinity have now been commercialized [88]. In route (3), highly oriented and highly crystallized graphite films are prepared under high-temperature and high-pressure conditions, and this has been commercialized [89,90]. It has been shown in the laboratory that the same materials, comparable to commercialized ones and even better, were able to be synthesized at a temperature above 3200 °C under simple mechanical constraint [29]. Since carbon nanotubes (CNTs) possess high κ along the axis, the use of CNT arrays as a thermal interface material (TIM) has attracted much attention, e.g. between a Si wafer and a substrate of another material [91].

Highly oriented graphite films show pronounced anisotropy: very high κ along the film, as explained above, but very low across the film, comparable to plastics, as shown in Table 16.1. So, these films are also used as heat-insulating materials and are particularly effective for high-temperature applications. Graphite sheet prepared from exfoliated graphite by compression has been shown to be effective as heat insulator in temperatures as high as 3200 °C by placing around a graphite heater [92]. Carbon foam prepared by compaction of exfoliated graphite, though complete compression results in highly oriented graphite film, is an excellent container of phase-change materials (PCMs) for storage of thermal energy, as explained in Chapter 9. Therefore, highly oriented and highly crystallized graphite has a high potential for thermal-management materials, i.e. not only as materials with high thermal conductivity but also as heat-insulating materials and heat-storage materials.

References

[1] Slack GA. Phys Rev 1962;127:694–701.
[2] Holland MG, Klein CA, Straub WD. J Phys Chem Solids 1966;27:903–11.
[3] Kaburagi Y, Yoshida A, Hishiyama Y, et al. TANSO 1995; No. 166: 19–27 [in Japanese].
[4] Inagaki M. New Carbons. Elsevier; 2000, 30–57.
[5] Inagaki M, Kang F. Carbon Mater Sci Eng 2011:307–42.
[6] Inagaki M, Kang F, Toyoda M. Chem Phys Carbon 2004;29:1–69.
[7] Moore AW. Chem Phys Carbon 1973;10:69–187.
[8] Ohnishi T, Murase I, Noguchi T, et al. Synth Met 1986;14:207–13.
[9] Murakami M, Watanabe K, Yoshimura S. Appl Phys Lett 1986;48:1594–6.
[10] Murakami M, Yoshimura S. Synth Met 1987;18:509–14.
[11] Shioya M, Shinotani K, Takaku A. J Mater Sci 1999;34:6015–25.
[12] Hishiyama Y, Yasuda S, Yoshida A, et al. J Mater Sci 1988;23:3272–7.
[13] Inagaki M, Harada S, Sato T, et al. Carbon 1989;27:253–7.
[14] Hatori H, Yamada Y, Shiraishi M. Carbon 1992;30:763–6.
[15] Yamashita J, Shioya M, Hatori H, et al. TANSO 2010; No. 245: 196–199 [in Japanese].
[16] Inagaki M, Takeichi T, Hishiyama Y, et al. Chem Phys Carbon 1999;26:245–333.
[17] Kaburagi Y, Hishiyama Y. Carbon 1995;33:773–7.
[18] Kaburagi Y, Yoshida A, Kitahata H, et al. TANSO 1996; No. 171: 24–29 [in Japanese].
[19] Hishiyama Y, Yoshida A, Kaburagi Y, et al. Carbon 1992;30:517–9.
[20] Murakami M, Nishiki N, Nakamura K, et al. Carbon 1992;30:255–62.
[21] Yoshida A, Kaburagi Y, Hishiyama Y. Carbon 2006;44:2333–5.
[22] Hishiyama Y, Kaburagi Y. Carbon 1992;30:483–6.
[23] Hishiyama Y, Yoshida A, Kaburagi Y. TANSO 2012; No. 254: 176–186.
[24] Smith AW, Rasor NS. Phys Rev 1956;104:885–91.
[25] Hove JE, Smith AW. Phys Rev 1956;104:892–900.
[26] Wagner P, Dauelsberg LB. Carbon 1968;6:373–89.
[27] Jamieson CP, Mrozowski S. Proceedings of the First and Second Conferences on Carbon. University of Buffalo: Waverly Press; 1956, 155.
[28] Wen CY, Huang GW. J Power Sources 2008;178:132–40.
[29] Kaburagi Y, Kimura T, Yoshida A, et al. TANSO 2012; No. 253: 106–115.
[30] Mason IB, Knibbs RH. Technical Report AERE-R 3973. United Kingdom Atomic Energy Authority; 1962.

[31] Nysten B, Issi JP, Shioyama H, et al. J Mater Res 1993,8.2299–304.
[32] Soule DE. Phys Rev 1958;112:698–707.
[33] Bonnissel M, Luo L, Tondeur D. Carbon 2001;39:2151–61.
[34] Yuan G, Li X, Dong Z, et al. Carbon 2012;50:175–82.
[35] Wang LW, Metcalf SJ, Critoph RE, et al. Carbon 2011;49:4812–9.
[36] Liu Z, Guo Q, Shi J, et al. Carbon 2008;46:414–21.
[37] Qiu HP, Song YZ, Liu L, et al. Carbon 2003;41:973–8.
[38] Liu Z, Guo Q, Shi J, et al. Carbon 2007;45:1914–6.
[39] Fukushima H, Drzal LT, Rook BP, et al. J Therm Anal Cal 2006;85:235–8.
[40] Kalaitzidou K, Fukushima H, Drzal LT. Carbon 2007;45:1446–52.
[41] Prieto R, Molina JM, Narcisoa J, et al. Scr Mater 2008;59:11–4.
[42] Zhou S, Chiang S, Xu J, et al. Carbon 2012;50:5052–61.
[43] Zhou S, Zhu Y, Du H, et al. New Carbon Mater 2012;27:241–9.
[44] Zhou S, Xu J, Yang Q-H, et al. Carbon 2013;57:452–9.
[45] Veca LM, Meziani MJ, WangW, et al. Adv Mater 2009;21:2088–92.
[46] Lavin JG, Boyington DR, Lahijani J, et al. Carbon 1993;31:1001–2.
[47] Adams PM, Katzman HA, Rellick GS, et al. Carbon 1998;36:233–45.
[48] Edie DD, Fain CC, Robinson KE, et al. Carbon 1993;31:941–9.
[49] Robinson KE, Edie DD. Carbon 1996;34:13–36.
[50] Gallego NC, Edie DD. Composites A 2001;32:1038–43.
[51] Gallego NC, Edie DD, Nysten B, et al. Carbon 2000;38:1003–10.
[52] Ma Z, Shi J, Song Y, et al. Carbon 2006;44:1298–301.
[53] Ma Z, Liu L, Lian F, et al. Mater Lett 2012;66:99–101.
[54] Kim YA, Kamio S, Tajiri T, et al. Appl Phys Lett 2007;90; 093125.
[55] Mahanta NK, Abramson AR, Lake ML, et al. Carbon 2010;48:4457–65.
[56] Berber S, Kwon YK, Tomanek D. Phys Rev Lett 2000;84:4613–6.
[57] Pop E, Mann D, Wang Q, et al. Nano Lett 2006;6:96–100.
[58] Hone J, Whitney M, Piskoti C, et al. Phys Rev B 1999;59:R2514–6.
[59] Hone J, Llaguno MC, Nemes NM, et al. Appl Phys Lett 2000;77:666–8.
[60] Kim P, Shi L, Majumdar A, et al. Phys Rev Lett 2001;87:215502.
[61] Shaikh S, Li L, Lafdi K, et al. Carbon 2007;45:2608–13.
[62] Jakubinek MB, White MA, Li G, et al. Carbon 2010;48:3947–52.
[63] Balandin AA, Ghosh S, Bao W, et al. Nano Lett 2008;8:902–7.
[64] Ghosh S, Calizo I, Teweldebrhan D, et al. Appl Phys Lett 2008;92:151911.
[65] Burgemeister EA. Physica B 1978;93:165–79.
[66] Anthony TR, Banholzer WF, Fleischer JF, et al. Phys Rev B 1990;42:1104–11.
[67] Nakamura K, Yamashita S, Tojo T, et al. Diam Relat Mater 2007;16:1765–9.
[68] Bertolotti M, Liakhou GL, Ferrari A, et al. J Appl Phys 1994;76:7795–8.
[69] Sukhadolau AV, Ivakin EV, Ralchenko VG, et al. Diam Relat Mater 2005;14:589–93.
[70] Wolter SD, Borca-Tasciuc DA, Chen G, et al. Diam Relat Mater 2003;12:61–4.
[71] Vlasov A, Ralchenko V, Gordeev S, et al. Diam Relat Mater 2000;9:1104–9.
[72] Kidalov SV, Shakhov FM, Vul AYa. Diam Relat Mater 2007;16:2063–6.
[73] Kidalov SV, Shakhov FM, Vul AYa. Diam Relat Mater 2008;17:844–7.
[74] Kidalov SV, Shakhov FM, Vul AYa, et al. Diam Relat Mater 2010;19:976–80.
[75] May PW, Portman R, Rosser KN. Diam Relat Mater 2005;14:598–603.
[76] Morath CJ, Maris HJ, Cuomo JJ, et al. J Appl Phys 1994;76:2636–40.
[77] Chen G, Hui P, Xu S. Thin Solid Films 2000;366:95–9.
[78] Shamsa M, Liu WL, Balandin AA, et al. Appl Phys Lett 2006;89:161921.

[79] Gray KJ. Diam Relat Mater 2000;9:201–4.

[80] Ekimov EA, Suetin NV, Popovich AF, et al. Diam Relat Mater 2008;17:838–43.

[81] Lee MT, Fu MH, Wu JL, et al. Diam Relat Mater 2011;20:130–3.

[82] Rosinski M, Ciupinski L, Grzonka J, et al. Diam Relat Mater 2012;27–28:29–35.

[83] Kim JW, Yang HS, Jun YH, et al. J Mech Sci Technol 2010;24:1511–4.

[84] Miranzo P, Osendi MI, Garcia E, et al. Diam Relat Mater 2002;11:703–7.

[85] Partridge PG, Lu G, May P, et al. Diam Relat Mater 1995;4:848–51.

[86] Norley J. Electron Cool Mag 2012; Sept: 1–5.

[87] GRAF Tech International, http://www.graftechaet.com/GRAFOIL/GRAFOIL-Products/ Single-Layer-Material.aspx.

[88] Minteq Pyrogenics Group, http://www.minteq.com/our-products/minteq-pyrogenics-group/ pyroid-pyrolytic-graphite/.

[89] Panasonic Electronic Device Co., Ltd, http://industrial.panasonic.com/www-data/ pdf/AYA0000/AYA0000CJ2.pdf [in Japanese].

[90] Kaneka Co., Ltd, http://www.elecdiv.kaneka.co.jp/graphite/index.html, "Graphinity" (2010) [in Japanese].

[91] Chung DDL. Carbon 2012;50:3342–53.

[92] Hishiyama Y. TANSO 2000; No. 193: 228–231 [in Japanese].

Isotropic High-density Graphite and Nuclear Applications

17

Most carbon materials are anisotropic owing to anisotropy in their basic structural units (BSUs), as represented by natural graphite flakes. However, some applications require isotropy for carbon materials; for example, the structural components in nuclear reactors. In order to realize isotropy in bulk carbon materials, anisotropic BSUs have to be randomly agglomerated into a block. Three fabrication processes have been employed for achieving isotropy in carbon materials:

1. The random aggregation of micrometer-sized particles, even though these particles are anisotropic.
2. The aggregation of spherical particles of carbon.
3. The random agglomeration of nanometer-sized crystallites in a block.

The first process has produced so-called isotropic high-density graphite blocks, where small-sized coke particles are molded under isostatic compression using pitch binder. The second process has been carried out using mesocarbon microbeads (MCMBs), micrometer-sized spherical particles with nanotexture of radial point orientation of anisotropic BSUs. The third process has been realized as glass-like carbons. The products of these methods can have different characteristics, as summarized in Table 17.1, showing a list of fabrication conditions: raw materials, formation and carbonization, and characteristics of the resulting carbon materials.

Process (1) is basically the same as the conventional one, using filler coke and binder pitch, for producing polycrystalline graphite blocks, except that filler coke particles must be fine and forming the carbon paste to the block has to be done under isostatic compression. For process (2), the raw material is limited to MCMBs (see Figure 4.5), conventional forming methods, extrusion and compression, can be applied, and no binder is necessary, because MCMBs are self-sintering. In process (3), thermosetting resin, such as furfuryl alcohol and phenol resin, has to be used and heating conditions for carbonization must be carefully controlled, i.e. heating rate should be very slow so as to complete the shrinkage of the resin during carbonization (see Chapter 6). Process (3) can confer isotropy on the product (glass-like carbons), but is not suitable for producing high-density graphite, because the products have a large number of closed pores, i.e. are low density, and non-graphitizing in nature.

Prerequisite for readers: Chapter 3.1 (Polycrystalline graphite blocks) in *Carbon Materials Science and Engineering: From Fundamentals to Applications*, Tsinghua University Press.

Table 17.1 Fabrication Processes of Isotropic Carbon Materials

Process	Raw Material	Formation	Carbonization	Characteristics
1	Filler coke particles and binder pitch	Cold isostatic pressing	Conventional carbonization and graphitization	High density
2	Mesocarbon microbeads (MCMBs)	Extruding or compressing	Conventional carbonization and graphitization	Self-sinterable, high density
3	Thermosetting resins (furfuryl alcohol, phenol resin, etc.)	Formation taking account of the shrinkage	Very slow heating	Low bulk density, gas impermeability, non-graphitizing

Isotropic high-density and high-strength graphite materials are needed for various applications, such as heating elements, crucibles and jigs for the synthesis and processing of semiconductors, electrodes in electric-discharge machining, and structural components in nuclear reactors, including high-temperature gas-cooled reactors.

In this chapter, isotropic high-density graphite produced mainly by process (1) is reviewed. After a brief description of production, the properties are summarized and discussed, focusing on their application in nuclear reactors, fission and fusion reactors.

17.1 Production

The industrial production process for isotropic high-density graphite is schematically shown in Figure 17.1 [1]. In order to achieve a high degree of graphitization, needle-like coke is usually selected as filler. Fine particles of filler coke are necessary to attain both isotropy and high density, in addition to isostatic compression for forming. Usually, particles of needle-like coke, sieved to between a few micrometers and a few tenths of micrometers, are mixed with a binder pitch by kneading to prepare carbon paste. The carbon paste is pulverized and sieved once again before sending to compression. Forming of the carbon paste into a block is performed by isostatic compression at room temperature (cold isostatic pressing; CIP). The principle of CIP is schematically shown in Figure 17.2. The compression is performed under a pressure of 100–200 MPa, using water as the pressurizing medium. Since the carbon paste is usually contained in a rubber case, the forming process is often called "rubber pressing." Formed products are subjected to carbonization (calcination) at around 1500 °C.

Binder pitch with a high carbonization yield must be selected to achieve high density in the final graphite blocks. Pitch is mainly composed of condensed aromatics with a wide range of molecular weights, so that binder pitch has been commonly characterized by fractionation to benzene-soluble, benzene-insoluble but quinoline-soluble,

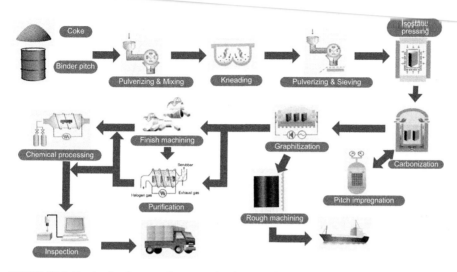

FIGURE 17.1 Production Process for Isotropic High-density Graphite

Courtesy of Toyo Tanso Co., Ltd

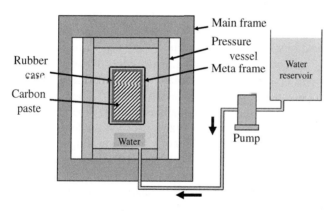

FIGURE 17.2 Cold Isostatic Pressing (CIP)

Courtesy of Toyo Tanso Co., Ltd

and quinoline-insoluble parts. The quinoline-insoluble part, called "α-resin" in the industry, is the component with highest molecular weight and low hydrogen content, H/C of less than 0.5, but also includes some solids, such as carbon blacks and coke particles. The benzene-insoluble but quinoline-soluble part, called "β-resin," is understood to be an important component for binding filler coke particles. The benzene-soluble part, called "γ-resin," has low molecular weight and a part of it evaporates even during kneading with the filler coke. The fractionation into more detailed fractions and the rheological analysis of each fraction have been performed to analyze the size and shape of pitch molecules [2–5]. Asphalt emulsion has been proposed for use

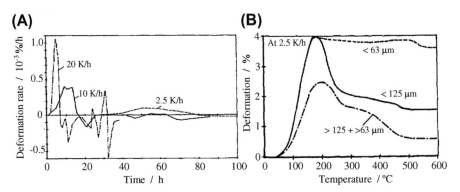

FIGURE 17.3 Deformation of Carbon that had been Compressed Under 150 MPa to a Diameter of 70 mm at Room Temperature, During Heating up to 600 °C

(A) Effect of heating rate, and (B) of particle size of filler coke

From [7]

as a binder, although this may not meet the requirement for a high carbon yield [6]. In industrial production of isotropic graphite, green coke, i.e. that before calcination to coke, is sometimes added because of its self-sinterability [1].

The heating procedure during carbonization has to be controlled carefully to avoid the formation of cracks in the carbonized body, particularly for the production of large-sized bodies. Upon heating, many kinds of gases are evolved, mainly owing to the pyrolysis of the binder pitch, which raises the pressure inside the block because the carbon paste becomes almost gas-impervious just after melting of the binder pitch. Accordingly, the paste block experiences expansion and shrinkage due to pyrolysis and carbonization of the binder pitch, which depend strongly on the particle size of the filler coke and the heating rate. In Figure 17.3A, the effect of heating rate is shown by plotting deformation rate for carbon paste formed at room temperature by compression [7]. The block heated at a high rate of 20 K/h showed marked expansion with subsequent complicated expansion and shrinkage behavior, leading to high risk for the formation of cracks. By slow heating at 2.5 K/h, deformation was greatly suppressed, i.e. almost no risk for cracking. As shown in Figure 17.3B, too small a grain size (smaller than 63 μm) led to considerably high dilatation; mixing of particles of different grain sizes reduced the dilatation during heating, leading to less risk for cracking. Slow heating and mixing of different particle sizes of the filler are therefore desirable, in order to avoid cracking during carbonization.

Since the carbonized block contains many pores, pitch impregnation and recarbonization are performed, even several times, to fill the open pores with carbon formed from impregnated pitch to increase the density. The carbonized blocks thus prepared are subjected to graphitization at around 3000 °C using conventional furnace and procedure. In some cases, pitch impregnation is carried out on graphitized blocks, followed by another cycle of carbonization and graphitization.

Table 17.2 Metallic Impurities in Graphite Block

	Mass Content (ppm)				Mass Content (ppm)		
Element	Highly Purified	Purified	Non-Purified	Element	Highly purified	Purified	Non-purified
Li	<0.001	<0.001	<0.03	V	<0.001	0.018	40
B	0.10	0.15	3	Cr	<0.004	0.006	<0.3
Na	<0.002	<0.002	<0.5	Mn	<0.001	<0.001	<0.2
Mg	<0.001	0.004	0.2	Fe	<0.02	0.06	26
Al	<0.001	0.012	14	Co	<0.001	<0.001	<0.2
Si	<0.1	<0.1	2	Ni	<0.001	0.006	4
K	<0.003	0.04	2	Cu	<0.002	<0.002	<1
Ca	<0.01	0.08	6	Zn	<0.002	<0.002	<0.6
Ti	<0.001	<0.001	33	Pb	<0.001	<0.001	<1

Courtesy of Toyo Tanso Co., Ltd

For many applications, high purity is required for graphitized blocks, in addition to isotropy, high density, and high graphitization degree. Therefore, purification is performed for the graphitized blocks in a flow of halogen gases at around 2000 °C. As an example, metallic impurities are listed for non-purified, purified, and highly-purified graphite blocks in Table 17.2. Most of the impurity metals can be excluded by evaporation as metal-chlorides at temperatures up to 1800 °C, but F_2 gas is needed for removing B. For the purification gas, Cl_2 and F_2 are often used, but halo-genized hydrocarbon gases, which give halogen gas at high temperatures, such as CCl_4 and CCl_2F_2, have also been selected. Heat treatment at high temperatures is performed under high vacuum in order to remove oxygen and hydrogen from graphitized blocks, hydrogen being necessary for removing sulfur.

By selecting MCMBs as the raw material, binder pitch becomes unnecessary because of their self-sinterability, and a conventional forming process, extrusion and compression, can theoretically be applied owing to their spherical morphology. However, isostatic pressing has been found to be preferable in practice. MCMBs are produced by the separation of mesophase spheres from mesophase pitches using solvents [8,9]. Mesophase pitch and green cokes are self-sintered without any binder, but the control of self-sinterability is not easy. Even though they are self-sintering, forming under isostatic pressing is preferable to attain isotropy in the final carbon materials. Sintering of MCMBs has been studied and discussed [10–12]. Figures 17.4A and 17.4B show changes in the density of the MCMB particles and the shrinkage of the compressed block with carbonization temperature [11]. Particle density increases and the block shrinks markedly in the range of 800–1200 °C (stage 2). Below 800 °C, particle density does not change and the block expands slightly (stage 1). Above 1200 °C, densification proceeds both in the particles and the block. Stage 2 is caused mainly by the pyrolysis of mesophase particles, and sintering both within

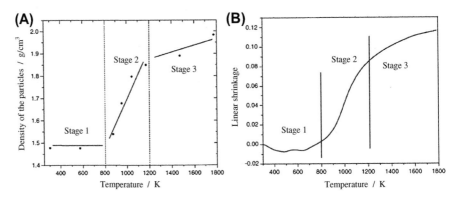

FIGURE 17.4 Sintering of MCMB Particles

(A) Changes in density of particles, and (B) shrinkage of the block formed from the particles with carbonization temperature

From [11]

and between particles occurs in the stage 3 owing to carbonization. The β-resin component in MCMBs has been pointed out to have an important role in sintering [13]. Oxidation at a temperature of 200 °C was effective to avoid the formation of cracks during sintering of the compacts prepared under 100 MPa, but fracture toughness of the resultant block after heat treatment at 2400 °C decreased, even though the bulk density did not change [12]. Oxidation of MCMBs has marked influence on their carbonization and graphitization behavior [14]. The main gases evolved during carbonization at 1000 °C were CO and H_2 for a slightly oxidized MCMB, but CO_2 for highly oxidized one. On the graphitized artifacts, cracks were mainly transgranular in the former, giving little chance for interconnection and high flexural strength, while cracks were mainly intergranular and could connect to form long cracks, giving low strength, in the latter.

17.2 Properties

The microtexture of isotropic high-density graphite prepared from fine coke particles via CIP is compared to conventional graphite prepared from coarse coke grains in Figure 17.5. Polarized-light micrographs reveal that the former has finer microtexture than the latter, much smaller anisotropic particles, and much smaller and fewer pores.

The properties of commercially available isotropic graphite are listed in Table 17.3. In the table, only the commercial-grade products of Toyo Tanso Co., Ltd, are listed, but other grades produced by other companies have very similar properties. They commonly have high bulk density of 1.7–1.9 Mg/m^3 and high mechanical properties with tensile strength of 20–55 MPa, higher than conventional graphite, but slightly higher electrical resistivity of 9–15 μΩ·m and a higher thermal expansion coefficient

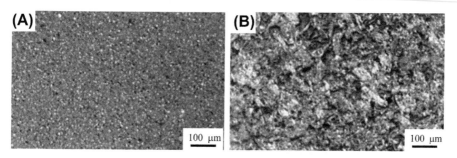

FIGURE 17.5 Comparison of Polarized-light Micrographs of Graphite Cross-section

(A) Isotropic high-density graphite, and (B) anisotropic conventional graphite

Courtesy of Toyo Tanso Co., Ltd

of about 5×10^{-6}/K. Temperature dependence of these properties is shown in Figure 17.6, including some of the grades of isotropic graphite listed in Table 17.3.

Temperature dependence of electrical resistivity, the thermal expansion coefficient, thermal conductivity, and strengths is common for all grades (Figure 17.6), and the same for other carbon materials. Hardness, mechanical strengths, Young's modulus, and thermal conductivity of these graphites are higher than those of conventional coarse-grained graphite for steel-refining electrodes, mainly because of the high bulk density, in the range of 1.75–1.90 Mg/m^3. Since fine-grained cokes are used as the filler, however, electrical resistivity is a little higher than coarse-grained graphite. Thermal conductivity is relatively high, close to metals, but cannot be as high as highly-oriented graphite films, discussed in Chapter 16, because of the random orientation of the crystallites.

The differences in properties among the grades of isotropic graphite are mainly caused by pore parameters; many properties are closely related to the pore parameters, as determined on the cross-section by image-analysis techniques [15]. Elastic modulus, bending strength, fracture toughness, and critical crack-opening displacement are related to the mean area of pores, as shown in Figure 17.7, revealing that many properties of high-density graphite depend strongly on microtexture. Similar analysis has been applied to the high-density graphite materials used for nuclear applications [16]. The distribution of pore area against pore size was expressed by a linear relationship on a log-log plot, the cross-section of pores was best fit by an ellipse with aspect ratio of around 2, and cracks attributed to shrinkage of filler cokes were observed.

Fracture behavior of isotropic graphite blocks has been studied in detail in relation to application as components of nuclear reactors, particularly of high-temperature gas-cooled reactors. The crack extension rate during fatigue testing was studied using a tapered double-cantilever beam specimen for fine-grained IG-11 [17]. Load was cyclically applied at a loading rate of 251 N/s in a range of stress ratio, R, of 0–0.8, where R is the ratio of the minimum stress intensity factor, K_{min}, to the maximum stress intensity factor, K_{max}. For a given value of R, crack extension

Table 17.3 Properties of Isotropic High-density Graphite

Grade Name	Bulk Density (Mg/m³)	Hardness	Electrical Resistivity (μΩ·m)	Flexural Strength (MPa)	Compressive Strength (MPa)	Tensile Strength (MPa)	Young's Modulus (GPa)	Thermal Expansion Coeffient (10⁻⁶/K)	Thermal Conductivity (W/m·K)
IG-11	1.77	51	11.0	39	78	25	9.8	4.5	120
IG-15	1.90	60	9.5	54	103	29	11.8	4.8	140
IG-45	1.88	55	9.0	60	110	40	12.0	4.9	140
ISEM-1	1.68	45	13.5	36	69	20	8.8	4.2	90
ISEM-3	1.85	60	10.0	49	103	29	11.8	5.0	130
ISO-63	1.78	76	15.0	65	135	46	12.0	5.6	70
ISO-68	1.82	80	15.5	76	172	54	13.2	5.6	70
TTK-4	1.78	72	14.0	73	135	49	10.9	5.0	90
TTK-50	1.80	70	13.0	60	130	40	11.5	5.1	100

Courtesy of Toyo Tanso Co., Ltd

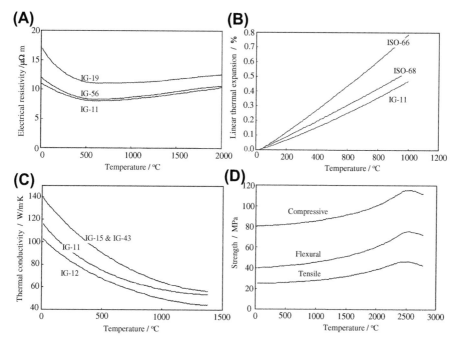

FIGURE 17.6 Temperature Dependence of Properties of Isotropic High-density Graphites

(A) Electrical resistivity, (B) linear thermal expansion, and (C) thermal conductivity for different grades of high-density graphite, and (D) strengths for the grade IG-11

Courtesy of Toyo Tanso Co., Ltd

rate was proportional to $(K_{max} - K_{min})$. The effects of R, irradiation, oxidation, specimen volume, and cumulative damage on fatigue behavior and fracture toughness were examined for nuclear-grade graphites IG-11 and IG-110 (purified IG-11), and medium-grained semi-isotropic PGX [18]. The fatigue strength of IG-11 decreased with decreasing R, and the crack extension rate was enhanced mainly by the increment of K_{max}. Fracture toughness of graphite was measured using the specimen shown as the insert in Figure 17.8, by changing specimen thickness and notch sharpness. In Figure 17.8, toughness, J_{iEP}, is shown as a function of specimen thickness, B, revealing that a constant value can be obtained at B of more than 20 mm. Since PGX is composed of coarse coke particles, J_{iEP} was a little lower than for IG-110, but no difference in J_{iEP} was detected in the two orthogonal directions, i.e. it is not anisotropic.

Fatigue strength has been studied in detail at room temperature [19] and at 1000 °C [20] for IG-11 and PGX under a biaxial stress state, axial tensile stress and hoop stress, using tubular specimens. Hoop stress was generated by increasing the internal gas pressure of the tubular specimen, and various ratios of hoop to axial stress, $\sigma_{hoop}/\sigma_{axial}$, in a range of 0 to 8.17 were applied [17]. The axial and hoop fracture stresses decreased in the biaxial stress state, especially when the ratio of the hoop

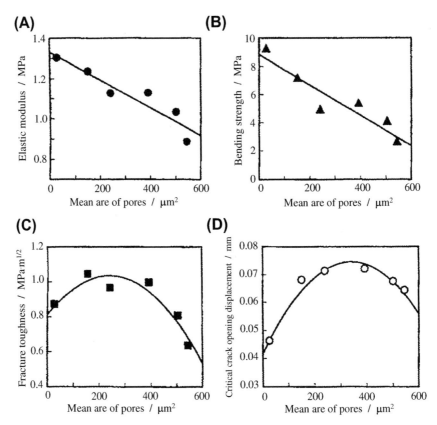

FIGURE 17.7 Relationships of Mechanical Properties to Mean Area of Pores for High-density Graphite Materials

(A) Elastic modulus, (B) bending strength, (C) fracture toughness, and (D) critical crack-opening displacement

From [15]

to axial stresses was around 1. In Figure 17.9, biaxial fracture probability is plotted against normalized stress (axial stress, σ_{axial}, or hoop stress, σ_{hoop}, normalized to the mean uniaxial tensile strength, σ_t) at different $\sigma_{hoop}/\sigma_{axial}$. The experimental points under each stress condition can be approximated by a straight line, suggesting a normal distribution of functions of σ_{axial}/σ_t and σ_{hoop}/σ_t.

Measurements of electrical resistivity during room-temperature fatigue testing suggested a possibility to predict fatigue failure [21]. Acoustic emission (AE) events during fatigue of high-density graphite were divided into low and high relative energies, the former due to cleavage between basal planes within the filler grains and the latter to microcracking in the binder region between the filler grains, also suggesting the possibility to predict fatigue failure [22]. AE onset stress was closely related to tensile and compressive strengths for four high-density graphites. The relationship

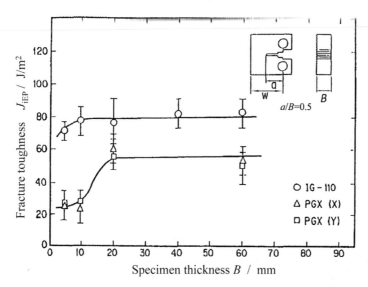

FIGURE 17.8 Facture Toughness, J_{iEP}, as a Function of Specimen Thickness for Isotropic Graphites: IG-110 and Two Orthogonal Directions of PGX

From [18]

between indentation load, L, and depth, h, during a dynamic micro-hardness test has been discussed for twelve nuclear-grade graphites and three carbon-carbon composites in relation to bending strength and Young's modulus [23]. The parameter determined as the slope of L/h vs h linear relationship on loading was proportional to bending strength.

Gas permeability of high-density graphite is relatively low in comparison with coarse-grained graphite. In Figure 17.10, dependence of He permeability on gas pressure is compared for two graphite blocks (A and B) fabricated by the conventional process using a binder and three graphite blocks (C–E) prepared by the self-sintering process, revealing that the permeability for He depends strongly on bulk density and fabrication process [24]. Permeability of gases, He and N_2, was expressed by the viscous flow permeability constant and Knudsen flow permeability constant, both constants being related to mean pore radius and open porosity of the block.

Behavior of oxidation by air at low temperatures of 400–650 °C is governed by the oxidation rate (reaction-controlled oxidation), and oxidation occurs uniformly in graphite block because oxygen can diffuse into the carbon matrix. At high temperatures above 750 °C, oxidation is controlled by diffusion of oxygen (diffusion-controlled oxidation) and so occurs preferentially at the surface of block. The reaction-controlled oxidation is markedly accelerated by metal impurity. Purification of isotropic high-density graphite was shown to be very important for suppressing oxidation in the reaction-controlled region. In Figure 17.11, Arrhenius plotting of oxidation rate is shown for two nuclear-grade graphites (IG-11 and PGX) before and after purification, revealing that

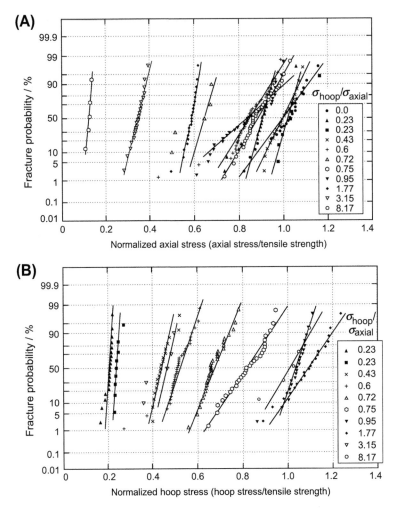

FIGURE 17.9 Biaxial Fracture Probability as a Function of Normalized Stresses for IG-11 Graphite at Room Temperature

(A) As a function of σ_{axial}/σ_t, and (B) of σ_{hoop}/σ_t

From [19]

oxidation rate is suppressed by purification, particularly for IG-11 [25]. Elimination of impurity metals, such as Co, Ni, and V, from graphite blocks was particularly effective for reducing oxidation rate. Oxidation rate of the carbonized blocks was 10–100 times higher than that of graphitized blocks. Oxidation behavior of purified isotropic graphite, IG-110, has been well characterized in the literature [26–30]. Four grades, including IG-110, were examined for oxidation behavior in a temperature range of 600–950 °C in a flow of dry air [30]. The graphites prepared from pitch coke were oxidized quickly, more rapidly than those prepared from petroleum coke, at 702 and

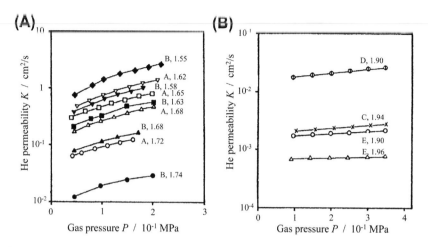

FIGURE 17.10 Changes in Permeability of He with Gas Pressure for Different Graphite Blocks

(A) Fabricated by the conventional process using a binder pitch, and (B) by using self-sintering raw materials. Bulk density is shown together with sample code (A–E)

From [24]

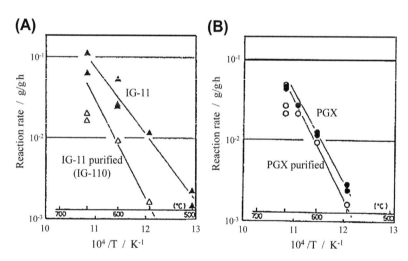

FIGURE 17.11 Arrhenius Plots of Oxidation Rate for Two Nuclear-grade Graphites Before and After Purification

(A) IG-11, and (B) PGX

From [25]

808 °C, although oxidation rates determined at 5–10% mass loss were nearly the same. Oxidation behavior at a temperature in the range 600–750 °C was studied for three commercial grades of isotropic graphite, and it was concluded that the difference in oxidation behavior depended strongly on the microtexture of the graphite, in particular the particle size and morphology of the filler coke [31].

17.3 Nuclear applications

17.3.1 Fission reactors

The high-temperature gas-cooled reactor (HTGR) has the capability to produce high temperatures of about 1000 °C that are expected to give efficient utilization of nuclear energy and to have inherent safety features. HTGRs have been developed and operated in many countries. The Japan Atomic Energy Research Institute (JAERI) designed and constructed the high-temperature engineering-test reactor (HTTR) in order to establish and upgrade the technology related to HTGR [32]. The HTTR is a helium-cooled and graphite-moderated HTGR with thermal power of 30 MW and a maximum reactor-outlet coolant temperature of 950 °C. The first criticality was attained on November 10, 1998, and the reactor achieved the full power of 30 MW with outlet coolant temperature of 850 °C on December 7, 2001. After a series of safety demonstration tests, it is planned to be used as the heat source of a hydrogen-production system by 2020. The major specifications of the HTTR are summarized in Table 17.4 [32]. High-temperature (950 °C) continuous operation of the HTTR was performed for 50 days from January to March in 2010. The results demonstrated the potential to supply stable heat of high temperature for hydrogen production [33].

The reactor internals consist of graphite core blocks, graphite core support structures, metallic core support structures, and other components. The graphite core support structures consist of reflector blocks, hot plenum blocks, core bottom structures, core support posts etc., as shown in Figure 17.12. The hot plenum blocks provide lateral and vertical positioning and support of the core array. The blocks contain flow paths, which guide the primary coolant from the outlet of the fuel columns and distribute it into the hot plenum beneath the hot plenum blocks. The core support

Table 17.4 Major Specifications of the HTTR

Thermal power	30 MW
Outlet coolant temperature	850 °C
Inlet coolant temperature	395 °C
Fuel	Low-enriched UO_2
Fuel element type	Prismatic block
Direction of coolant flow	Downward flow
Pressure vessel	Steel
Number of main cooling loops	1
Heat removal	Intermediate heat exchanger Pressurized water cooler
Primary coolant pressure	4 MPa
Containment type	Steel containment
Plant lifetime	About 20 years

From [32]

posts are designed so as to support the core and hot plenum block arrays, which form the hot plenum. The active core is surrounded by replaceable reflector blocks and permanent reflector blocks. The permanent reflector blocks are fixed tightly by the core restraint mechanism. The active core consists mainly of hexagonal fuel blocks, control-rod guide blocks, and replaceable reflector blocks, and is 2.9 m in height and 2.3 m in diameter, containing 30 fuel columns and seven control-rod guide columns. Each column is made up of five fuel blocks and four replaceable reflector blocks above and below the fuel blocks.

In the graphite components, key-keyway structures are adopted to maintain the array during earthquakes, as shown in Figure 17.13 [34,35]. The key-keyway structure is crucial for assessing the integrity of the components against the aseismic characteristics of the assembly and the dynamic load caused by earthquakes [36]. The equivalent stiffness affected by the contact behavior between the key and the keyway, and the stress distribution around the keyway were experimentally studied using a scale model, where the plenum block was made of PGX and the key of IG-11 [35].

The fuel assembly of the HTTR is shown in Figure 17.14. It has three dowel pins on the top and three mating dowel sockets at the bottom to align the fuel assemblies. Fuel particles with UO_2 kernel (about 6 mass% enrichment and 600 μm in diameter),

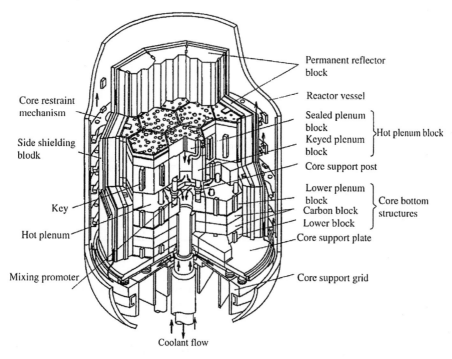

Core restraint mechanism

Side shielding blodk

Key

Hot plenum

Mixing promoter

Permanent reflector block

Reactor vessel

Sealed plenum block
Keyed plenum block } Hot plenum block

Core support post

Lower plenum block
Carbon block } Core bottom structures
Lower block

Core support plate

Core support grid

Coolant flow

FIGURE 17.12 Core Structure of the HTTR

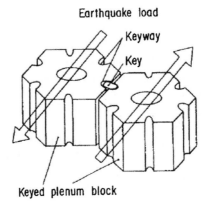

FIGURE 17.13 Key-keyway Structure Between Keyed Plenum Blocks of Graphite

From [35]

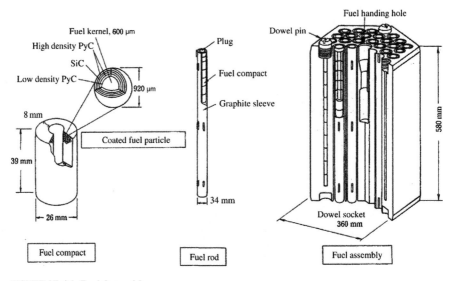

FIGURE 17.14 Fuel Assembly

From [32]

each being coated by pyrolytic carbons (PyC) and SiC, are dispersed in the graphite matrix and sintered to form a fuel compact. Fuel compacts are contained in a fuel rod and inserted into vertical holes in the graphite block to construct the fuel assembly. Helium gas coolant flows through gaps between the holes and the rods.

Isotropic graphite blocks with high strength and high purity developed by industry have been examined for use as structural components of HTTR in detail: fine-grained IG-110 for replaceable reflector blocks and medium-grained PGX for permanent reflector blocks. Structural design guidelines for graphite have been established.

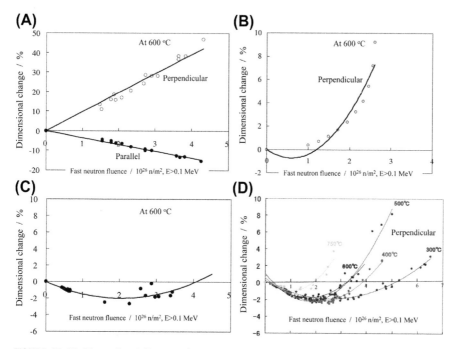

FIGURE 17.15 Dimensional Change with Fast Neutron Irradiation for Nuclear-grade Graphites

(A) HOPG, (B) PGA, (C) IG-110, and (D) ATR-2E

From [52]

Many Japanese companies have participated in the JAERI project since the research and development stage of the HTTR [37].

For the design of the core graphite component of the HTTR, comprehensive measurements of various properties—bulk density, thermal conductivity, thermal expansion, elastic modulus, and mechanical strengths (tensile, bending, and compressive strengths)—were carried out for IG-110 using various test specimens [38]. Uniaxial mechanical strengths of IG-110 were statistically assessed on the basis of seven tests from 1981 to 1989 on different blocks of IG-110, comprising more than 350 measurements [39]. The strengths data were approximated by a normal distribution, making it easier to make a reliability-oriented definition of the specified minimum ultimate strengths. The anisotropy in tensile strength was concluded to be practically negligible.

Nondestructive evaluation methods using ultrasonic waves and micro-indentation were developed for evaluation of the effects of oxidation damage on the mechanical properties of the graphite components in HTGRs [40]. Ultrasonic wave velocities at 1 MHz were empirically correlated to burn-off by exponential formulas. It was possible to evaluate the porous state of the oxidized graphite with wave propagation analysis using a wave-pore interaction model. It was important to consider the non-uniformity of pore distribution in the oxidized block. The micro-indentation method was expected to determine local oxidation damage.

The effects of neutron irradiation on various properties—dimensional change, thermal conductivity, Young's modulus, strengths, creep, and thermal expansion—was examined for commercially available nuclear-grade graphite materials, including IG-110, PGA, H-451, and NBG-10 [41–53]. Results of irradiation were reviewed and analyzed, focusing on IG-110 [52]. In Figure 17.15, dimensional change of three nuclear-grade graphites, IG-110 [46], PGA [44], and ATR-2E [52], by fast neutron irradiation at 600 °C is compared with highly oriented pyrolytic graphite (HOPG) [41,52]. For ATR-2E, the results at different irradiation temperatures are shown. HOPG shows large expansion perpendicular to the plate and large shrinkage along the plate with irradiation, showing marked anisotropy in dimensional change. Nuclear-grade graphites give much smaller dimensional change (the scale of the ordinate in Figures 17.15B–17.15D is much smaller than that in Figure 17.15A) and isotropic change. The changes exhibit a turn-around from shrinkage to expansion at a certain dose, depending on graphite grade and irradiation temperature. Among the three graphites, IG-110 showed the smallest dimensional change and the highest turn-around dose [46,52]. These dimensional changes were simulated by a quadratic equation of irradiation fluence [53].

For IG-110, the effect of neutron irradiation at 600 °C was measured on volume change, electrical resistivity, Young's modulus, and strength [46]. Relative changes of these parameters with irradiation fluence are shown in Figure 17.16, all parameters showing a turn-around at about 15 dpa (**d**isplacements **p**er **a**tom), the same influence as observed for dimensional change in Figure 17.15C (15 dpa corresponds to about 2.2×10^{26} n/m^2).

The accumulated data of neutron irradiation of nuclear-grade graphite materials were shown to be possible to generalize by interpolation and extrapolation for the application of isotropic high-density graphite to the main components of a very-high temperature reactor [52].

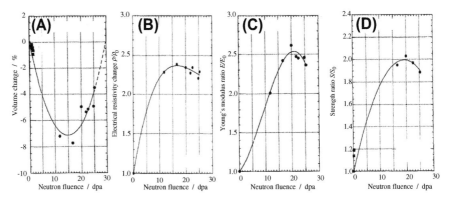

FIGURE 17.16 Changes in Properties of IG-110 with Neutron Irradiation at 600 °C

(A) Volume, (B) electrical resistivity, (C) Young's modulus, and (D) strength

From [46]

The effects of neutron irradiation on creep rate were measured on medium-grained semi-isotropic graphite H451, which was produced by extrusion from particles of 0.5 mm mean size and had a bulk density of about 1.75 g/cm³ [49]. Compressive creep stresses of 13.8 MPa at 600 °C and 20.7 MPa at 900 °C, or tensile creep stress of 6 MPa at 900 °C, were applied. The experimental results suggest that the creep model has to be modified before and after the turn-around of volume change.

17.3.2 Fusion reactors

Designing the advanced experimental tokamak nuclear fusion reactor ITER (International Thermonuclear Experimental Reactor) began in 1992 under an international nuclear fusion research and engineering project [54], and the construction of ITER started in 2005 at Cadarache, France. The cross-sectional view and the structure of the divertor cassette of the ITER are shown in Figure 17.17. The joint program between Japan and Europe is planning to construct the JT-60SA (an upgraded device based on the JT-60 tokamak, with fully superconducting coils) for early realization of fusion energy based on the tokamak concept. The materials for the plasma-facing components placed at the inner and outer targets and at the dome in the divertor cassette are being carefully studied, because the material is required to have high heat transfer and high resistance to thermal shock. Materials used for the vacuum vessel and in-vessel components have been summarized [55].

For fusion reactors, carbon and graphite materials, including isotropic high-density graphite and carbon fiber/carbon composites (CFCs), have been examined as one of the candidates for the first wall facing to plasma [56]. In ITER, however, tungsten has been

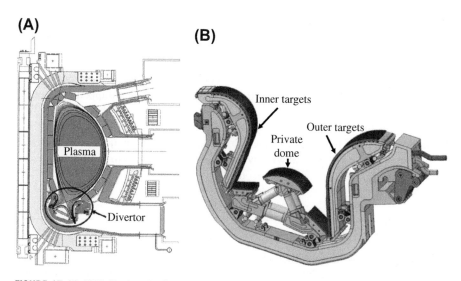

FIGURE 17.17 ITER Nuclear Fusion Reactor

(A) Cross-sectional view, and (B) structure of the divertor cassette selected as the plasma-facing material mainly because of its low sputtering yield and good thermal properties. Isotropic high-density graphite materials are mostly used as a substrate for tungsten deposition [57–59]. High heat flux experiments using electron-beam irradiation have been performed for tungsten deposited on CFC and isotropic graphite with multilayered diffusion barriers of rhenium and tungsten, showing effectiveness of the barrier to the diffusion of carbon into the tungsten layer at 1300 °C [60]. In JT-60SA, the divertors, consisting of brazed CFC mono-block targets as shown in Figure 17.18, are under examination [61,62]. Improvement in thermal conductivity even at 1200 K was observed for isotropic high-density graphite and CFC impregnated with Cu (10–18 vol%) containing a small amount of Ti (0.5–0.8 vol%) [63]. Improvement of the heat-transfer characteristics of mechanically joined CFC and Cu blocks has been attempted by inserting graphite sheets with high thermal conductivity (Chapter 16) [64].

Divertor plates have been examined for installation in the vacuum vessel of the large helical device (LHD) for nuclear fusion research in Japan [65,66]. Isotropic high-density graphite is used together with highly-oriented graphite film of high thermal conductivity to connect to Cu cooling pipes [67–70]. One of the divertor plates studied is shown in Figure 17.19 [68]. Two isotropic high-density graphite armor tiles sandwich a stainless-steel cooling pipe, and are tightly fixed with two stainless-steel bolts. To improve thermal contact between the armor tiles and the cooling pipe, graphite sheet with 0.1 mm thickness is used.

Understanding the retention and recycling of hydrogen isotopes (protium, deuterium, and tritium) for the plasma-facing components is essential for the design of fusion reactors [71–73]. Hydrogen isotope absorption and transport of hydrogen in isotropic high-density graphite and CFC have been actively studied [74–77]. The amount of hydrogen retained depends strongly on the graphite material [74,75]. The existence of two hydrogen-trapping sites in polycrystalline graphite has been proposed, one at the interstitial cluster loop edge (trap 1) and the other at the edge surface of a crystallite (trap 2) [76,77], as schematically shown in Figure 17.20A. In trap 1, hydrogen atoms are bonded to carbon atoms of interstitial clusters formed in the interlayer spaces of the crystallite by ion or neutron irradiation, and have a high absorption enthalpy of about 4.4 eV. In trap 2, hydrogen atoms are trapped at the edge carbon atoms of the crystallite with a low absorption enthalpy of about 2.6 eV. The latter performs the principal part of hydrogen retention, as shown in Figure 17.20B. When the surface of the graphite or CFC was sputtered, the majority of sputtered carbon atoms were co-deposited with hydrogen isotopes on the surrounding surfaces, to increase the retention of hydrogen isotopes.

17.4 Concluding remarks

Industrial production of isotropic high-density graphite materials has been established using CIP, followed by carbonization and graphitization. In order to attain high density, a cycle of pitch impregnation, carbonization, and graphitization is often applied. Fundamental investigation of filler cokes and binder pitches still continues in industry,

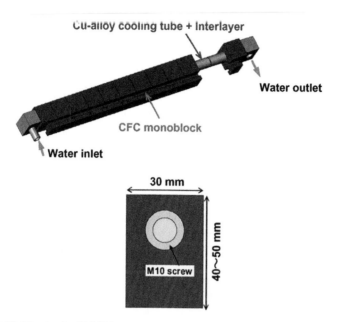

FIGURE 17.18 Divertor for JT-60SA

From [61]

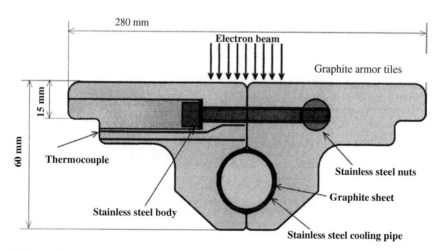

FIGURE 17.19 Structure of a Surface-boltless Mechanically Attached Module for Divertor Plates of LHD

SS, stainless steel

From [68]

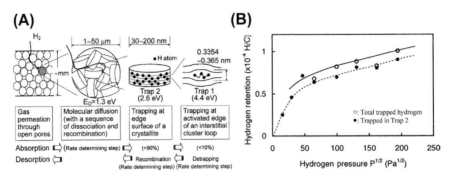

FIGURE 17.20 Hydrogen Trapping Sites and Trapped Amounts in Polycrystalline Graphite

(A) Proposed model of two trapping sites, trap 1 and trap 2, and (B) trapped hydrogen in the two sites as a function of hydrogen pressure

From [73]

FIGURE 17.21 A Large-sized Isotropic High-density Graphite Block Exhibited at an International Conference in 1982

Courtesy of Toyo Tanso Co., Ltd

although the results are rarely published. For industrial production of large-sized graphite blocks, the long-term accumulation of technological experience and efforts was required. In 1982, a large block was exhibited for the first time at the International Symposium on Carbon held in Toyohashi, Japan, as shown in Figure 17.21. Now, large blocks with a size of $0.79 \times 1.22 \times 1.275$ m^3 (2.2 T) are industrially produced.

The application of isotropic high-density graphite has also been established in a wide range of industrial fields, for manufacturing different semiconductors as heating elements, crucibles, susceptors, etc.; for hot-pressing ceramics as heaters, dies, sleeves, spacers, trays, etc.; for electric-discharge machining (EDM) as electrodes; for nuclear fission reactors as structural components; and for nuclear fusion reactors as the first wall of divertors. In this chapter, nuclear applications have been briefly reviewed. The future development of nuclear-related industries is not guaranteed, particularly after the Fukushima disaster in 2011. It has to be mentioned, however, that research activities on isotropic high-density graphite materials, which have been done with the aim of applying them to HTGR, have made great contributions to developments in industrial production and in materials quality and reliability.

References

[1] Tojo J. TANSO 2008; No. 234: 234–243 [in Japanese].
[2] Sakai M, Inagaki M. Carbon 1981;19:37–43.
[3] Sakai M, Yoshihara M, Inagaki M. Carbon 1981;19:83–7.
[4] Sakai M, Sasaki K, Inagaki M. Carbon 1983;21:593–6.
[5] Sakai M, Sogabe T, Kitagawa H, et al. Carbon 1983;21:601–3.
[6] Born M, Seichter A, Starke S. Carbon 1990;28:281–5.
[7] Shen K, Huang ZH, Yang J, et al. Carbon 2013;58:238–41.
[8] Yamada Y, Imamura T, Kakiyama H, et al. Carbon 1974;12:307–19.
[9] Honda H. Mol Cryst Liquid Cryst 1983;94:97–108.
[10] Gao Y, Song HH, Chen XH. J Mater Sci 2003;38:2209–13.
[11] Norfolk C, Mukasyan A, Hayes D, et al. Carbon 2004;42:11–9.
[12] Zhou C, McGinn PJ. Carbon 2006;44:1675–81.
[13] Martinez-Escandell M, Carreira P, Rodriguez-Valero MA, et al. Carbon 1999;37:1662–5.
[14] Shen K, Huang ZH, Yang J, et al. Carbon 2011;49:3200–11.
[15] Oshida K, Ekinaga N, Endo M, et al. TANSO 1996; No. 173: 142–147 [in Japanese].
[16] Kane J, Karthik C, Butt DP, et al. J Nucl Mater 2011;415:189–97.
[17] Ishiyama S, Eto M, Oku T. J Nucl Sci Technol 1987;24:719–23.
[18] Ishiyama S, Oku T, Eto M. J Nucl Sci Technol 1991;28:472–83.
[19] Eto M, Ishiyama S. J Nucl Sci Technol 1997;34:476–83.
[20] Eto M, Ishiyama S. J Nucl Sci Technol 1998;35:808–15.
[21] Eto M, Konishi T. TANSO 1999; No. 186: 30–35.
[22] Ioka I, Yoda S, Konishi T. Carbon 1990;28:879–85.
[23] Oku T, Ohta S, Eto M, et al. TANSO 1993; No. 156: 15–21 [in Japanese].
[24] Kamiyama M, Sogabe T. TANSO 1992; No. 151: 8–12 [in Japanese].
[25] Kawakami H. TANSO 1986; No. 124: 26–33.
[26] Takahashi M, Kotaka M, Sekimoto H. J Nucl Sci Technol 1994;31:1275–86.
[27] Fuller EL, Okoh JM. J Nucl Mater 1997;240:241–50.
[28] Kim ES, Lee KW, No HC. J Nucl Mater 2006;348:174–80.
[29] Kim ES, No HC. J Nucl Mater 2006;349:182–94.
[30] Chi S-H, Kim G-C. J Nucl Mater 2008;381:9–14.
[31] Contescu CI, Guldan T, Wang P, et al. Carbon 2012;50:3354–66.

[32] Shiozawa S, Fujikawa S, Iyoku T, et al. Nucl Eng Des 2004;233:11–21.
[33] Takamatsu K, Ueta S, Sumita J, et al. JAEA Res 2010:2010–38 [in Japanese].
[34] Saito S, Tanaka T, Sudo Y. Nucl Eng Des 1991;132:85–93.
[35] Futakawa M, Takada S, Takeishi H, et al. Nucl Eng Des 1996;166:47–54.
[36] Iyoku T, Futakawa M, Ishihara M. Nucl Eng Des 1994;148:71–81.
[37] Minatsuki I, Tanihira M, Mizokami Y, et al. Nucl Eng Des 2004;233:377–90.
[38] Arai T, Sato S, Oku T, et al. J Nucl Sci Technol 1991;28:713–20.
[39] Arai T, Ioka I, Ishihara M. JSME Int J 1991;34:470–6.
[40] Shibata T, Tada T, Sumita J, et al. J Solid Mech Mater Eng 2008;2:166–75.
[41] Kelly BT, Brocklehurst JE. Carbon 1971;9:783–9.
[42] Oku T, Eto M, Ishiyama S. J Nucl Sci Technol 1987;24:670–1.
[43] Maruyama T, Harayama M. J Nucl Mater 1992;195:44–50.
[44] Brocklehurst JE, Kelly BT. Carbon 1993;31:155–78.
[45] Kelly BT, Burchell TD. Carbon 1994;32:119–25.
[46] Ishiyama S, Burchell TD, Strizak JP, et al. J Nucl Mater 1996;230:1–7.
[47] Neighbour GB, Hacker PJ. Mat Lett 2001;51:307–14.
[48] Burchell TD, Snead LL. J Nucl Mater 2007;371:18–27.
[49] Burchell TD. J Nucl Mater 2008;381:46–54.
[50] Snead LL, Burchell TD, Katoh Y. J Nucl Mater 2008;381:55–61.
[51] Chi S-H, Kim G- C. J Nucl Mater 2008;381:98–105.
[52] Kunimoto E, Shibata T, Shimazaki Y, et al. JAEA Res 2009; 2009-008 [in Japanese].
[53] Shibata T, Kunimoto E, Eto M, et al. J Nucl Sci Technol 2010;47:591–8.
[54] Aymar R, et al. J Nucl Mater 1998;258–263:56–64.
[55] Kalinin G, Barabash V, Fabritsiev S, et al. Fusion Eng Des 2001;55:231–46.
[56] Yamashina T, Hino T. J Nucl Sci Technol 1990;27:589–600.
[57] Tokunaga K, Yoshida N, Noda N, et al. J Nucl Mater 1999;266–269:1224–9.
[58] Tokunaga K, Yoshida N, Kubota Y, et al. Fusion Eng Des 2000;49–50:371–6.
[59] Tokunaga K, Kubota Y, Noda N. Fusion Eng Des 2006;81:133–8.
[60] Tokunaga K, Matsubara T, Miyamoto Y, et al. J Nucl Mater 2000;283–287:1121–7.
[61] Higashijima S, Sakurai S, Suzuki S, et al. Fusion Eng Des 2009;84:949–52.
[62] Barabaschia P, Kamadab Y, Ishida S. Fusion Eng Des 2011;86:484–9.
[63] Oku T, Kurumada A, Sogabe T, et al. J Nucl Mater 1998;257:59–66.
[64] Masaki K, Miyo Y, Sakurai S, et al. Fusion Eng Des 2010;85:1732–5.
[65] Motojima O, Akaishi K, Fujii K, et al. Fusion Eng Des 1993;20:3–14.
[66] Iiyoshi A, Komori A, Ejiri A, et al. Nucl Fusion 1999;39:1245–55.
[67] Kubota Y, Noda N, Sagara A, et al. Fusion Eng Des 1998;39–40:247–52.
[68] Kubota Y, Masuzak S, Morisaki T, et al. Fusion Eng Des 2005;75–79:297–301.
[69] Masuzaki S, Shoji M, Tokitani M, et al. Fusion Eng Des 2010;85:940–5.
[70] Tokitani M, Yoshida N, Masuzaki S, et al. J Nucl Mater 2011;415:587–91.
[71] Zakharov AP, Gorodetsky AE, VKh Alimov, et al. J Nucl Mater 1997;241–243:52–67.
[72] Causey RA. J Nucl Mater 2002;300:91–117.
[73] Atsumi H. J Vac Soc Jpn 2006;49:49–55 [in Japanese].
[74] Atsumi H, Iseki M, Shikama T. J Nucl Mater 1992;191–194:368–72.
[75] Atsumi H, Iseki M, Shikama T. J Nucl Mater 1994;212–215:1478–82.
[76] Kanashenko SL, Gorodetsky AE, Chernikov VN, et al. J Nucl Mater 1996;233–237:1207–12.
[77] Atsumi H, Tauchi K. J Alloys Comp 2003;356–357:705–9.

Index

Note: Page numbers followed by "*f*" denote figures; "*t*" tables; "*b*" boxes.

Printed and bound by CPI Group (UK) Ltd, Croydon, CR0 4YY

08/05/2025

01864900-0004